Sounds of Our Times

Robert T. Beyer

Sounds of Our Times

Two Hundred Years of Acoustics

Robert T. Beyer
Department of Physics
Brown University
Providence, RI 02912
USA

With 261 illustrations.

Library of Congress Cataloging-in-Publication Data
Beyer, Robert T. (Robert Thomas), 1920–
    Sounds of our times : two hundred years of acoustics / Robert T.
  Beyer
      p. cm
    Includes index and bibliographical references.
    ISBN 0-387-98435-6 (alk. paper)
    1. Sound—History—19th century.   2. Sound—History—20th century.
  I. Title
  QC221.7.B49   1998
  534—dc21   98-9607

Printed on acid-free paper.

Production coordinated by Brian Howe, and managed by Steven Pisano; manufacturing
    supervised by Jacqui Ashri.
Typeset by Best-set Typesetter Ltd, Hong Kong.
Printed and bound by Maple-Vail Book Manufacturing Group, York, PA.
Printed in the United States of America.

9 8 7 6 5 4 3 2 1

ISBN 0-387-98435-6 Springer-Verlag New York Berlin Heidelberg   SPIN 10659851

To all acousticians, both those who are mentioned in this book and those who are not. Together, they have constructed the vast edifice of sound and vibration.

Preface

> But at my back I always hear
> Time's wingèd chariot hurrying near.
> Andrew Marvell (1621–1678)

In writing this book, I acknowledge two important predecessors. In 1930, Professor Dayton C. Miller, of Case Institute, wrote his *Anecdotal History of Acoustics*, which followed mainly the individual work of a hundred or so acousticians, up to about 1930. And in 1978, *Origins in Acoustics* by Professor F.V. ("Ted") Hunt of Harvard appeared. Professor Hunt had not completed his book at the time of his death in 1972, but portions of it were published posthumously under the editorship of Professor Robert Apfel of Yale.

This book begins roughly where the published portions of Hunt's work left off—the period at the beginning of the nineteenth century—and moves forward into the modern era. Because of the vast amount of acoustical research in this period, the personal and anecdotal style of Professor Miller's book did not seem appropriate. On the other hand, my age suggested to me that I might have not have the time for pursuing the intense but time-consuming scholarship of Professor Hunt (note the quotation above). I have therefore tried to steer a middle course between Miller and Hunt, and have also relied more on secondary sources.

Before going on with the text, I must also cite two remarks from the Preface to Norman Davies' *God's Playground: A History of Poland* that have served as guiding principles to this neophyte in the writing of history:

Like all history books [this volume] had to be written through the distorting medium of the mind of the historian. . . . All the historian can do is to recognize the particular distortions to which his work is inevitably subject and to avoid the grosser forms of retouching and excision. [1]

In view of the unfamiliarity of Polish History to most British and American readers, the treatment of weighty events has been sweetened with a mild infusion of anecdotes, epitaphs, *bons mots* . . . and antiquarian curiosities—in short, with a selection of all those historical whimsicalities which scientific scholars judge unworthy of their genius. [2]

The resultant history of acoustics now lies before you.

Notes and References

[1] Norman Davies, *God's Playground: A History of Poland* (paperback), Columbia University Press, New York, NY, 1982, 2 vols., vol. 1, p. vii.
[2] *God's Playground: A History of Poland*, p. xiv.

Acknowledgments

Among those who have helped me I would like to express my thanks to the Honorable Guy Strutt of Terling, Essex, England, for contributing photographs of his grandfather, Lord Rayleigh. Thanks also to Professor William Wooten and the Brown Psychology Derpartment for the picture of Professor S.S. Stevens, to the Khokhlov family for the photograph of Academician Khokhlov, to Professor V.A. Krasilnikov for the photograph of Academicians Kolmogorov and Obukhov, to Dr. Dieter Guicking for the photograph of Professor Erwin Meyer, and to Elaine Moran of the Acoustical Society of America for supplying photographs of a number of departed acousticians, and much assistance otherwise. I am also indebted to Joe Anderson and Michele Zeccardi of the Niels Bohr Library of the American Institute of Physics for assistance on birth and death dates, to the librarians of the Bell Telephone archives, and to the various librarians of Brown University for their unfailing assistance. Thanks are also due to President Vartan Gregorian of Brown University for suggestions of people and sources that might be useful to me, and to various colleagues at Brown for their help, and to Burt Hurdle, Murray Korman, Werner Lauterborn, Konstantin Naugolnykh, Jim West, and many others in the Acoustical Society of America for useful conversations.

Special thanks are due to Fred Fisher, Katherine Harris, and Charles Schmid, who reviewed the text in detail and gave me many variable suggestions and corrections. The errors that may remain are, of course, my own.

Thanks also to my original publisher, AIP Press, and its editors, Maria Taylor and Charles Doering, for the many courtesies and encouragements that they have provided during the course of preparation of this text.

Appreciation is also extended to Springer-Verlag, who have taken over the AIP books program, making it possible for me to change publishers "in mid-stream" without any loss in forward progress. Thanks also to my secretary, Jane Martin, for much help in the final preparation of the text.

And as always, my acknowledgment to and appreciation of my wife. At a time when I was supposed to be retired, I seem to have come up with a full-time, four-year occupation that has allowed little time to pause and smell the roses.

Robert T. Beyer

Contents

1
The State of Acoustics in 1800

> The soughing of the wind,
> the tuneful noise of birds in the spreading branches,
> the measured beat of water in its powerful course,
> the harsh din of the rocky avalanches,
> the invisible swift course of bounding animals,
> the roaring of the savage wild beasts,
> the echoes rebounding from the cleft in the mountains.
>
> *The Book of Wisdom*, Chapter 17

1.1. Introduction

Two of the commonest words describing sound are acoustics and phonics, deriving, respectively, from the Greek *akouein*, to hear, and *phonein*, to speak, and with only these two words, we can conjure up the beginning of the science of acoustics with one human being communicating with another. This must have been followed by the discovery that pleasing sounds could be made by vibrating strings, by blowing across open tubes or by striking stretched membranes, i.e., the birth of music. This appreciation of music has been a hallmark of humanity almost from the beginning.

To limit the scope of this volume, the writer has thought it appropriate to choose the turning of the eighteenth century into the nineteenth as a convenient point at which to begin. Professor Frederick Vinton Hunt (1905–1972), in his admirable *Origins in Acoustics* [1], had covered the origins of the subject from the earliest days through those of Newton (1642–1727). There are a few further flashes in that book that illuminate certain portions of acoustics in the eighteenth and nineteenth centuries, but little coverage of later periods [2]. Both Hunt and his editor have also acknowledged the excellent treatment of eighteenth century acoustics in the prefaces written by Clifford Truesdell (1919–   ) that appear in the reprinting of the complete works of Leonhard Euler (1701–1783) [3]. We shall therefore begin our story with a review of acoustics as it existed in and around 1800, with a few flashbacks of our own into earlier times, to provide some continuity with the published work of Hunt and Truesdell.

The choice of the year 1800 is good in another sense. The dying decades of the eighteenth century marked the fading of the *ancien régime*. The new age, enflamed by the American and French Revolutions, saw the struggling development of democracy, the ascendancy of Napoleon and, in science and engineering, the beginnings of modern chemistry, steam power, mass production, and a growing knowledge of optics and electricity. All these events marked the period as one of tremendous change. Since it has been widely recognized that acoustics lives by its interaction with the other sciences and engineering, as well as with society in general, we shall be examining the literature for any corresponding surge in the field of acoustics.

Chladni and Young

We are greatly aided in this task by two texts that appeared in the first decade of the new century. In 1802, Ernest F.F. Chladni (1756–1827) (Fig. 1-1) [4] published *Die Akustik* [5]. Chladni himself translated the book into French (a book "in which I have abridged, changed and added a great deal" [6]). He published this volume in 1809, dedicating it to Napoleon (a wise choice, no doubt, during that "sun of Austerlitz," especially since the French government contributed funds to support the translation and revision). By just glancing through this volume, we can see several features that distinguish the book from a modern one. The first is the almost complete absence of mathematics. Acoustics, as it was studied at the time, at least, in the mind of Chladni, and aside from music and vibrating structures, was largely one of observations and descriptions. Magnificent mathematics had been developed by Euler, Jean LeRond d'Alembert (1717–1783), and

(a)   (b)

FIGURE 1-1. (a) Ernst F.F. Chladni (1756–1827) and (b) Thomas Young (1773–1829). (From D. Miller [4].)

Joseph Louis Lagrange (1736–1813) in the eighteenth century [3] and applied by them to acoustical problems, but Chladni clearly passed over the details of this mathematics in writing his treatise.

A second difference is the emphasis on vibrations. If one had any doubt that vibrations have long been recognized as an integral part of acoustics, a reading of Chladni would eliminate that misconception. Of the four sections of Chladni's book, the one devoted to vibrations comprises more than 60% of the text. Propagation of sound through air and other gases covers about 15%, while the remaining 25% of the volume is divided among propagation in liquids and solids, musical scales, speech and hearing. (Parenthetically we might note that, even though the period around 1800 was far from a quiet time in the world, there is virtually nothing in the book on noise.)

In the French edition, Chladni identified himself as the son of a Professor of Law at Wittenberg. The younger Chladni was evidently heavily under his father's thumb, and wrote ruefully of his lack of freedom as a youngster and of his attendent unhappiness. He studied at both Wittenberg and Leipzig Universities, taking a degree of Doctor of Law from the latter, and had every expectation of following in his father's footsteps. The death of his father in this period apparently changed everything for him. Chladni tells us that he then gave up jurisprudence and turned his attention to the study of music. There he found that "the theory of sound was more neglected than the several other branches of physics, which gave birth in me of a desire to remedy this defect [7]."

Our second author was the English physician and polymath, Thomas Young (1773–1829) (Fig. 1-1 [4]). Young was a master of many languages, both contemporary and ancient, a practicing physician, and a student of Egyptology (he pioneered in translation work on the hieroglyphics of the famous Rosetta stone before Champollion). He was also a first class physicist, especially in optics and elasticity. He was elected a Fellow of the Royal Society of London at the age of 21, primarily for his work on the eye. Early in his medical education career, a course that took him from London to Edinburgh to Göttingen, and, finally, to Cambridge, Young made a bet with his fellows that "Young will produce a pamphlet or paper on the theory of sound more satisfactory than any thing that has already appeared, before he takes his Bachelor's Degree" (March 14, 1799) [8]. He did publish a paper on the "Theory of Sound and Light" in 1800 [9], but history records that the decision of his peers was against him and he lost the bet. His biographer Alexander Wood observed in 1954 "It is doubtful whether a competent tribunal would now uphold the decision on appeal [10]."

In 1807, Young published a two-volume text, entitled *A Course of Lectures on Natural Philosophy and the Mechanical Arts* [11]. These lectures were given at the Royal Institution in London, and the books contain three chapters on sound that give an good summary of the then current knowledge of acoustics. The book apparently proved popular; a second edition appeared in 1845, long after Young's death.

In passing, it is interesting to note how many famous acousticians actually did major work in other fields—Young in optics; Helmholtz in optics, electricity, and thermodynamics; Rayleigh in many fields, including gases, electricity, and optics. And, very often, scientists, world-famous in other fields, have published one or two papers in acoustics. We shall remark on this attraction of the field as the book progresses, noting the contributions to acoustics from such great names as Einstein, Debye, Schrödinger, and Schwinger, to name a few.

In his book, Chladni divided the subject of acoustics into sources of sound, the passage of sound through matter, and its reception. We shall indeed adopt this grouping of the subject for the current chapter: sound is produced, sound travels, sound is observed. However, as Young noted in his text, the state of knowledge of the subject in his time was such that it was more convenient to begin with the subject of sound propagation and, in particular, the study of the velocity of sound, rather than beginning with the source. Hence we shall first discuss sound propagation, then the production of sound, and, finally, its reception.

In dealing with the origins of various portions of acoustics, the author of this book recalls a cautionary saying of his own, "nothing was ever invented or discovered or written for the first time; somebody else always did it earlier." While this is an obvious oversimplification, not to say a contradiction, it is a good guiding principle. In covering a period of two hundred years, there is no doubt that some historical circumstances will be incorrectly cited in this text. While every effort has been made to give proper credit, apologies are made in advance for any errors [12].

1.2. Sound Propagation

Perhaps the first problem of sound propagation was the question of whether or not air (or other material) was necessary for sustaining its propagation. By 1800, this was thought to have been long settled. One of the oldest and most frequently repeated experiments in acoustics is the use of a bell or other mechanical source of sound in a chamber that had been evacuated to some extent. This experiment was first carried out by Gianfrancisco Sagredo in 1615 [13], and repeated by a number of distinguished scientists over the next two hundred years [14], with the conclusion, as of 1800, that it had been proven that sound could not travel through a vacuum.

Sound Velocity in Air

The next question was that of the velocity of propagation in air. By 1800, accurate measurements of the velocity of sound in air had existed for more than one hundred and fifty years. Robert Bruce Lindsay (1900–1985) points out in his essay on "The Story of Acoustics" [15] that by 1738 the measurements in air were within 1% of the best present-day measurements. He wrote

This (the measurement of the velocity of sound in air) is a tribute to the care with which those Paris academicians carried out their work. Actually, very few early physical measurements have stood the test of time as well as these of the velocity of sound in air. [16]

However, the theoretical basis for calculation of the sound velocity in gases still remained a puzzle. For its calculation, the scientists of the day went back to Sir Isaac Newton (1642–1728). Newton had worked on this problem some one hundred years earlier, and arrived at a value from intricate geometric considerations [17]. He summed up this analysis by saying that, if the Earth's atmosphere were considered homogeneous, it would have a certain height, and the velocity of sound in air is equal to the velocity acquired by a heavy body when it falls through one-half this height! The reader can easily verify that this statement is equivalent to saying that the sound velocity is equal to the square root of the ratio of the atmospheric pressure to the density. And, in fact, a few pages earlier, Newton had written

The velocities of pulses propagated in an elastic fluid are in a ratio compounded of the subduplicate [i.e., square root of (the)] ratio of the elastic force directly, and the subduplicate ratio of the density inversely. [18]

Chladni, using Newton's method, cited velocities for a number of gases, such as oxygen and carbon dioxide, that come close to currently accepted values, but was far off in his estimate of the value in hydrogen—680–810 m s^{-1} as against today's accepted value of 1240 m s^{-1}.

This analysis in effect assumed sound propagation to be an isothermal process, and therefore gave too low a value for the sound velocity in air, a difference that Newton accounted for by what Hunt called "a monstrous exhibition of teleological data manipulation [19]." The interesting history of this problem is described in Hunt's book, and we shall not repeat it here. What we are concerned with is that the problem was still not entirely settled in 1800. It must be remembered that much of our understanding of the behavior of gases, and of thermodynamics, stems from work done in the first quarter of the nineteenth century. Further progress in our knowledge of the velocity of sound had to wait for such a development (see Chapter 2).

The effect of the temperature on the velocity of propagation of sound was known qualitatively in 1800, but the precise connection also had to await a better knowledge of the thermal properties of gases. Young [20] describes the experiments of Count Giovanni Bianconi (1717–1781) as demonstrating that the velocity of sound in air increased with the temperature. Lindsay [21] attributes the date of this work to 1740.

In his lectures, published in 1807, Young describes again the peculiar method of arriving at the sound velocity given by Newton [22]. Young did hold out hope for the future by noting "the happy suggestion of [Pierre Simon] Laplace (1749–1827) [23] that an increase in temperature accompanies the condensation in the sound wave, and a decrease accompanies the rarefaction." However, the detailed solution still had to wait.

Another student of sound propagation in the eighteenth century was William Derham (1657–1738). In 1708, he reported to the Royal Society [24] on the effect of wind, barometric pressure, temperature, and humidity on sound propagation. While he was mostly correct in his conclusions, as was pointed out by Hunt [25], he was under the impression that rain and fog reduced the transmission of sound, a conclusion that was accepted until Tyndall found it to be otherwise (see Chapter 3), more than 150 years later. Derham was also among the first to note that the distance of an observer from a lightning flash could be measured by the time delay for the arrival of the thunder.

Sound Velocity in Liquids and Solids

In 1800, there was virtually no knowledge of sound transmission in liquids. Chladni makes the curious remark that "The propagation of sound in water may be concluded from the fact that aquatic animals are also possessed with organs of hearing [26]." However, he does repeat the seventeenth century observation of Francis Hauksbee (died 1713) [27] that "when underwater, one can hear sounds that are produced in the air, but that one hears more strongly the sounds that are also produced underwater." Indeed, it is reported that Benjamin Franklin (1706–1790) held his head under water while an associate clicked two stones together (under water) at some distance, and that he heard them perfectly. (Was there anything that Franklin didn't try?) [28]

The velocity of sound in water had not yet been measured. In fact, Chladni remarked that "the velocity with which sound is propagated in water or in other liquids is completely unknown [29]." However, a rather good theoretical value for this velocity is given by Young. He noted that the elasticity (we would call this the bulk modulus) of water had been measured by John Canton (1718–1772) in 1762 [30] and found to be 22,000 times that of air. Young then resorted to the Newtonian method of calculation:

It (the elasticity) is therefore, measured by the height of a column which is in the same proportion to 34 feet, that is 750,000 feet, and the velocity corresponding to half this height is 4900 feet in a second. [31]

Using the elasticity data of Canton, Chladni was able to determine the sound velocity in a number of liquids with accuracy similar to that obtained for water.

A measurement of the velocity of sound in solids did exist in 1800, and is described by Chladni in his book [32]. He compared the musical pitch emanating from a struck solid bar (undergoing longitudinal oscillations) with the pitch of the (standing) wave in a closed, air-filled pipe of the same length. Arguing that the difference is due to the difference in the two sound velocities, he came up with values of the sound velocity. Thus, for tin, the pitch of the bar was higher than that of air by two octaves and a major seventh, from which he deduced that the sound velocity in air "will be surpassed by that in tin by about $7\frac{1}{2}$ times," while the pitch in copper was higher by about three octaves and a fifth, or about 12 times that of air. If we take the value of the sound velocity in air at $20°$C to be $343\,\mathrm{m\,s}^{-1}$, this gives us $2573\,\mathrm{m\,s}^{-1}$ for tin and $4116\,\mathrm{m\,s}^{-1}$ for copper, against tabulated values of 3320 and $5177\,\mathrm{m\,s}^{-1}$, respectively [32], [33]. Chladni's wording is a bit obscure; he might have meant that the actual values of the sound velocity were $8\frac{1}{2}$ and 13 times that of air, which would still leave his values as too low by nearly 15%.

Nevertheless, he did demonstrate that sound velocity is considerably higher in solids than in gases or liquids. A more accurate measurement was developed in the early 1800s, but we shall leave that to Chapter 2.

Diffraction, Reflection

By 1800, it was well established that sound coming from a small source traveled in all directions (spherical waves), but sound could be constrained to travel in tubes in one direction only (mainly plane waves), although the terms spherical and plane were not yet used. Young came close in his essay in noting that

when a sound is transmitted in a fluid . . . the impulse spreads in every direction, so as to occupy at any one time nearly the whole of a spherical surface. But it is impossible that the whole of this surface should be affected in a similar manner by any sound originating from a vibration confined to a certain direction, since the particles behind the sounding body must be moving towards the centre, whenever the particles before it are retreating from the centre. [34]

Clearly, Young was missing the distinction between a monopole and a dipole, and the role of wavelength in the whole matter had not yet been broached.

The law of reflection was also well understood and in fact its role in so-called whispering galleries was mentioned by Young. Young was also aware that sound, emitted at one focus of an ellipse, would travel to the walls of the ellipse and be reflected to the second focus. This led him to the idea that a parabola being the limiting case of an ellipse, could be used to collect sounds from distant points at the focus of the parabola (or a paraboloid of revolution in three dimensions). He comments

It appears, therefore, that a parabolic conoid is the best form for a hearing trumpet, and for a speaking trumpet; but for both purposes the parabola ought to be much elongated, and to consist of a portion of the conoid remote from the vertex. [35]

Young also observed that a trumpet of such shape is very similar to a cone, and that "conical instruments are found to answer sufficiently well for practice." In these statements, he is anticipating the megaphone and horn loudspeakers, as well as providing a clue for the stethoscope of later development. The use of cylindrical pipes was also discussed by Chladni [36], who called them "porte-voix," and attributes the first use to Sir Samuel Morland (1615–1695) in 1671, with later work by Lambert (1728–1777) and Jean Henri Hassenfratz [37] (1755–1827). A text of the period by Robison (1822) discusses "speaking trumpets" and asserts that Alexander the Great used one to address his army (fourth century B.C.) [38].

Echoes

The term echo goes back to Greek mythology, where she was the nymph who loved Narcissus, but the physical nature of an echo was reasonably well understood in this period. In his discussion of the subject, Chladni notes that it is possible (by having two reflecting surfaces facing one another) to have multiple echoes. Chladni remarks that the ear is capable of distinguishing eight or nine different sounds in a single second [39]. Therefore, when these repeated echoes take place more rapidly than eight or nine per second, he states that the phenomenon is known as resonance. Young provides a very similar description of the sounds from a footstep in a narrow corridor [40]. This mixing of repeated echoes with resonance will be reflected later in the discussion regarding Tartini tones. While we would define resonance somewhat differently today than Chladni did, he gave a good qualitative description of what is taking place when one passes through a resonance.

Sound Intensity

In considering sound intensity, Chladni [41] noted that it depended on: the size of the sonorous body, the intensity of the vibrations of that body, the frequency of the vibrations, the distance at which the sound is heard, the density of the air, the direction in which the sound is heard, and the direction of the wind. We would recognize today that these dependences are a mixture of what we call the intrinsic intensity ($\frac{1}{2}\rho c v_0^2$, where ρ is the local gas density, c the speed of sound, and v_0 the amplitude of the displacement velocity), and such related quantities as the strength of the source, the directionality of the source, and the attenuation due to geometric spreading and sound absorption. All these quantities were not well sorted out in 1800.

1.3. Sound Production

In 1800, the available means for the production of sound were the human voice, musical instruments, cannons and other explosive devices, and natural phenomena such as animal sounds, thunder, etc. It is not surprising, therefore, that Chladni (and others of the time) used music as the basis on which to build almost all of acoustics. When dealing with vibrating strings, they were concerned with stringed musical instruments. Vibrating air columns were of interest because of organ pipes, and also various musical horns, while stretched membranes were related to drums. Almost every subject in Chladni's text is studied from the point of view of music. And Chladni was not alone. A somewhat earlier text by Matthew Young (1750–1800), Bishop of Clonfort, made the same emphasis [42]. The great advances of theoretical acoustics in the eighteenth century were perhaps due to the common interests of music patrons, researchers, and the listening audiences.

The reason for this substantial interest in music at that time can be understood in historical terms. The art of music itself underwent prodigious growth in the eighteenth century, with the development and improvement of musical instruments, leading to the construction of the greatest violins ever, the development of chamber and symphonic music and of the opera, and with such composers as Bach, Haydn, and Mozart. This in turn stimulated interest in the mathematical and physical rules that undergirded such marvelous sounds and combinations of sounds. Today's researchers have marked the changes in the character of composed music over the ages as the churches, chambers, and concert halls have varied in design. As Christopher Herr and Gary Siebein put it, "there has been a clearly discernible connection between the development of music, musical instruments, ensembles, and the design of spaces for musical performances [43]."

Measurement of Sound Frequency

Well before 1800, the understanding of music led to two great contributions to the science of sound. First, it emphasized the importance of ratios for different tones. The simple ratios appropriate for all the notes on the diatonic scale were known, and musicians with trained ears could easily identify the pitches of the various notes, starting from some accepted standard, such as middle C or, more commonly, the A above middle C. At the same time, it was also known that the pitch of a musical note was measured by its frequency of oscillation.

Marin Mersenne (1588–1648) was apparently the first to evaluate the frequency corresponding to a given pitch. By working with a very long rope, he was able to verify the dependence of the frequency of a standing wave on the length, mass, and tension of the rope. He then used a short wire under tension and matched it with an organ pipe that emitted the same tone as the wire. From his formula, Mersenne was then able to compute the frequency of oscillation [44].

The ideas behind Mersenne's measurements were correct, but the accuracy of his measurements was low [45]. His results were greatly improved in the following century by Joseph Sauveur (1653–1716) [46]. Sauveur combined a knowledge of the ratios of musical tones with the recently observed phenomenon of beats to make it possible to determine the actual frequencies of the tones. Sauveur considered two organ pipes whose rather low tones differed by a half-tone, standing in the ratio of 15 to 16. Sauveur was able to count six beats when the two pipes were sounded simultaneously and therefore assigned them the frequencies of 90 and 96 oscillations per second. (The two tones correspond to F$^\sharp$ and G in the second octave below middle C.) He then worked his way up the musical scale, using his knowledge of beats, so that he could evaluate the frequencies of all the musical notes. To Sauveur we also owe the first use of the scale that establishes the frequencies of the various C's as powers of 2 (often called the physicists' scale). Thus, middle C was assigned the frequency of 256 vibrations per second (2^8). The G of 96 oscillations per second corresponds to this scale. We note, however, that the determination of six beats as the difference must have been approximate. Since we do not know his accuracy, we might guess that the correct value lay between 5.5 and 6.5 beats, which would allow the G to be in the range from 88 to 104 and the A above middle C to lie between 391 and 462 cycles per second. As we note in the paragraph below, organs in western European churches used virtually any value within this range for their basic pitch.

Because of these earlier observations by Mersenne and Sauveur, Chladni was able to base his discussion of musical scales on the grounds of a knowledge of the frequencies involved in the tones of Guido's scale [47], even though there remained some uncertainty as to what was the appropriate standard for middle C. In his discussion, Chladni cited Euler's value [48] for the C below middle C as 118 vibrations per second in a paper in 1739, and 125 in a later paper in 1772, and Giuseppe Sarti (1729–1802) gave a value of 131 for the same tone, but then opted for the simpler number of 128, as assigned by Sauveur. Nearly one hundred years later, Alexander John Ellis (1814–1890) [49] reported the wide variation that existed in standards of pitch on the organs of various churches in Europe, giving a tabulated list of such pitches, ranging from 370 to 516 for the A above middle C—a wider range than the uncertainty attributed above to the measurements of Sauveur [50]. One must note, however, that Chladni devoted most of his attention, in musical matters, to discussing ratios of frequencies of the different tones, where the quantities are far more accurate, rather than the absolute values of these frequencies.

The Human Voice

Both Chladni and Young provided brief descriptions of the operation of the human voice system. Chladni described the origins of the human voice as follows:

The voice of men and animals is formed in the same manner. It is found in the larynx, two approximately semi-circular membranes, which together form a circular surface. The circumference of these membranes, which is called the ligaments of the glottis, is attached to the walls of the larynx, and their straight edges can either be joined following the diameter of the circle, or form a lenticular window, which is called the glottis. If this opening is sufficiently large, the air passes without producing a sound, as is the case in ordinary breathing; but if it is contracted, the air coming out of the lungs through the larynx is thrown against these two membranes and produces some rapid vibrations, which are communicated to the current of outcoming air. To this current of vibrating air, which is called the voice, the other organs of the mouth oppose different obstacles and form very different openings, each of which articulates the voice in a different manner. The more the glottis is contracted, the sharper is the sound. All possible varieties are produced by changes in the opening, whose extremes differ by $\frac{1}{10}$ of an inch. [51]

Thomas Young was peculiarly qualified to be interested in the human voice from his work as a physician, his profound knowledge of different languages, and his studies of physical phenomena. In his thesis at Göttingen, Young had tried to enumerate the number of different sounds of which the human voice was capable (his number was 47), in order to define a kind of international alphabet of sounds [52].

In his lectures on natural philosophy [53], Young devoted several paragraphs to the human voice. He noted

The human voice depends principally on the vibrations of the membranes of the glottis, excited by a current of air, which they alternately intercept and suffer to pass; the sounds being also modified in their subsequent progress through the mouth.

Drawings of the glottis given by Young are shown in Fig. 1-2. Of the glottis, he wrote:

there are two ligaments on each side but it is not fully understood how they operate. [54]

The name "vocal cords" was given to these ligaments by Antoin Ferrein (1693–1769), a French physician, who compared them with the strings of a violin [55].

Young also described briefly the character of the production of vowels, "chiefly formed by this apparatus in the glottis and modified either in their origin or in their progress by the various arrangements of the different parts of the mouth." What these arrangements were, Dr. Young forbore to say.

As far as consonants were concerned, Young observed that

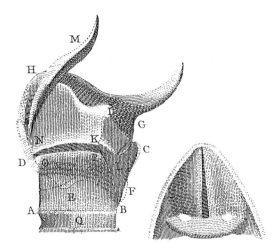

FIGURE 1-2. Views of the glottis. (From T. Young [54].)

the perfect consonants may be either explosive, susurrant or mute. The explosive consonants begin or end with a sound formed in the larynx, the others are either whispers, or mere noises, without any vowel sound. [56]

Artificial Voice

The idea that vowel sounds were bursts of air modified by closing or partial closing of the tube-like interior of the throat and mouth caused experimenters of the day to attempt the construct voice machines, starting the long trail of research that led to the synthetic speech of our own day [57]. The best known of these machines were those designed by Christian Gottlieb Kratzenstein (1723–1795) and Wolfgang von Kempelen (1734–1804) [58].

In 1791, The Imperial Academy of Sciences at St. Petersburg had offered a prize for answering the questions: (1) "What is the nature and character of the sounds of the vowels a, e, i, o, u (that make them) so different from one another? and (2) Can an instrument be constructed like the vox humana pipes of an organ which shall accurately express the sounds of the vowels?" The German Kratzenstein won this prize, publishing a Latin essay on the subject and constructing a set of resonators (shown in Fig. 1-3) [59]. These were equipped with a vibrating reed, similar to that in a mouth organ. Kratzenstein served as a Professor at Halle, then at the Imperial Academy in St. Petersburg and, finally, at the University of Copenhagen.

Von Kempelen was an Austrian, a minor nobleman of the Holy Roman Empire, who did architectural and engineering work for the Hapsburg monarchs in Vienna. He improved upon Kratzenstein's device [59]. A model of his instrument, which was reconstructed by Sir Charles Wheatstone [60] (1802–1875), is pictured in Fig. 1-4.

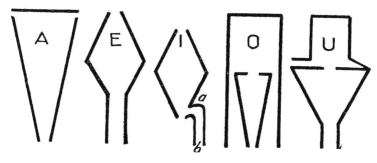

FIGURE 1-3. Shaped resonance chambers for the different vowels, for Kratzenstein's talking machine [59].)

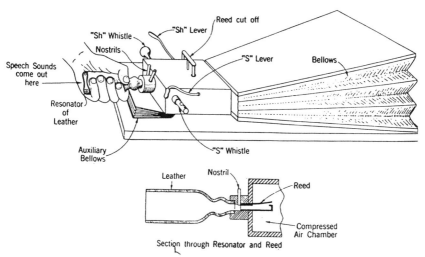

FIGURE 1-4. Wheatstone's reconstruction of the von Kempelen talking machine. (From H. Dudley and T. Tarnoczy [59].)

Vibrating Strings

The vibrating string was perhaps at the very origin of music, since interest in this phenomenon goes back at least to King David around 1000 B.C. [61], and scientific interest to Pythagoras [62] in the sixth century B.C. By 1800, the mathematical developments of the eighteenth century had had their impact on the subject. D'Alembert [63] had developed the wave equation and applied it to the vibrating string, obtaining the solution that the vibrations actually consist of two waves traveling in opposite directions along the string, forming standing wave patterns. This indeed was the vital link between vibrations and acoustics. The sonorous body (a popular term in those days) underwent its vibrations and could be studied in its own right, but somehow, during these vibrations, the air particles that immediately abutted the body were themselves put into vibration and could pass it on to their neighbors. Just how this was done was still something of a mystery, but the two phenomena had to have inseparable links.

Another advance of this same period was that made by Daniel Bernoulli (1700–1782) [64] who showed that an individual string could actually contain many different vibrations (what we today would call simple harmonic vibrations) at the same time. These vibrations (or the traveling waves of which they are composed) exist in the string independently of one another, and the total effect at any point would be the algebraic sum of the individual motions. Thus was born the principle of superposition [65]. At the same time, it was realized that sound waves themselves involved very small disturbances. Young estimated that the amplitude of the displacement velocity was as small as a hundredth of an inch per second. We might note that a velocity amplitude of $1 \, \text{in} \, \text{s}^{-1}$ would correspond to a sound pressure level 26.4 dB above the threshold level of hearing.

Musicians have long recognized that the vibrating string alone did not produce much in the way of sound unless it was reinforced by the presence of some surface, such as the faces of a violin or the soundboard of a harpsichord. Before the study of such two-dimensional vibrations was perfected, however, great attention was paid to the vibratory motions of air in confined tubes or pipes. The work of Euler in 1727 [66] and Lagrange in 1759 [67] in the eighteenth century made it possible to predict the approximate harmonic frequencies of open and closed pipes, and to understand the characteristics of overtones. The major unsolved problem was the correct estimate of end corrections, i.e., the distinction between the effective length of the pipe and its actual physical length.

Chladni was also aware that the human ear could hear tones as low as $30 \, \text{s}^{-1}$ and up to "8,000 to 12,000 vibrations" [68]—which goes considerably beyond the range of frequencies achieved by most musical instruments.

Vibrating Plates

Somewhere between the production and detection of sound is Chladni's own work on vibrating plates, since the study involved not only the production of sound by the vibrating plates, but also the experimental technique of identifying the vibrations. Chladni was well aware of the vibration of strings, and of the localization of nodal points in a standing wave, and the theoretical work of Lagrange and others gave a strong foundation to the subject. There was, however, no theory of vibrating plates.

Chladni was drawn to a study of the vibrations of plates from work done by Georg Christophe Lichtenberg (1742–1799) [69], who scattered "electrified powder over an electrified resin-cake, the arrangement of the powder revealing the electric condition of the surface [70]." In his first work, reported in 1787 [71], Chladni held fixed one or more points on a plate and stroked the side of the plate with the bow of a violin. In order to render the effect of the vibrations visible, he placed a little sand on the plate. The sand "was thrown aside by the trembling of the vibrating parts (of the plate) and accumulated on the nodal lines [72]." Chladni must have been fascinated by the patterns taken on by the sand particles, a fascination that these "Chladni figures" continue to generate [73]. In those days before photography, he included hundreds of drawings of different modes of excitation for triangular, circular, square, and even elliptical plates, in his 1809 book (Figs. 1-5 and 1-6). In this preface to the latter, Chladni wrote

When I applied the (violin) bow to a round plate of yellow copper, attached at its center, it yielded different sounds which, compared with one another, were equal to the squares of 2, 3, 4, 5, etc.; but the nature of the movements to which these correspond, and the means of producing each of the movements separately were still unknown to me. [74]

The solution to this problem will be discussed in Chapter 2.

There was, by 1800, a new device that was both a source of sound and an instrument for frequency measurement. This was the tuning fork. The inventor of this device (in 1711) was John Shore, a trumpeter in the band of Queen Anne in England and for the composer Handel. [According to Dayton Clarence Miller (1866–1941), Shore liked to make the pun, "I don't have my pitch-pipe with me to tune my instrument, but I have something just as good, a pitch-fork [75]." The use of the tuning fork in research, however, did not occur until a much later time (see Chapter 5).

Singing Flames

In 1802, an Irish born physician, Dr. Bryan Higgins (1737–1820) [76] reported a curious phenomenon—that of the singing flame. He produced a flame by burning hydrogen gas, and placed glass tubes of various diameters over it, sealed at the far end. If the tube diameters were too small, the flame would be extinguished, but as he proceeded to larger and larger diameters, he produced "several sweet tones, according to the width, length and thickness of the glass jar or sealed tube [77]." When the diameters became too large, the sounds grew fainter and ultimately ceased. Higgins noted that he first observed the effect in 1777 and cites (without reference) a similar experiment by Volta.

Many theories were advanced to explain this effect, both at the time and in the ensuing years, but they were largely incorrect. A good discussion of them is given in the book by Arthur Tabor Jones (1876–1950) [78]. Further discussion of the effect and of other effects involving flames will be deferred until Chapters 3 and 6.

FIGURE 1-5. Chladni figures. (From E.F.F. Chladni [6].)

FIGURE 1-6. Chladni figures. (From E.F.F. Chladni [6].)

1.4. Sound Reception

The Human Ear

The only instrument available for sound reception in 1800 was the ear. By the time of Chladni, the structure of the exterior and middle portions of the ear were quite well understood (Fig. 1-7). Chladni [79] recognized that the impulses of sound received in the outer ear were transmitted through the small bones (ossicles) of the middle ear, more or less faithfully, to the cochlea. While the gross features of the cochlea were well described by him, its role, so far as he was concerned, was very much that of a "black box." The sound impulses impinged on one end, and, somehow or other, the sensations were picked up by the auditory nerve at the other and transmitted to the brain. Chladni offered the (incorrect) opinion that the signals from the ossicles affected the cochlea as a whole rather than locally. Further developments in this area had to wait another century.

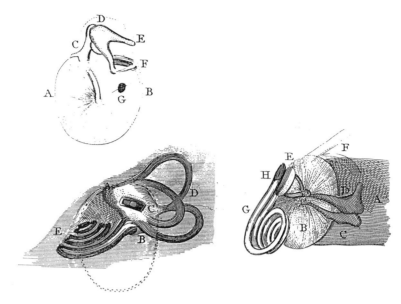

FIGURE 1-7. The ear as observed in 1800. (From T. Young [11].)

Chladni did not mention some major discoveries on the ear that had already been made. In 1775, Dominico Cotugno (1736–1822) described the basilar membrane for the first time. In describing in some detail the passages in the inner ear, he drew the conclusion that, contrary to the then prevailing opinion, the inner ear was filled with a liquid, and not with air [80]. Shortly after Cotugno's work, the anatomist P.F. Meckel provided convincing proof of this by a simple but ingenious experiment. He let an intact fresh cadaver ear be frozen one very cold night, and the next morning, when he broke open the ear, he found that it was solidly packed with ice [81].

At the time of Chladni's book (1807), Ludwig van Beethoven (1770–1827) was gradually losing his hearing, perhaps the most celebrated case of deafness in the western world. It is now believed [82] that Beethoven was suffering from otosclerosis, a hardening of the tissues surrounding the ossicles of the middle ear, preventing these bones from transferring the sound to the inner ear. Today, this ailment can be remedied by surgery, but not in Beethoven's time. However, there was knowledge at that time of bone conduction, i.e., the fact that sound could get to the inner ear by conduction through the bones of the head. Beethoven was able to hear some sound by placing a piece of wood between his teeth, with the other end resting on the sounding board of the piano. Alas that acoustical research did not proceed fast enough!

The ability of the ear to discern the direction of a distant sound was also known. Chladni pointed out that, if both ears are open, one can determine the direction by the inequalities of the effect on one ear or the other, except when, the position of the listener remaining the same, the sound is produced in front or in the rear, which cannot be distinguished. It appears that animals sometimes turn their ears toward different sides in order to learn the direction of the sound. [83]

Teaching of the Deaf

There is a long history connecting our knowledge of speech and hearing with the education of the deaf, so it is not inappropriate to consider the historical development of this subject here. Our knowledge of the teaching of the deaf both by signs and by lip-reading, goes back into the sixteenth century [84], so that, by 1800, there was considerable literature on the subject. There is a long record among the Spanish and French Catholic clergy of such teaching, and a somewhat later one of German Lutheran pastors, all of whom worked (primarily in one-on-one instruction), with love, care, and ingenuity. Often, however, their techniques were not handed down and valuable information was lost. By 1800, however, two schools of thought had emerged as to how the deaf should be educated.

The first of these was pioneered by a French priest, Abbé Charles Michel de l'Epée (1712–1789), who opened the first public school for the deaf in Paris in 1755 [85]. He emphasized sign language at this school, with little attention paid to lip-reading (or speech-reading, as it is now called). His associate, Abbé Roche Ambroise Sicard (1742–1822) (who had several close escapes from being murdered by the Paris mob during the revolution), continued this emphasis, so that the French school became almost completely one of manual or sign training.

This almost exclusive use of signs was challenged by a German school, beginning with Samuel Heinicke (1727–1790), a contemporary of de l'Epée, who opened a school for the deaf in 1778 in Leipzig, using oral methods exclusively. He was supported by John Graser (1766–1841), who required that only oral methods be used in the instruction in his school at Bayreuth. As we shall see in later chapters, this split has remained to the present day. The supporters of signs would argue that they are attempting to develop the most out of the individual child, while the oralists would maintain that they are attempting to bring the child into normal living with the nondeaf society. In many respects the arguments parallel those on the use of bilingual education for immigrant children as opposed to English only. In fact, one can see, in almost any area of human activity, the conflict between those who would organize their group on the basis of their common origin, race, religion, or ethnic origins, and those who would endeavor to integrate that group within the larger framework of society. There is, perhaps, no winner, but one hopes for an increased awareness of and tolerance for the two viewpoints.

Tartini Tones

Virtually all of acoustics studied before 1800 was linear acoustics, i.e., the differential equations used were all linear. That meant, of course, that the principle of superposition applied to any combination of waves or vibrations occurring at the same point at the same time. However, a phenomenon was observed in the eighteenth century that was later demonstrated to be nonlinear, and to have important consequences in physical acoustics, although this identification and application did not occur until much later. This is the phenomenon of Tartini tones.

Like many discoveries, there is considerable dispute as to the original discoverer. In 1748, Georg Sorge (1703–1778) [86] noted that, when two musical sounds of different pitch were sounded simultaneously and loudly, a tone was heard with a pitch equal to the difference in the pitches of the two tones. In 1754, the Italian violinist, Giuseppe Tartini (1692–1770) [87] noted the same result when playing two different tones loudly on the violin. Tartini also claimed that he had heard this difference tone, as it was later called, as early as 1714. To muddy the water still further, Jean Baptiste Romieu [88] reported that he had heard this "low harmonic" in 1751. Whatever the truth of the various French, German, and Italian claims, the sounds have since come to be known as Tartini tones or the Tartini pitch. Some typical cases are shown in Fig. 1-8 [89].

During the next half-century Lagrange [90], Chladni [91], and Young [92] studied the problem and concluded that it was a form of beats. Under the usual description of beats, when the difference in pitch is small (5–10 beats per second), we can distinguish them clearly. When the number of beats per second increases, we first distinguish the unpleasant sounds of dissonance, but as the number gets very large, they argued, we ultimately hear the pure tone of the difference frequency. While this interpretation was later proved to be incorrect, it satisfied the acoustics community for the next half-century.

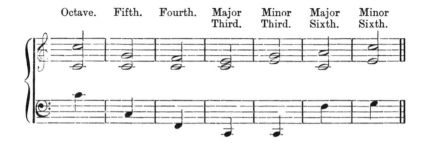

FIGURE 1-8. Tartini tones. (From H. von Helmholtz [83].)

Overview

In looking back over the period before 1800, one is humbled by the accomplishments of acousticians who worked with an almost complete lack of anything we would call apparatus. What they had were the human voice and ear, musical instruments, bells and tuning forks, vibrating strings and plates, the basic equations developed over the years, and a great deal of ingenuity and resourcefulness. In a day when we can scarcely add numbers without a calculator, or perform an experiment without vast arrays of electronic equipment, we can only marvel at the success of their ingenuity and resourcefulness.

Notes and References

Because of their frequent citing in the references, the books in the series *Benchmark Papers in Acoustics*, R.B. Lindsay, series editor. Dowden, Hutchinson, and Ross, Stroudsburg, PA (except for volume 13, for which the publisher is Van Nostrand, Reinhold, New York, NY) are listed below and will be cited in the chapter references *only* by "*Benchmark*" plus the volume number.

1. *Underwater Sound*, V. Albers (ed.).
2. *Acoustics*, R.B. Lindsay.
3. *Speech Synthesis*, J. Flanagan and L. Rabiner (1973).
4. *Physical Acoustics*, R.B. Lindsay.
5. *Musical Acoustics, Violin*, Part I. C.M. Hutchins.
6. *Musical Acoustics, Violin*, Part II, C.M. Hutchins.
7. *Ultrasonic Biophysics*, F. Dunn and W. O'Brien.
8. *Vibration*, A. Kalnins and C. Dym (1976).
9. *Musical Acoustics: Piano and Wind Instruments*, E. Kent.
10. *Architectural Acoustics*, T. Northwood.
11. *Speech Recognition*, M. Hawley (1977).
12. *Disc Recording*, W. Roys.
13. *Psychological Acoustics*, E. Schubert (1979).
14. *Acoustic Transducers*, I. Groves.
15. *Physiological Acoustics*, J. Tonndorf.
16. *Acoustical Measurements*, H. Miller.
17. *Noise Control*, M. Crocker.
18. *Acoustics of Bells*, T. Rossing.
19. *Nonlinear Acoustics in Fluids*, R.T. Beyer.

The author has endeavored to identify the first and last page of every article cited. Where he has failed, there is only the first page given. If the article is known to be only one page in length, the page reference is followed by a period. The reader can also mark the author's success (or lack thereof) in the citation of birth and death dates. Corrections and additions in either of these categories would be appreciated.

[1] F.V. Hunt, *Origins in Acoustics*, Yale University Press, New Haven, CT, 1978. Reprinted by the Acoustical Society of America, Woodbury, NY, 1993. Only part of Professor Hunt's manuscript was published at the time. It is the hope of the author of this present work to prepare the remaining portions for publication in the near future.

[2] Professor Hunt's original intention was to produce four Chapters, covering acoustics up to the present era. His sudden death in 1972 prevented this, but the incomplete material for Chapters 3 and 4 has been made available to the author through the courtesy of Professor Robert Apfel of Yale University, who edited the 1978 publication. Excerpts from this material will be used at various points in this book.

[3] C. Truesdell, *The Rational Mechanics of Flexible or Elastic Bodies* 1638–1788. Second Series, vol. XI, second section, *Leonhardi Euleri Opera Omnia*. Lausanne, 1960; op. cit., final volume, Lausanne, 1955.

[4] D.C. Miller, *Anecdotal History of the Science of Sound*, Macmillan, New York, NY, 1935, opposite pp. 24, 51.

[5] E.F.F. Chladni, *Die Akustik*, Breitkopf & Hartel, Leipzig, 1802. The name Chladni "(klad'-nee)" is Slavic in origin.

[6] E.F.F. Chladni, *Traité d'Acoustique*, Courcier, Paris, 1809.

[7] E.F.F. Chladni, *Traité d'Acoustique*, pp. v, vi. Thus, the paranoia that acousticians often suffer relative to the other fields of physics has been around for a long time.

[8] Alexander Wood, *Thomas Young, Natural Philosopher*, Clarendon Press, Oxford, UK, 1954, p. 61.

[9] Thomas Young, *Phil. Trans. Roy. Soc. (London)* **90**, 106–128 (1800). Reprinted in Ref. [10, pp. 531–554].

[10] Alexander Wood, *Thomas Young, Natural Philosopher*, p. 62.

[11] Thomas Young, *A Course of Lectures on Natural Philosophy and the Mechanical Arts*, Joseph Johnson, London, 1807, 2 vols.

[12] It goes almost without saying that the author would appreciate receiving word of any such errors.

[13] G. Sagredo, a letter to G. Galilei, 11 April 1615.

[14] A detailed account of these experiments, including their final interpretation in the twentieth century by Bruce Lindsay, is given in F.V. Hunt, Ref. [1, pp. 112–121].

[15] R.B. Lindsay, *J. Acoust. Soc. Am.* **39**, 629–644 (1966). Robert Bruce Lindsay, a Brown graduate and an MIT Ph.D., served long as Professor of Physics and Departament Chair at Brown and later as Dean of its Graduate School, but is best remembered by acousticians from his tenure as Editor-in-Chief of the *Journal of the Acoustical Society of America* from 1957 to 1985.

[16] R.B. Lindsay, *J. Acoust. Soc. Am.* **39**, 636 (1966).

[17] I. Newton, *Principia Mathematica* (*Mathematical Principles of Natural Philosophy*), Sec. VIII. Reprinted in *Benchmark/2*, pp. 75–86. See also, Thomas Young, Ref. [10, vol. 1, p. 370].

[18] Ref. [12, p. 84].

[19] F.V. Hunt, *Origins in Acoustics*, p. 151. Newton blamed it all on the "crassitude" of the air particles.

[20] T. Young, *A Course of Lectures on Natural Philosophy*, vol. 1, p. 371.

[21] R.B. Lindsay, *J. Acoust. Soc. Am.* **39**, 635 (1966).

[22] T. Young, *A Course of Lectures on Natural Philosophy*, vol. 2, p. 635. Bianconi's original work appeared in *Comm. Bonon*, **ii**, I, 265. See also E. Cherbuliez, *Mitt. Naturforsch. Ges. Bern*, 1870, p. 141ff.; 1871, p. 1ff; J.M.A. Lenihan, *Acustica*, **2**, 205–212 (1952).

[23] T. Young, *A Course of Lectures on Natural Philosophy*, vol. 1, p. 370.

[24] W. Derham, *Phil. Trans. Roy. Soc. (London)* **26**, 2–35 (1708).

[25] F.V. Hunt, *Origins in Acoustics*, pp. 111–112. See also D.C. Miller, *Anecdotal History of the Science of Sound*, pp. 37–38.

[26] E.F.F. Chladni, *Traité d'Acoustique*, p. 311.

[27] F. Hauksbee, *Phil. Trans. Roy. Soc. (London)*, No. 521. See also *J. Scavans*, 1678, p. 278.

[28] B. Peirce, *An Elementary Treatise on Sound*, James Munroe, Boston, MA, 1836, p. 44.

[29] E.F.F. Chladni, *Traité d'Acoustique*, p. 313.

[30] J. Canton, *Phil. Trans. Roy. Soc. (London)* **52**, 640–642 (1762); **54**, 261–262 (1764).

[31] Thomas Young, Ref. [10, vol. 1, p. 372].

[32] E.F.F. Chladni, *Traité d'Acoustique*, pp. 318–319.

[33] *American Institute of Physics Handbook*, McGraw-Hill, New York, NY, 3rd ed., 1972, pp. 3–104.

[34] T. Young, *A Course of Lectures on Natural Philosophy*, vol. 1, pp. 373–374.

[35] T. Young, *A Course of Lectures on Natural Philosophy*, vol. 1, p. 375.

[36] E.F.F. Chladni, *Traité d'Acoustique*, p. 284ff. See also Samuel Morland, *Phil. Trans. Roy. Soc. (London)* (1671), No. 79, VI 3056; J.H. Lambert, *Mém. de l'Acad. Berlin* (1763), p. 87.

[37] J.H. Hassenfratz, *Nicholson's Journal* **11**, 127–132 (1805); *Ann. Chemie* **53**, 64–75 (1805).

[38] J. Robison, *Encyclopaedia Britannica*, 3rd ed., 1799.

[39] E.F.F. Chladni, *Traité d'Acoustique*, Ref. 6, p. 293ff.

[40] One of the best examples of this "chatter" is provided in a side entry way to the Prado Museum in Madrid, where the chatter continues for a considerable period of time.

[41] E.F.F. Chladni, *Traité d'Acoustique*, Ref. [6, p. 276].

[42] Matthew Young, *Enquiry into the Principal Phenomena of Sounds and Musical Sounds*, G. Robinson, Dublin, 1784.

[43] Recent researches have tracked changes in the character of composed music over the ages as the churches, chamber, and concert halls have varied in their design. See Christopher Herr and Gary W. Siebein, *J. Acoust. Soc. Am.* **99**, 2530 (A) (1996).

[44] M. Mersenne, *Harmonicorum Libri XII*, G. Baudry, Paris, 1635; *Harmonie Universelle*, S. Cramoisy, Paris, 1636. The latter work was translated into English by J. Hawkins, *General History of the Science and Practice of Music*, A. Novell, London, 1853, pp. 600–616, p. 650ff.

[45] [Ref. 1, pp. 89–94].

[46] J. Sauveur, *Mém. Acad. Roy. Sci.* 297–300, 347–354 (1701). Translated into English by R.B. Lindsay, *Benchmark/1*, pp. 87–94.

[47] F.V. Hunt, *Origins in Acoustics*, pp. 59–60.

[48] E.F.F. Chladni, *Traité d'Acoustique*, Ref. [6, p. 7].

[49] This material appears as an Appendix to H. von Helmholtz, *On Sensations of Tone*. First (German) edition, Heidelberg, 1862. English translation by A.J. Ellis, reprinted by Dover, New York, NY, 1962. See also Alexander J. Ellis and Arthur Mendel, *Musical Pitch*, Frits Knuf, Amsterdam, 1968.

[50] Clearly they all could have used the services of the Standards Secretariat of the Acoustical Society of America. Standards in terminology and measurement of acoustical quantities has become an important feature of recent work in acoustics, and an elaborate structure of committees exists within the Acoustical Society of America (and also within the American Society of Mechanical Engineers) for the development and promulgation of such standards.

[51] E.F.F. Chladni, *Traité d'Acoustique*, pp. 67–68.

[52] Alexander Wood, *Thomas Young, Natural Philosopher*, p. 118.

[53] Thomas Young, *A Course of Lectures on Natural Philosophy*, Lecture XX, XIV, p. 312.

[54] Thomas Young, *A Course of Lectures on Natural Philosophy*, p. 313.

[55] A.T. Jones, *Sound*, p. 359.

[56] Thomas Young, *A Course of Lectures on Natural Philosophy*, p. 313.

[57] A discussion of some of these early machines is given by J.L. Flanagan, *J. Acoust. Soc. Am.* **51**, 1375–1387 (1972); reprinted in *Benchmark/3*, pp. 9–21.

[58] Both cited by Thomas Young. C.G. Kratzenstein, *J. Phys.* **21**, 358; *Acad. Petr.* 1789, **4**, II, H. 16. W.v. Kempelen, *Ueber den Mechanismus der Menschlichen Sprache*, J.V. Degen, Vienna, 1791. This volume was reprinted by Friedrich Frommann Verlag, Stuttgart, Germany, 1970. Its new preface makes reference to the work of Flanagan and, so a search on the part of the current author, extending over volumes from two centuries, led him back to a volume in this own office! See also R. Willis, *Trans. Cambridge Philos. Soc.* **3**, 131 (1830).

[59] See J. Flanagan, *J. Acoust. Soc. Am.* **51**, 1375–1387 (1972). According to Flanagan (loc. cit.), Kratzenstein was apparently the inventor of the mouth organ. It was said that his talking machine imitated the five vowels with "tolerable accuracy." Flanagan also noted that, while von Kempelen's machine was a serious scientific effort, he was not taken too seriously, for it was known that he had already perpetrated a major hoax in exhibiting his automatic chess player, which had a legless player concealed inside. The work of von Kempelen is also discussed in a paper by Homer Dudley and Thomas Tarnoczy, *J. Acoust. Soc. Am.* **22**, 151–166 (1950).

[60] Wheatstone's model was observed by Melville Bell, the father of Alexander Graham Bell, and the boy's interest was stimulated into trying to duplicate and improve upon the device. See Chapter 5.

[61] There are numerous references in the first book of Samuel to the playing of stringed instruments by King David.

[62] R.B. Lindsay, *J. Acoust. Soc. Am.* **39**, 630 (1966).

[63] J. Le R. D'Alembert, *Recherches sur le Courbe que Forme une Corde Tendue Mise en Vibration*, Royal Academy, Berlin, 1747, p. 214ff.

[64] D. Bernoulli, *Réflexions et Éclaircissements sur les Nouvelles Vibrations des Cordes, Exposées dans les Mémoires de l'Academie de 1747 et 1748*, Royal Academy, Berlin, 1755, p. 147ff.

[65] The principle of superposition states that the sum of any two solutions of a linear differential equation (such as the wave equation), say of different

frequencies, is also a solution of that equation. As such it provides a firm basis for the work of Fourier in analyzing functions (Chapter 2) and indeed of much of modern physics.

[66] L. Euler, *Dissertatio Physica de Sono*, E. and J.R. Thurnisis, Basel, 1727; *Opera Omnia*, series III, B. Teubner, Leipzig, 1926, vol. 1, pp. 182–196.

[67] J. Lagrange, *Oeuvres de Lagrange*, Gauthier-Villars, Paris, 1867, Vol. 1, p. 39ff.

[68] Ref. [5, p. 6].

[69] G.C. Lichtenberg, *Mem. Roy. Soc. Göttingen* (mentioned by Chladni, but no further reference given).

[70] J. Tyndall, *The Science of Sound*, 3rd ed., 1875. Reprinted by Citadel Press, New York, NY, 1964, p. 178.

[71] E.F.F. Chladni, *Entdeckungen über die Theorie des zu Klanges*, Weidman, Werben, and Reich, Leipzig, 1787.

[72] Ref. [6, p. 121].

[73] The strength of this fascination in the popular mind is reflected in the Civil War essay by Dr. Oliver Wendell Holmes *My Hunt After the Captain* (the search for his son, the future Supreme Court Justice, who had been wounded at the battle of Antietam); reprinted in Vol. VIII of *Holmes's Works*, Houghton Mifflin, Boston, MA, 1891, p. 19: "my thoughts . . . arranging themselves in curves and nodal points, like the grains of sand in Chladni's famous experiment."

[74] Ref. [6, p. vi].

[75] Ref. [4, p. 39]. See also A.T. Jones, *Sound*, p. 170.

[76] B. Higgins, *Nicholson's Journal, London*, pp. 129–131 (1802). The note also includes a comment by William Nicholson.

[77] B. Higgins, *Nicholson's Journal, London*, p. 130.

[78] A.T. Jones, *Sound*, pp. 223–227. Jones was a professor of physics at Smith College and carried out the task of producing the section "Contemporary References to Acoustics" of *JASA* until his sudden death in 1950. The author of this text began his association with the Acoustical Society of America by succeeding Professor Jones in this role.

[79] Ref. [6, p. 334].

[80] Dominico Cotugno, *De Aquaeductibus Auris Humanae Iternae Anatomica Dissertatio*, Typographia Simoniaa, Naples, 1761, pp. 18–20. The article has been translated into Italian by M. Mirolo, University of Bari, Italy, 1951, 70pp.

[81] A brief excerpt of Cotugno's paper has been published in English in *Benchmark*/15, along with a discussion of the work of both Cotugno and Meckel.

[82] Cooper, *Beethoven*, Appendix A, Beethoven's Illnesses, by Edward Larkin.

[83] Ref. [6, p. 343].

[84] Ruth E. Bender, *The Conquest of Deafness*, The Press of Western Reserve University, Cleveland, OH, 1960, This book covers the early history of teaching of he deaf in considerable detail.

[85] Richard Winefield, *Never the Twain Shall Meet*, Gallaudet University Press, Washington, DC, 1987. An historical background of the teaching of the deaf is given in this book, largely from the "signer's" viewpoint.

[86] G. Sorge, *Vorgemach der musicalische*, 1745–1747. An interesting history of the early studies of the phenomenon is given by A.T. Jones, *Am. J. Phys.* **3**, 49 (1935).

[87] G. Tartini, *Trattato di Musica*, Padua, 1754.

[88] J.B. Romieu, Assemblée publique de la Société Royale des Sciences tenue dans la grande salle de l'Hôtel de Ville de Montpelier, 16 Déc. 1751.

[89] H. von Helmholtz, *On Sensations of Tone*, p. 154.

[90] J. Lagrange, *Misc. Taurnens*, vol. 1, Sec. 64; *Ouevres de Lagrange*, Paris, 1867, vol. 1, p. 142.

[91] E.F.F. Chladni, Ref. [5, p. 253].

[92] T. Young, *Phil. Trans. Roy. Soc. (London)*, Jan. 16, 1800.

2
Acoustics 1800–1850

> There was a celebrated Fourier in the Academy of Sciences whom posterity has forgotten, and an obscure Fourier in I don't know what garret that the future will remember.
>
> Victor Hugo, *Les Miserables*

The lack of apparatus for acoustical research, mentioned in Chapter 1, continued well into the nineteenth century, since the field had to wait for the great development of electrical apparatus that began near the middle of that century. It is difficult, in our time, surrounded as we are by electric, electronic and optical devices, to appreciate the problems, both perceptual and experimental, faced by the acousticians of the early nineteenth century. They had few devices to produce sound other than the voice and musical instruments, almost no direct ways of visualizing the waveform of these sounds, and few mechanisms of reception or analysis other than the ear. But, as each advance in electricity and optics was made, acoustical applications could be perceived, and the field moved forward.

There were three major problems faced by acousticians of the early nineteenth century. Sound sources were needed whose frequencies could be simply controlled and measured. Then there was the need for identifying the detailed time dependence of these sounds and how they traveled through the acoustical medium. Finally, there was the need for instruments that could detect the sound in a consistent and measurable way. We shall see in this chapter what significant progress was made during this period.

2.1. Sound Production

Chladni Figures

Chladni figures were discussed in Chapter 1. The theoretical basis for these vibrations had not yet been established in 1800, and a prize was established by the French government for the person who could first work out the

FIGURE 2-1. A bust of Sophie Germain by Z. Astruc. (From Ref. [1].)

differential equation governing the behavior of these plates. This was won in 1815 by Sophie Germain (Fig. 2-1) (1776–1831) [1], who deduced the fourth-order differential equation involved. Mile. Germain was a French mathematician, who corresponded with the great scientists of her day, but who had no public career. She belonged to the same generation as Jane Austen, when public recognition of a woman was unlikely, if not impossible in those days. It should be noted that she is the first woman to be mentioned in this history; regrettably, there will be few others until we reach the later part of the twentieth century.

When Mlle. Germain sent some of her work to Gauss (when she was 21), that celebrated mathematician responded, writing

When a woman, because of her sex, or customs and prejudices, encounters infinitely more obstacles than men in familiarizing herself with their knotty problems, yet overcomes these fetters and penetrates that which is most hidden she doubtless has the most noble courage, extraordinary talent, and superior genius. [2]

Not a bad letter of recommendation! Mlle. Germain did make an error in setting up the boundary conditions for the vibrating plate, so that the correct solution of the problem was not obtained at that time. This had to wait until 1850, when Gustav Kirchhoff produced the correct boundary conditions and the solution [3].

Chladni's first observations on vibrating plates were made with sand, and he found that the particles accumulated on the nodal lines. Later, however, he found that, when he used very fine particles (shavings of hairs from the violin bow), these did not fall at the nodal lines, but were bounced about in the air, and came to rest (when the sound ceased) in the middle of the spaces between the nodal lines, i.e., in the vicinity of the antinodes.

The reason for this curious behavior was established by Michael Faraday (1791–1867) [4] in 1831. As he wrote in his paper,

When a plate is made to vibrate, currents are established in the air lying upon the surface of the plate, which pass from the quiescent lines towards the centres or lines of vibration . . . and then proceeding outwards from the plate to a greater or smaller distance, return towards the quiescent lines.

Thus, Faraday, whose name is mostly associated with electrical currents and electromagnetics, made his contribution to acoustics on the subject of air currents.

Faraday noted that the velocity of motion of these currents and how far they rise above the plate depend on the intensity of the vibration, the nature of the medium and other circumstances. Faraday had observed an effect that has since been called variously hydrodynamic flow, acoustic streaming and even quartz wind, during the century and a half that have followed.

The art of violin making as practiced by the craftsmen of Cremona has been a wonder of the musical world for more than three hundred years. As Carleen Hutchins (1911–   ) has written, "The master violin makers of the seventeenth and eighteenth centuries found by experience what sounds should be evoked when tapping different areas of a top or backplate when held at some prescribed point [5]." The later work of Chladni on making vibrations of a plate visible permitted the French scholar Felix Savart (1791–1841) to construct a simplified trapezoidal violin, with flat face and back plates, so that he could obtain Chladni figures for both of them. He also worked with some of the famous violins of Stradivarius and Guarnerius and, in the opinion of Carleen Hutchins, came close to employing the technical methods used for developing new violins today [6]. He also deduced the correct function of the sound post [7].

2.2. Artificial Sources of Sound

Before 1800 there was apparently little interest in discerning the way in which sounds were produced by the vocal cords, even though people had been peering down throats, with the aid of mirrors, for a considerable time. Major attention had been paid to the position of the mouth and lips during the production of vowels, and to the possible creation of machines that would produce such sounds such as those of Kratzenstein and von Kempelen, described in Chapter 1. In attempting to strike out in new directions, Robert Willis (1799–1878) wrote in 1830,

These considerations soon induced me, upon entering this investigation, to lay down a different plan of operations; namely, neglecting entirely the organs of speech, to determine, if possible, by experiments upon the usual acoustic instruments, what forms of cavities or other conditions are essential to the production of these sounds, after which, by comparing these with the various positions of the human organs, it might be possible, not only to deduce the explanation and reason of their various positions but to separate these parts and motions which are destined for the performance of their other functions from those which are immediately peculiar to speech. [8]

In 1799, John Robison developed a controllable sound source. Tyndall described Robison's work as follows:

A stop-cock was so constructed that it opened and shut the passage of a pipe 720 times in a second. The apparatus was fitted to the wind-chest of an organ . . . the sound g in alt was most smoothly uttered equal in sweetness to a clear female voice. When the sound was reduced to 360 . . . the sound was more mellow than any man's voice of the same pitch. [9]

Hunt observes [10] that the manner in which Robison was able to control the stopcock was perhaps the most interesting part of the experiment. It was suggested by Benjamin Peirce [11] in 1836 that the stopcock must have been driven by a high-speed rotator, perhaps similar to the one used by Savart [12]. It was, in effect, the first siren.

The name siren was applied to a sound source developed 20 years later by Charles Cagniard de la Tour (1777–1859) [13]. He prepared two disks, each perforated with equally spaced holes that formed concentric circles. He apparatus is described by Dayton Miller

One disk formed the top of the wind-chest while the other rotated close to the first, interrupting the flow of air from all of the holes in one circle at the same time, producing a smooth musical sound. The holes in the two disks were bored with opposite slants, so that the air pressure caused the moveable disk to rotate; the apparatus included a counting device to register the number of turns of the disk. [14]

Some early forms of the siren are shown in Fig. 2-2 [15].

It is a characteristic of technology that, once a device has been invented, there is a rapid succession of improvements. Such was the case with the siren. In 1840, August Seebeck made some improvements [16], and about the same time, Heinrich Wilhelm Dove (1803–1879) [17] invented the siren that bore his name, providing the rotating disk with four annular rows, each containing a different number of holes, and four independent nozzles. In 1862, Helmholtz (1821–1894) [18] arranged two Dove sirens on a single spindle, and added other improvements that made the device useful not only for determining pitch, but also for studying beats and other combinations of musical tones. Thus, the first of the three problems outlined at the start of this chapter had a solution. Sound sources with variable, known frequencies were now available.

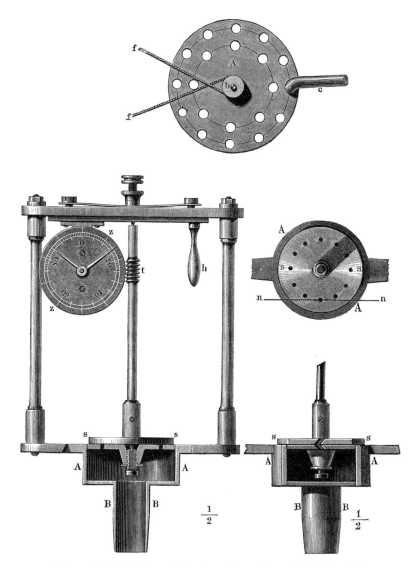

FIGURE 2-2. Early forms of the siren. (From H. von Helmholtz [15].)

2.3. The Human Voice

It was noted in Chapter 1 that both Chladni and Thomas Young were able to describe the larynx, and give a rough description of the origins of sound. This was put on a more systematic basis by the experiments of Johann Müller (1801–1858), Professor of Anatomy and Physiology at the University of Berlin [19]. After making various experiments with sound production in tubes and musical instruments, Müller performed experiments both on cadavers and living subjects, and concluded that "the sound of the voice is generated at the glottis, and neither above nor below this point [20]." He carefully removed the larynx from human cadavers, and by controlling both the tensions in the various muscles and the opening of the glottis, he was able to reproduce the different vowel sounds by blowing air from the trachea through the isolated larynx. Unfortunately, his careful and detailed work was largely neglected, and had to be rediscovered at a much later time.

2.4. The Tuning Fork

The tuning fork, which in this period became of very widespread use, was also used for determining pitch. In particular, Johann H. Scheibler (1777–1837) developed a tuning-fork "tonometer" [21] which consisted of some 56 tuning forks. One was adjusted to what he regarded as the mean pitch of the A above middle C of the concert pianos used in Vienna at the time (1834). A second fork was adjusted, no doubt by ear, to be one octave below the pitch of the first fork. The others were then adjusted, to differ successively by four vibrations per second above the lower A. Thus, he divided the octave into 55 intervals, each of about four vibrations per second. He then measured the number of beats in each interval as accurately as he could. The sum total of such beats thus give him the absolute frequency of the lower A. (This is in principle the same method used earlier by Sauveur.) As a result of his efforts, Scheibler determined the lower A to be 220 vibrations per second, with the higher A therefore at 440. He then recommended this pitch as the standard for musical pitch. It has been widely used as such, and is known as the Stuttgart pitch.

Moving in a somewhat different direction, F. Savart (1791–1841) used a toothed wheel [22], a device first suggested by Robert Hooke (1635–1703) [23]. A wheel with 600 teeth could be rotated 40 times per second, with the teeth striking against a card or wooden wedge, thus producing a sound of 2400 Hz. Savart used his device to attempt to measure the lower and upper frequency thresholds of hearing and arrived at values of eight vibrations/s^{-1} and 24,000 vibrations/s^{-1} respectively.

2.5. Sound Propagation

As noted in Chapter 1, the correct accounting for the velocity of sound in gases was still an unsolved problem in 1800. Laplace had suggested that there was a temperature variation in the wave, with an elevated temperature occurring at the points of condensation and a decrease in temperature at the points of rarefaction but the concept of adiabaticity had not yet been developed.

The Adiabatic Sound Velocity

What was needed soon came, in a great avalanche of publications. As early as 1699, Guillaume Amontons (1663–1705) had noted that

unequal masses of air increase equally in pressure with equal degrees of heat, and contrariwise, [24]

but this was not cast in the form of a general law relating the pressure with the temperature. In 1801, however, the English chemist John Dalton (1766–1844) [25] reported that

all elastic fluids under the same pressure expand equally by heat. [26]

In the following year, a young Frenchman Joseph Louis Gay-Lussac (1778–1850), published a report on the expansion of gases with increase in their temperature [27]. He observed that this phenomenon had been observed 15 years earlier by J.A.C. Charles (1746–1823), who had, however, not published his results. Gay-Lussac carefully dried nine different gases (Dalton had dried only his air sample), and concluded that, over the temperature range from 0° to 100°C, all the gases expanded by 1/266.66 of their volume per degree Celsius at 0°C. Thus, the volume could be expressed as

$$\frac{V(T)}{V_0} = 1 + \beta t, \tag{2.1}$$

where t is the temperature in degrees Celsius and β is the factor common to all gases, given by Gay-Lussac as 1/266.6, but now determined to be 1/273.16.

The next step came almost immediately. Jean-Baptiste Biot (1774–1862) [28] expressed the ambient density in the form $\rho' = \rho(1 + s)$, where s is the fractional change in the density (what we would today call the condensation). If the process in question was an isothermal one, the corresponding pressure change could be written as $P' = P(1 + s)$. Biot argued that the variation in temperature in a sound wave suggested by Laplace would alter the pressure, increasing it by a factor $(1 + ks)$, where k was an undetermined parameter. In his *Caloric History of Gases* [29], Robert Fox points out that this idea could have been connected with the proposal of Lagrange [30], some 40 years before Biot, that the pressure in the sound wave was proportional to a higher power of the density than one, i.e., $P < \rho^m$, where m, in Lagrange's estimate, was equal to about $1\frac{1}{3}$. Biot did not avail himself of this idea but, so long as s was very small, the two approaches were the same. Using his own expression for $P' = P(1 + s)(1 + ks)$, and neglecting higher powers of s, Biot arrived at the value

$$v = \left[\frac{P_0(1 + k)}{\rho_0}\right]^{1/2} \tag{2.2}$$

for the velocity of sound. Here we have introduced modern notation. In Biot's expression, P_0 was written $gHn\rho$, where g is the acceleration due to gravity, H the height of the mercury column of the barometer, and n the specific density of mercury. Working from the best available data for v and n, Biot arrived at a value $k = 0.392$ [31]. However, an independent determination of $1 + k$ (which was, in fact, the ratio of specific heats) had to wait for its experimental determination by Clément and Desormes in 1819 [32].

For our acoustical story, there is another important point. In a series of experiments, Dulong and Petit [33] established the relation

$$\frac{P(T)}{P_0} = 1 + \beta t, \tag{2.3}$$

where β is the same factor as in Eq. (2.1). If one writes

$$\frac{PV}{P_0 V_0} = 1 + \beta t \tag{2.4}$$

it can be noted that both Eqs. (2.1) and (2.3) are satisfied. It is pointed out by Bailyn [34] that the idea that $PV/P_0 V_0$ could be expressed as a function of temperature was apparently first suggested by André Marie Ampère (1775–1836) [35]. Thus, the structure of the ideal gas law approached its final form. From the acoustical point of view, the important fact is that the relationship of Eq. (2.4) could be used to convert the expression $v = [\gamma P_0/\rho_0]^{1/2}$ to the form

$$v = [\text{const} \times T]^{1/2}, \tag{2.5}$$

where $T = t + 1/\beta$ is the absolute temperature, demonstrating that the sound velocity in air could be expressed as a function of the temperature alone.

It was in the 1820s that the direct measurement of the velocity of sound in water was first made. Theoretical calculations by Laplace in 1816 suggested that the velocity in fresh water was 1525.8 m s^{-1} and in sea water 1620.9 m s^{-1} [36]. The fresh water value was similar to that obtained by Chladni slightly earlier (1494 m s^{-1}), but now it was time for an experimental measurement. In 1826, the Swiss scientist Jean-Daniel Colladon (1802–1893) undertook, along with the French mathematician Charles Sturm (1803–1855), to measure the sound velocity in the waters of Lake Geneva [37]. The description of Colladon's adventures with the customs agents, in his efforts to bring some explosive charges into Switzerland for the experiment, makes very amusing reading. In the actual experiment, Colladon's father lowered a bell into the water from one boat (Fig. 2-3(a)), and

FIGURE 2-3. Colladon's measurement of
the sound velocity in Lake Geneva. (a)
Sound receiver; and (b) sound source.
(From Colladon and Sturm [37].)

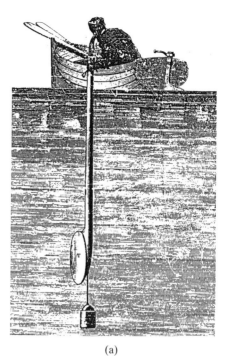

(a)

(b)

Colladon himself listened with an underwater listening tube at the other (Fig. 2-3(b)). The explosive, by this time successfully gotten across the border (with some degree of subterfuge), was placed on a plate and connected with the striking mechanism, so that, when the bell was sounded, there was a flash of explosive, observed by Colladon the younger in the second boat, and the time of transmission was measured from the instant of sighting of the flash. Colladon's final result was 1437.8 m s^{-1}. Lindsay observes in his editorial comment accompanying the reprinting of Colladon's description [38] that, with a lake temperature of 8° C, modern compressibility data yield a value of $v = 1438.8$ m s^{-1}, so that Colladon's results were quite accurate.

Some interesting remarks about diffraction were made at this time. It was well known that sound appeared to bend around corners and other obstacles in air. However, Colladon, in his work on the sound velocity in water just cited, noted that, if his hearing pipe was located in the water behind an object or wall screening it from the direct line to the source the received sound intensity was greatly diminished. The reason for this—the much longer wavelength of the sound wave in water—was not yet appreciated.

2.6. Velocity of the Sound in a Solid

The velocity of sound in a solid was measured directly for the first time by Biot in 1808 [39]. He used an iron water pipe in Paris. The sections of the pipe were attached together and formed a path about 1000 m in length. Biot sounded a bell at one end and had an observer at the other measure the time difference between the arrival of the sound through the metal and its arrival in air. Since the velocity of sound in air was accurately known, and also the length of the pipe, the time of flight could be measured accurately, and thus the sound velocity could be determined. Since the time of flight of the sound through the air measured close to 3 s, while the time difference between the two sounds was in excess of $2\frac{1}{2}$ s, the difference between the two times, which amounted to about 0.3 s, was subject to the usual inaccuracy involved in taking the difference between two large numbers, so the resultant measurement was rather approximate. But he did make the measurement. Apparently Biot never attempted to fill the pipe with water and thus achieve a direct measurement of the sound velocity in that medium.

In 1821–1822, Claude L.M. Navier (1785–1836) [40] and Augustin Louis Cauchy (1789–1857) [41] developed the equations of elasticity and in 1830, Siméon Denis Poisson (1781–1840) [42] used them to show that there can exist just two fundamental and independent modes of wave propagation through the interior of a homogeneous solid. One of these waves is a compressional wave (similar to a sound wave in a fluid and involving a volume change), while the second is a shear wave, and involves no change in the volume. The velocity of propagation of the compressional wave is greater than that of the shear wave, so that it arrives first in any situation. It is therefore called the primary or P wave, while the shear wave is known as the secondary or S wave. As we shall note in later chapters, these waves form the basis for the study of the seismic waves in the Earth.

2.7. Sound and Vibrations in Two Dimensions

In his 1800 paper, Thomas Young had discussed the state of affairs when two sound beams cross one another ("the coalescence of musical sounds" [43]). He criticized the work of an earlier author, one Smith, who maintained that the two intersecting beams crossed one another "without affecting the same individual articles of air by their joint forces [44]." Young adds, "undoubtedly they cross without disturbing each other's progress; but this can be no otherwise effected than by each particle's partaking of both motions." This, of course, follows from the principle of superposition, but that principle had not yet been formulated.

The idea of a particle being acted upon by two oscillating forces at right angles to one another caught the interest of researchers in this period. These included James Dean (1776–1849), a Professor at the University of Vermont [45], Hugh Blackburn (1823–1909) [46], Professor of Mathematics at the University of Glasgow, Nathaniel Bowditch (1773–1838) [47], the famed compiler of astronomical tables, Charles Wheatstone [48], and, finally, J. Lissajous (1822–1889) [49], who, more than 40 years after the first discovery, repeated the experiment and received the credit for finding what have since been known as Lissajous' figures (Fig. 2-4). In Bowditch's experiment, a pendulum was made of a bob, suspended from a cord supported at two points. By appropriate initial displacement the bob could be made to execute motions in a plane in the two directions that were in the ratio of two small numbers, the bob could execute simple two-dimensional patterns, which Bowditch reported.

2.8. The Stokes–Navier Equation

In his classic paper of 1845, Sir George Gabriel Stokes (1819–1903) (Fig. 2-5) [50] noted that, in the usual analysis of a sound wave, treated as a one-dimensional problem by Simon Denis Poisson (1781–1840) and others, attention was paid to forces acting on the bulk of the medium, in the direction of motion, but no attention was paid to any forces that act tangential to the direction of motion. Such forces are those provided by the shear viscosity.

The three-dimensional equation of motion of a viscous fluid is known as the Stokes–Navier equation [51]. Its history, of course, goes right back to

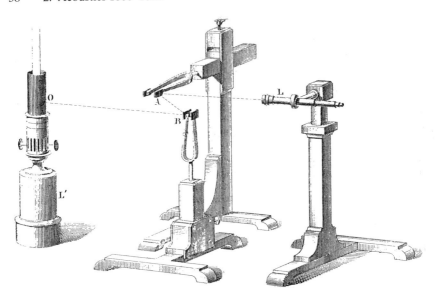

FIGURE 2-4. Lissajous' method of obtaining his figures. Two tuning forks, with mirrors attached, are arranged perpendicular to one another, providing displacements in two directions [49].

Newton, who observed in is *Principia* that "the resistance arising from the want of lubricity in the parts of a fluid, is ... proportional to the velocity with which the parts of the fluid are separated from one another [52]." In 1743, Bernoulli introduced the concept of internal pressure in this problem [53], and in 1752, Euler stated the principle of linear momentum: "the total

FIGURE 2-5. Sir George G. Stokes (1819–1903). (From D. Miller [77].)

force on a body is equal to the rate of change of the total momentum of the body [54]," where the word "body" refers to any element of a continuous medium that we care to examine in detail.

From this principle, Euler was able to state the equation of motion for an inviscid fluid. In 1822, Cauchy [55] added the concept of the stress tensor, which gave a framework for developing the equations of motion of any continuous medium. As Acheson points out, it only remained to add the constitutive relation describing the physical properties of the system in order to have the complete equations of motion. But this did not occur finally until Stokes wrote them down in 1845.

Actually, the progress was made on two fronts. Horace Lamb (1849–1934) [56] noted that

(the) dynamical equations were first obtained by Navier [57] and Poisson [58] on various considerations as to the mutual action of the ultimate molecules of the fluids. The method adopted above (considering the normal and tangential forces acting, in the case of a viscous fluid, on a rectangular element of that fluid), which does not involve any hypothesis of this kind, appears to be due in principle to de Saint-Venant [Adhémar de Saint Venant, 1797–1886] [59] and Stokes. [60]

In its general form, the Stokes–Navier equation involved two viscosity coefficients λ and μ, where μ is the ordinary shear viscosity. If we introduce a bulk viscosity coefficient $\mu_b = \lambda + 2\mu/3$ (which would represent the total viscous effect of a pure dilatation), then the Stokes–Navier equation for one dimension can be expressed as

$$\frac{\rho Du}{Dt} = -\frac{\partial p}{\partial x} + \rho g + \left(\mu_b + \frac{4\mu}{3}\right)\left(\frac{\partial^2 u}{\partial x^2}\right). \tag{2.6}$$

It is at this point that Rayleigh remarks in his book,

it has been argued with great force by Prof. Stokes, that there is no reason why a motion of dilatation uniform in all directions should give rise to viscous force. [61]

And since the time of Stokes, this viscous effect of a pure dilatation, the so-called bulk viscosity coefficient, has been set equal to zero, i.e., $\lambda = -2\mu/3$.

2.9. Sound Absorption in Fluids

Starting with this modified equation (2.6), Stokes was able to derive an expression for the sound absorption coefficient in a fluid. The resultant values were extremely small; Stokes had, perhaps, the advantage of not having his theoretical results compared with experiment until well after his death. It was then found that the experimental values were always larger than the theoretical, often enormously so. The Stokes assumption had therefore to be reexamined (see Chapter 7).

2.10. Finite-Amplitude Sound

At almost the same time as the appearance of the French edition of Chladni's book, an important paper was written by Poisson [62] in which he argued that the velocity of propagation of a longitudinal wave would be affected by the local particle velocity (often called the displacement velocity), so that the general form of a plane wave particle velocity u should not be $u = F(x \pm c_0 t)$, where c_0 is the ordinary velocity of propagation, but rather $u = F[x \pm (c_0 + u)t]$. That is, when the local particle velocity has the same direction as the sound wave, the propagation velocity is increased, and, conversely, it is decreased when the local particle velocity is in the direction counter to that of the sound wave.

This work by Poisson was not exploited at the time, but in 1848, John Challis (1803–1881) [63] pointed out a problem that the logical application of Poisson's result would produce. Since the portions of the wave with the high positive velocity would be moving more rapidly than those portions in front of it with smaller (or even negative) particle velocities, a point would be reached at which the equation would become multivalued and therefore, in Challis' view, nonsensical. This problem, called a "difficulty" by Stokes [64], was pictured by him as shown in Fig. 2-6. However, Stokes also pointed out that the theoretical development of Poisson assumed an ideal, lossless medium, and suggested that the presence of sound absorption (as we have seen, Stokes was also working on this problem) would be enough to prevent the discontinuity (and the "difficulty") from arising.

In a paper published in the following year, Sir George Airy (1801–1892) [65] noted that the differential equation governing the motion was a nonlinear one, and in the case of water, the "difficulty" would lead to the appearance of higher harmonics of the fundamental frequency of the wave, and, eventually, to the phenomenon in tidal motion known as a "bore," which was well known at the time. Thus, in the early days of the study of acoustics, the sophisticated problem of nonlinear propagation and the possible creation of discontinuities in the wave were given their initial considerations. No solutions were worked out, but the studies were under way.

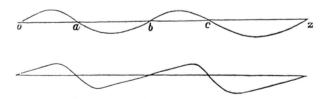

FIGURE 2-6. Steepening of the wavefront due to finite amplitude of signal. (From Stokes [64].)

2.11. Sound Reception

Work of Charles Wheatstone

The name of Charles Wheatstone (later Sir Charles, 1802–1875) is usually associated with the electrical bridge that bears his name [66], but he did considerable work in acoustics. He was the son of a well-known musician of the period, and started his career as the maker and seller of musical instruments in London. It was during this period that he concocted a demonstration called the "Enchanted Lyre," in which he arranged two sounding boards in different rooms, and connected them with a wooden rod. When a musical instrument was placed in contact with one of the boards and played, the music sounded in the other room. Wheatstone had a flair for showmanship and constructed the visible sounding board in the form of a lyre, hence its name (Fig. 2-7) [67].

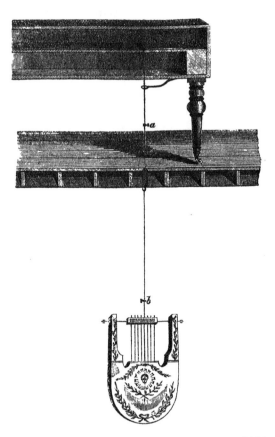

FIGURE 2-7. Wheatstone's lyre. (From Wheatstone [67].)

Wheatstone was soon writing papers on acoustics, studying the passage of sound through solids, including the observation of the phenomenon of polarization of sound (1823) [68] and transmission of sound through the bone of the skull, in which paper he developed the use of the word "microphone" for a device that resembled the stethoscope [69]. He then studied a number of musical instruments, including the jew's-harp and the mouth organ and, in 1829, he invented the concertina, which is similar to the accordion. In 1833, he gave an extended study of the phenomenon of Chladni figures, and, broadening his interests, measured the speed of propagation of electricity. In these two papers, he established himself as a scientist of significance, and in 1834 he was appointed as Professor of Experimental Philosophy at King's College, London, a position he held for the rest of his life.

One of his most useful contributions to acoustics, however, was the introduction of the rotating mirror to study periodic events. We mentioned the phenomenon of the singing flame in Chapter 1, a phenomenon that Faraday thought arose from a series of explosions of the hydrogen gas used to fuel the flame [70]. One could, in fact, sweep one's eyes rapidly past the flame and see that the height of the flame is changing regularly during the process. Wheatstone then proposed the use of a mirror such as that shown in Fig. 2-8 [71]. By rotating the mirrors at appropriate speeds, one could, by following the reflections, see the flame rise and fall.

Such a rotating mirror was later given widespread use in optics, in measuring the velocity of light in water by Foucault in 1850, but it was rapidly replaced in acoustics by the techniques introduced by August Toepler (1836–1912) (see Chapter 6).

The Stroboscope

In 1836, a Belgian scientist, Joseph Plateau (1801–1883), reduced to practice an idea that had floated around since antiquity, that of viewing something at regularly spaced short intervals, and thereby produced stroboscopic observation for the first time [72]. According to Miller [73], the idea goes back to the Roman poet Lucretius (94–55 B.C.). There are two ways of employing the effect: one can illuminate the object periodically, or use a sectored, rotating disk (as Plateau did) and get periodic short glimpses of the object. The technique was put to acoustic use in the 1860s by Toepler and von Helmholtz in the visualization of sound.

2.12. The Doppler Effect

Acoustics profited much from gains in optical spectroscopy in this period. Studies of double stars by astronomers in the early 1800s had indicated that the characteristic lines of a discrete spectrum emitted from such double

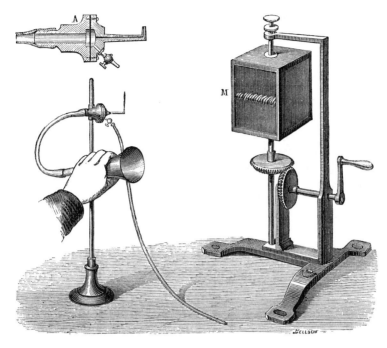

FIGURE 2-8. Apparatus for production of images of vowels, Koenig's version. (From Koenig [71].)

stars were displaced in the spectrum by a small amount relative to the same characteristic lines measured in the laboratory on Earth. Thus, the frequency of the lines from one of the stars became higher and from the other, lower. In 1842, an Austrian, Christian Doppler (1803–1853) (Fig. 2-9) published a short book on the light from such double stars, in which he presented the conclusion that the perceived frequency of the light from the distant star would be shifted from its objective value if the star and the observer were moving relative to one another. Hence, if the double star was rotating as a unit, one-half would be approaching the terrestrial observer, while the other half would be receding. Doppler then showed that the total relative shift in frequency would be equal to $2v/c$, where v was the speed of approach (or recession) and c the speed of light [74]. In his book, Doppler pointed out that this shift was a characteristic of wave theory and should apply equally well to sound waves.

This idea was quickly taken up by a prominent Dutch meteorologist, C.B.J. Buys Ballott (1817–1890) [75]. In what was perhaps the first application of the steam engine to a problem of acoustics, Buys Ballott obtained the services of a locomotive of the Rhine Railroad and carried out a series of what must have seemed very bizarre tests on a portion of the track

FIGURE 2-9. Christian Doppler [74].

between Utrecht and Maarsen in 1845. First, trumpeters were stationed on a car accompanying the locomotive, while other musicians stood at fixed points along the track. The musicians on the ground estimated the shift in pitch of the trumpets as the train approached and then receded from the observation point. The speed of the locomotive was recorded (from 30 to 45 miles per hour), and the pitch change was estimated in eighths or sixteenths of a tone. Buys Ballot was careful to avoid the change in the angle of approach when the train was near the observer (so that the component of the velocity in the direction of the observer was reduced substantially by a cosine factor) by restricting all his measurements to distances greater than 20 m from the stationary point of observation.

In a second set of experiments, the trumpeters stood on the ground and the observers rode on the train. The results of both experiments fully confirmed Doppler's theory. While the method of observing the frequency shift seems extremely approximate when compared with present-day methods, Buys Ballott found consistency among the measurements and his results were quickly accepted as proof of the Doppler effect in acoustics.

2.13. Fourier Analysis

In 1822, Baron J.B.J. Fourier (1768–1830) (the "celebrated" Fourier in the Hugo quotation at the beginning of this chapter, Fig. 2-10), published his book on *Théorie Analytique de la Chaleur* (Analytic Theory of Heat) [76], in which there was no acoustics, but which, nevertheless, has had a profound effect on our knowledge of the field. In it, he demonstrated that a finite and continuous periodic motion can always be decomposed into a

FIGURE 2-10. J.B.J. Fourier (1768–1830). (From D. Miller [77].)

series of simple harmonic motions of suitable amplitudes and phases. This idea was soon picked up by those in the acoustics community. In 1843, Georg Simon Ohm (1789–1854) stated, as a law of acoustics [78], that all musical tones are periodic functions of the time and that the ear is capable of analyzing any musical tone into its separate component simple tones, each of which can be distinguished by the ear. In a sense, the ear is its own Fourier analyzer. While it was not yet possible in 1844 to portray the waveform of a complex tone, Ohm and Fourier both indicated that these components exist and could be sorted out if the actual waveform were somehow known or measured. In his paper, Ohm considered a sound pulse impinging on the ear. He then represented this impulse, à la Fourier, in a series of sines and cosines whose arguments contain a base frequency (corresponding to the pitch of the musical tone) and its multiples. Thus, the pulse consists of a fundamental frequency and a collection of higher harmonics, all of which are multiples of the fundamental.

Lord Rayleigh, in his book *The Theory of Sound*, remarked

The reason of the preeminent importance of Fourier's series in Acoustics is . . . that, in general, simple harmonic vibrations are the only kind that are propagated through a vibrating system without suffering decomposition. [79]

In any event, we can safely say that Hugo's deprecating remarks about "our" Fourier have proved to be greatly in error.

2.14. The Stethoscope

In was noted in Chapter 1 that horns of various shapes had long been used for projecting a beam of sound in a specific direction, as an aid to the deaf, and also for listening to faint sounds. A simple application of this latter use was made in 1816 by a French physician, René Laennec (1781–1826). In his paper, Laennec wrote

In 1816, I was consulted by a young woman laboring under severe symptoms of diseased heart, and in whose case percussion and the application of the hand were of little avail on account of the great degree of fatness. . . . I happened to recollect a simple and well-known fact in acoustics and fancied, at the same time, that it might be turned to some use on the present occasion. The fact I allude to is the augmented impression of sound when conveyed through certain solid bodies. . . . Immediately, on this suggestion, I rolled a quire of paper into a sort of cylinder and applied one end of it to the region of the heart and the other to my ear, and was not a little surprised and pleased, to find that I could thereby perceive the action of the heart in a manner much more clear and distinct than I had ever been able to do by the immediate application of the ear. [80]

Laennec started out with the idea of using a more or less solid cylinder, but quickly switched to a hollow tube and, indeed, attached a funnel-shaped end to the part closest to the heart, and this instrument he called the "cylinder" or the "stethoscope." From such a simple beginning developed what must be the physician's most identifying piece of equipment.

2.15. Hearing

Modern texts on hearing make very few references to the work done in this period. We have already mentioned one important contribution, that of Georg Ohm (1843) on the ability of the ear to analyze a complex sound into its component simple tones. Even earlier (1826), M. Delezenne [81] studied in detail the ability of listeners to distinguish the pitch of musical tones. In Delezenne's experimental approach, he would present his listeners with a pair of tones (on a single string sonometer), and ask his subjects "to judge whether the interval was sharp or flat as he moved the sonometer bridge (presumably randomly) from the position specified for the mathematically defined position for the given interval [82]." Delezenne's experimental procedure was well ahead of his time. In limiting his listeners to a simple "sharp or flat" response, he anticipated twentieth-century researchers, and in giving the results of his listener's sense of mistuning (of octaves, fifths, etc.), in terms of log units of the musical comma, he anticipated the work of Fechner and his associates at the end of the century (see Chapter 6).

Other advances were made in physiological acoustics. To assist the reader in understanding the development of research on the structure of the ear, we supply two relatively recent drawings of the ear structure. Figure 2-11 is von Békésy's representation (1960) of von Helmholtz's drawing of the ear (1863), a two-dimensional picture that gives an accurate, although simplified, picture [83]. The drawing by Max Brödel (Fig. 2-12) is more vivid, but gives no hint as to the structure of the cochlea [84]. We shall refer to

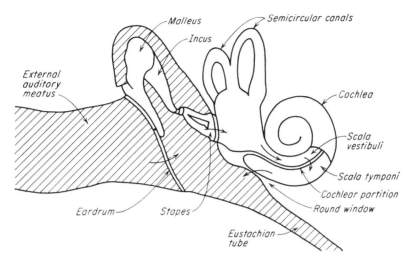

FIGURE 2-11. The ear. The arrows indicate the paths of a sound. (From G. von Békésy [83].)

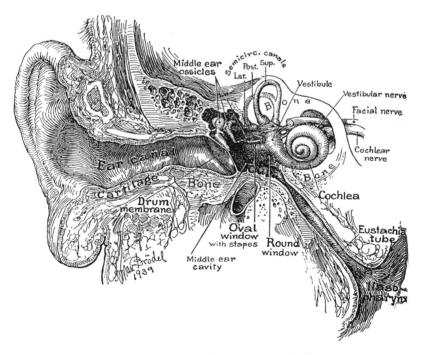

FIGURE 2-12. The ear. (From M. Brödel [84].)

these drawings in discussing the work, not only of the period 1800–1850, but of later periods.

In 1824, Flourens found that the vestibular receptors (the structure along the walls of the vestibula, Fig. 2-11) take no part in the auditory process, but assist in the sense of balance [85]. Edouard F. Weber (1834) and H. Huschke (1844) advanced the knowledge of the structure of the cochlea, spotting a kind of teeth on its walls, but made no further connection of this structure with the hearing process [86].

In 1837, Carl Gustav Lincke made the drawings of the cochlea shown in Fig. 2-13 [87]. The first (a) is a cross section of the entire unit, showing the various cross section of the cochlea structure from one turn to the next, while the second (b) shows the cochlear partition. In each turn of the cochlea, the basilar membrane is evident, but further detail was not observed. von Békésy noted that type of picture shown in Fig. 2-13(b) results from having the fluid escape when the cochlea is opened, with the result that the tissues dry out [88]. The technique of fixation of the tissue sample had not yet been developed in the early 1800s.

2.16. Room Acoustics

The only apparent advance in room acoustics in the first half of the nineteenth century (except for the work of Joseph Henry to which we shall come in Chapter 3.) was provided in a small pamphlet by H. Matthews in 1826 [89]. Its author found extensive fault with the acoustics of churches, courts, and auditoria. He accurately described the interfering role of what would later be called reverberation. He notes the ability of a flat wall to reflect the sound, and if the sound strikes the wall obliquely, that it is possible for the listener to hear sound from two sources: the direct sound and that which has been reflected from the wall, which arrives later. He then expanded on this theme:

This (the phenomenon just described) is the principle which destroys the distinctness of sound in churches, chapels and all large edifices; the high side walls, the dome or flat ceiling and the broad flat fronted galleries, being as close to the speaker, echo does not politely wait, like as for the distant rock, until the speaker has done; but the moment he begins, and before he has finished a word, she mocks him as with ten thousand tongues. That part of the gallery, which is nearest, commences the mockery first . . . (then) the nearest part of the ceiling begins; then the more distant parts claim to be heard; the walls next . . . so that at the time the real voice is sounding the letters at the end of the word, some of these echoes are sounding those of the beginning and the middle. [90]

Matthews recognized that sound is little absorbed by the air during its passage through a room or hall, although it is absorbed by rugs and drapes. He also knew that the sound path could be altered by temperature changes

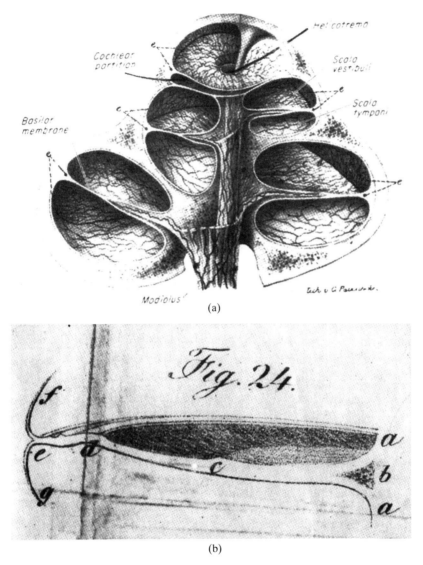

FIGURE 2-13. Drawings of the cochlea. (a) Central cross section; and (b) the cochlear partition. (From Lincke [87].)

and air currents (the heat radiated by the congregation warms the air, causing the air to rise, and with it the sound moves over the heads of the congregation (and with it, no doubt, the sermon) [91]. His suggested solution of the problem, involving special shapes of the ceiling and balconies, and special air currents to move the sound toward the audience, did not in fact, provide a successful design. The problem was correctly stated; the solution remained unknown.

2.17. Teaching of the Deaf

In 1815, a student at the Andover Theological Seminary, Thomas Hopkins Gallaudet (1787–1851) was sent abroad by Boston businessmen to learn the best techniques for teaching the deaf [92]. in order to establish a school in the United States for such purposes. He first went to England to the Braidwood School which followed the oralist philosophy, but he was not well received there (teaching methods were apparently a trade secret), and he went on to Paris, where he was welcomed by Abbé Sicard. Gallaudet spent 3 months in Paris, absorbing the sign language methods, and brought one of the French teachers, Laurent Clerc, back to America with him, to teach in the American Asylum for the education of the Deaf and Dumb, which was begun in Hartford, CT, in 1817. Clerc became a vigorous proponent of sign language, as was Gallaudet, who passed this point of view on to his son, Edward Miner Gallaudet (1837–1917). Edward Gallaudet founded the National Deaf–Mute College in Washington, with federal funding. Later, he changed the name to Gallaudet College (now Gallaudet University) in honor of his father.

It is of interest that two other prominent Americans, Horace Mann (1796–1859) and Samuel Gridley Howe (1801–1876) [93] went to Germany in 1843 to study techniques for teaching the deaf. They returned enthusiastic supporters of the oralist methods. There was some delay before institutions were developed, but the New York Institution (now the Lexington School for the Deaf) and the Clarke Institution for Deaf Mutes, in Northampton, MA, were established in the 1860s. These two were both oralist institutions. Thus, the signing and the oralist camps each had their own institutions, and conducted their own private civil war with one another.

Overview

From the modern viewpoint, the work in acoustics in the first half of the nineteenth century was scattered, and with little coherence of one area with another. Little of the research was at academic or industrial institutions, but the work of Airy, Stokes, and others at the University of Cambridge was leading the way to the future, and such towering figures as Henry, von Helmholtz, and Tyndall were just beginning their great careers.

Notes and References

[1] Louis L. Bucciarelli and Nancy Dworsky, *Sophie Germain*, Reidel, Boston, MA, 1980.
[2] Letter to Sophie Germain, dated 30 April 1807. Translation appears in Louis L. Bucciarelli and Nancy Dworsky, *Sophie Germain*, p. 25.

[3] G. Kirchhoff, *Ann. Physik* **81**, 258–263 (1850).

[4] M. Faraday, *Phil. Trans. Roy. Soc. (London)*, 299–318 (1831).

[5] *Benchmark/6*, p. 12.

[6] *Benchmark/6*, p. 13.

[7] Neville H. Fletcher and Thomas D. Rossing, *The Physics of Musical Instruments*, Springer-Verlag, New York, NY, 1991, p. 336.

[8] R. Willis, *Trans. Cambridge Philos. Soc.* **3**, 233–234 (1830).

[9] J. Robison, *Encyclopaedia Britannica*, 3rd ed., 1799; cited by J. Tyndall, *On Sound*, D. Appleton, New York, NY, 1867. The "windchest" was the source of compressed air.

[10] F.V. Hunt, unpublished manuscript, p. 3/29.

[11] B. Peirce, *An Elementary Treatise on Sound*, James Munroe, Boston, MA, 1836, p. 67.

[12] F. Savart, *Ann. Physik* **20**, 290–304 (1830).

[13] C. Cagniard de la Tour, *Ann. Chim. Phys.* **12**, 167–171 (1819).

[14] D.C. Miller, *Anecdotal History of Sound*, Macmillan, New York, NY, 1935, p. 54.

[15] H. von Helmholtz, *On Sensations of Tone*, 3rd ed., Longmans, Green, London, 1875. Reprinted by Dover, New York, NY, 1954, pp. 11, 12.

[16] August Seebeck, *Ann. Physik* **53**, 417–436 (1841).

[17] H.W. Dove, *Ann. Physik* **82**, 596–598 (1851). Dove was a German; his last name has two syllables (doh-vay).

[18] H. von Helmholtz, On *Sensations of Tone*, pp. 11, 13, 162, 182, 372.

[19] Johann Müller, *Elements of Physiology*, translated from the German by William Baly, Taylor and Walton, London, 1842, vol. 2, p. 1002ff.

[20] Johann Müller, *Elements of Physiology*, p. 1003.

[21] Johann Haunch Scheibler, *Der physikalische und musikalische Tonmesser*, G.D. Bädeker, Essen, 1834. His work is cited in H. von Helmholtz, *On Sensations of Tone*, pp. 199, 443.

[22] Ref. [7].

[23] Cited by Lindsay in his essay "The Story of Acoustics," *J. Acoust. Soc. Am.* **39**, 629–644 (1966). This article is reproduced in *Benchmark/2*, pp. 5–21.

[24] Quoted in F.V. Hunt, *Origins in Acoustics*, Yale University Press, New Haven, CT. 1978. Reprinted, Acoustical Society of America, Woodbury, NY, 1994, p. 55.

[25] J. Dalton, *Mem. Proc. Manch. Lit. Phil. Soc.* **5**, part 2, 595–602 (1802). The paper was actually read before the society in Oct. 1801.

[26] J. Dalton, *Mem. Proc. Manch. Lit. Phil. Soc.*, p. 602.

[27] J.L. Gay-Lussac, *Ann. Chim.* **43**, 137–175 (1802). There appears to be no evidence that Gay-Lussac was aware of the work by Dalton. See the excellent account of the work on gases in this period by Robert Fox, *The Caloric History of Gases, from Lavoisier to Regnault*. Clarendon Press, Oxford, UK, 1971, especially Chapters 1–5.

[28] J.B. Biot, *J. Phys.* **55**, 173–182 (1802).

[29] R. Fox, *The Caloric History of Gases, from Lavoisier to Regnault*, p. 83.

[30] J. Lagrange, *Mélang. Phil. Math. Soc. R. Turin* **2**, 153–154 (1760–1761).

[31] J.B. Biot, *Traité Phys*, vol. 1, p. 140.

[32] N. Clement and C.B. Desormes, *J. Phys.* **89**, 321–346, 428–455 (1819). There is in the literature some amusing counterplay regarding the names of these

authors. In the French *Nouvelle Dictionnaire Biographique*, Didot, Paris, 1860, vol. 10, p. 794, it is stated that Clement-Desormes was one person, but that an earlier index had unfortunately divided him into two. The apparent solution is given in the book by Fox, Ref. [6]. It seems that Clement worked with Desormes, then married the daughter of Desormes and, in the later part of his career, hyphenated his names, so that Clement, Desormes, and Clement-Desormes all existed at one time or another!

[33] P.J. Dulong and A.T. Petit, *Ann. Chem.* **2**, 240–263 (1819).

[34] M. Bailyn, *A Survey of Thermodynamics*, AIP Press, Woodbury, NY, 1994, p. 35. This book gives an excellent account of the historical development of our ideas of heat and temperature. Another important text on this subject is that of D.S.L. Cardwell, *From Watt to Clausius*, Cornell University Press, Ithaca, NY, 1971, especially Chapter 5.

[35] A.M. Ampère, *Ann. Chim.* **94**, 145–169 (1815).

[36] P.S. Laplace, *Ann. Chim. Phys.* **111** (1816). See also P.S. Laplace, *Collected Works*, vol. 14, p. 293.

[37] J.-D. Colladon and J.K.F. Sturm, *Ann. Chim. Phys.* **36**, 113–159, 225–257 (1827). See also J.-D. Colladon, *Souvenirs et Mémoires: Autobiographie de J. Daniel Colladon*, Aubert–Schuchardt, Geneva, 1893.

[38] *Benchmark/2*, pp. 194–201.

[39] J.B. Biot, *J. Mines* **24**, 319–320 (1808). There is some confusion here. Miller lists as a reference, Biot. *Ann. Chim. Phys.* **13**, 5 (1808); and Lindsay also cites this same reference. However, there is no such article in that journal in the period 1800–1820. The note in the *Journal des Mines* is very brief, but it covers the same subject and gives the basic results. (The Royal Society *Catalogue of Scientific Literature* incorrectly lists the pages of this latter article as 310–320.)

[40] C.L.M. Navier, *Mém. Acad. Sci. Paris* **7** (1827).

[41] A.L. Cauchy, *Bull. Sci. Soc. Philomathique*, 1823; *Exercices de Mathematique*, 1827, 1828.

[42] S.D. Poisson, *Mém. Acad. Sci. Paris* **10**, 578–605 (1831). An excellent historical coverage of this early work on elasticity is found in A.E.H. Love, *A Treatise on the Mathematical Theory of Elasticity*, 3rd ed., Cambridge University Press, Cambridge, UK, 1920, Chapter 1.

[43] Thomas Young, *A Course of Lectures on Natural Philosophy and the Mechanical Arts*, Joseph Johnson, London, 1807, p. 544.

[44] Smith, cited by T. Young, loc. cit.

[45] J. Dean, *Mem. Amer. Acad. Arts Sci.* **3**, 241 (1815).

[46] This is the origin of the Blackburn pendulum, which is cited by Thomas Young, loc. cit., as well as by A.T. Jones, *Sound*, p. 128.

[47] N. Bowditch, *Mem. Amer. Acad. Arts Sci.*, Ser. 1, **3**, 33–37 (1815).

[48] C. Wheatstone, *The Scientific Papers of Sir Charles Wheatstone*, The Physical Society, London, 1879.

[49] J. Lissajous, *Ann. Chim. Phys.* **51**, 147–231 (1857).

[50] G.G. Stokes, *Cambridge Philos Soc. Trans.* **8**, 287–325 (1848).

[51] For the fundamentals of the Stokes–Navier equation see D.J. Acheson, *Elementary Fluid Dynamics*, Oxford University Press, Oxford, UK, 1990, Chapter 6; and also C.-S. Yih, *Fluid Dynamics*, West River Press, Ann Arbor, MI, 1977, Chapter 2.

[52] I. Newton, *Principia Mathematica*, translated from the Latin by Andrew Motte, H.D. Symonds, The Physical Society, London, 1803, vol. 2, p. 146. The original Latin read

Resistentiam quae oritur ex defectu lubricitatis partium fluidi caeteris paribus, proportionalem esse velocitati, qua partes fluidi separantur ab invicem.

[53] D. Bernoulli, *Hydrodynamica sive viribus et motibus fluidorum commentarii.* Argent, 1738.

[54] L. Euler, *Hist. Acad. Berlin*, 1755. Translation and commentary by C. Truesdell, Editor's Introduction to *Euleri Opera Omnia*, Ser. II, vol. 12, Orell Füssli, Zurich, 1954.

[55] A. Cauchy, *Mém. Acad. Roy. Sci.* **1** (1827).

[56] H. Lamb, *Hydrodynamics*, 6th ed., Dover, New York, NY, 1945, p. 577.

[57] C.L. Navier, *Mém. Acad. Sci. Paris* **6**, 389 (1822).

[58] S.D. Poisson, *J. École Polytech.* **13**, 1 (1829).

[59] J. de Saint-Venant, *Compt. Rend.* **17**, 1240–1243 (1843).

[60] G.G. Stokes, *Trans. Cambridge Philos. Soc.* **8**, 287 (1845).

[61] Rayleigh, *The Theory of Sound*, vol. 2, p. 314.

[62] S.D. Poisson, *J. École Polytech.* **7**, 364–370 (1808). The part of the paper discussed here appears in English translation in *Benchmark*/19, pp. 23–28.

[63] J. Challis, *Phil. Mag.*, Ser. 3, **32**, 494–499 (1848).

[64] G.G. Stokes, *Phil. Mag.*, Ser. 3, **33**, 349–356 (1848). As the modern expression goes, "every difficulty is an opportunity."

[65] G.B. Airy, *Phil. Mag.*, Ser. 3, **34**, 401–405 (1849).

[66] The Wheatstone bridge was invented by a man named Christie, as Wheatstone always pointed out, but the name of Wheatstone has been attached to it ever since. Geoffrey Hubbard, *Cooke and Wheatstone and the Invention of the Electric Telegraph*, Routledge and Kegan Paul, London, 1965, p. 98.

[67] G. Hubbard, *Cooke and Wheatstone*, p. 15. Hubbard remarks that Wheatstone "was able to give the impression of a high-minded attachment to science while at the same time seldom doing anything that was not to his own financial advantage. The attachment to science was genuine and substantial. So were the financial advantages" (pp. 15–16). This demonstration was repeated by Tyndall in his lectures at the Royal Institution (see Chapter 3) and is often associated with him.

[68] C. Wheatstone, *Ann. Phil.* **6**, 87–90 (August, 1823).

[69] G. Hubbard, *Cooke and Wheatstone*, p. 17.

[70] M. Faraday, *Quart. J. Science Arts*, **5**, 174 (1818).

[71] Illustration is from R, Koenig, *Quelques Expériences* d'Acoustique, A. Lahure, Paris, 1882, p. 57. See also C. Wheatstone, *Phil. Trans. Roy. Soc.* **124**, 586 (1834).

[72] J.A. Plateau, *Bull Acad. Roy. Sciences Belles-Lettres Bruxelles*, **3**, 365 (1836). The term stroboscopic was apparently first used by Simon Stampfer [*Jahb. Kaiserlichen Königlichen Polytech Inst. Wien* **18**, 237 (1834)]. Others who thought of the idea are mentioned by A.T. Jones in his book (*Sound*, Van Nostrand, New York, NY, 1937, pp. 436–437), but he assigned the credit for the actual device to Plateau.

[73] D.C. Miller, *Amecdotal History of Sound*, pp. 52–53. The early work is also discussed in F.V. Hunt, Ref. [5, p. 3/36] and E.G. Richardson, *Sound*, Edward Arnold, London, 1940, pp. 79–84.

[74] C. Doppler, On the Colored Light of Double Stars and Other Stars of the Heavens. *Abh. Böhmischen Ges. Wiss.* 5th ser., vol. 3 (1842).

[75] C.B.J. Buys Ballott, *Ann. Physik* **66**, 321–350 (1845).

[76] J. Fourier, *Theorie Analytique de la Chaleur*, Didot, Paris, 1822. English translation, *The Analytic Theory of Heat*, Cambridge University Press, Cambridge, UK, 1878.

[77] D.C. Miller, *An Anecdotal History of Sound*, pp. 50, 62.

[78] G.S. Ohm, *Ann. Physik* **59**, 497–568 (1843).

[79] Rayleigh, *Theory of Sound*, 2nd ed. Reprinted by Dover, New York, NY, 1945, Vol. 1, p. 25.

[80] R. Laennec, "Traité de la ausculation médiate et les maladies des pommons et du coeur." English translation by John Forbes, M.D., is extracted in R.B. Lindsay (ed.), *Acoustics. Benchmark Papers in Acoustics*, Dowden, Hutchinson, and Ross, Stroudsburg, PA, 1972, pp. 166–172.

[81] M. Delezenne, *Recl. Trav. Soc. Sci. Agric., Arts de Lille* **8**, 1–57 (1826–1827). This paper is discussed in some detail in *Benchmark/*13, pp. 122–123.

[82] *Benchmark/*13, p. 123.

[83] G. von Békésy, *Experiments in Hearing*, translated and edited by E.G. Wever, McGraw-Hill, New York, NY, 1960. Reprinted by the Acoustical Society of America, Woodbury, NY, 1989, p. 11.

[84] Max Brödel, *Three Unpublished Drawings of Anatomy of the Human Ear*, W.B. Saunders, Philadelphia, PA, c. 1946.

[85] M.J.P. Flourens, *Acad. Roy. Sci.* 36–42, 52–58 (1824).

[86] E.H. Weber, *De Pulsu, Resorptione, Auditu et Tactu, Annotationes, Anatomicae et Physiologicae*, C.F. Kohler, Leipzig, 1834; E. Huschke, in *Vom Bau des menschlichen Körpers*, edited by S.T. von Sömmering, Leipzig, 1844.

[87] Carl Gustau Lincke, *Handbuch der teoretischen und praktischen Ohrenheilkunde*, I. Hinrichs, Leipzig, 1837.

[88] G. von Békésy, *Experiments in Hearing*, Mc Graw-Hill, New York, 1968, p. 11.

[89] Henry Matthews, *Observations on Sound*, Sherwood, Gilbert & Piper, London, 1826.

[90] Henry Matthews, *Observations on Sound*, pp. 8–9.

[91] The author recalls a college friend who sat looking up at the ceiling during a mathematics class. When asked why, he replied, "I'm watching the lecture go over my head."

[92] Richard Winefield, *Never the Twain Shall Meet*, Gallaudet University Press, Washington, DC, 1987, p. 6.

[93] Ref. [92, pp. 7–8]. Mann (1819) and Howe (1821) had been undergraduates together at Brown University.

3
von Helmholtz and Tyndall

In the present work an attempt will be made to connect the boundaries of two sciences, which, although drawn towards each other by many natural affinities, have hitherto remained practically distant—I mean the boundaries of physical and physiological acoustics on the one side, and of musical science and esthetics on the other.

H. von Helmholtz [1]

In the following pages I have tried to render the science of Acoustics interesting to all intelligent persons, including those who do not possess any special scientific culture.

J. Tyndall [2]

The development of acoustics in the early part of the nineteenth century was dominated by the work of French scholars. After the middle of the century, however, the tide seemed to have turned to British and German scientists. Three valuable books appeared in this later period: *On Sensations of Tone*, by Hermann von Helmholtz, 1862 [3]; *On Sound*, by John Tyndall, 1867 [4], and *The Theory of Sound*, by John William Strutt, later 3rd Baron Rayleigh, 1877 [5]. All three of these remarkable books are still in print and used today. We shall spend this chapter and the next, looking at them and at the contributions of their authors to the field of acoustics.

3.1. Hermann L.F. von Helmholtz (1821–1894)

It is perhaps not borrowing too much from optics to remark that two brilliant stars lit up the acoustical firmament in the second half of the nineteenth century—von Helmholtz and Rayleigh. It is difficult to imagine where the field of acoustics would be without either of them (or, for that matter, where the rest of physics would have been, since acoustics was in the case of each of them only a small part of their scientific contribution).

Hermann Ludwig Ferdinand von Helmholtz was born in 1821 in Potsdam, Prussia, the only son of a Gymnasium teacher in that city [6]. He had a sickly childhood and was late in entering into schooling [7]. Once in school, however, he made up for lost time and was ready for the university at the age of 17. He wanted to study physics, but his father could not afford to support him in that enterprise. However, the Prussian government did offer financial support for students of medicine, who would sign up for an extended period of service with the military, so Helmholtz entered a medical institute in Berlin, and received his medical degree in 1842 at the age of 21.

He was then assigned a post with a Guards regiment in Potsdam, but had space to set up his own laboratory in "physics and physiology." There he began a variety of experiments in optics and nerve impulses. In addition, in spite of a very full-time position as army surgeon and his work in the laboratory, von Helmholtz was able to devote thought to establishing a firm basis for the concept of conservation of energy, which he affirmed in a celebrated lecture to the Physical Society of Berlin in 1847. In was in this paper that he introduced the concept of potential energy. von Helmholtz was later to give full credit to the pioneering work of the obscure German physician Julius Robert Mayer (1814–1878) on the law of energy conservation, work that was carried out before the work of von Helmholtz [8], but it was von Helmholtz who was able to establish a mathematical proof of the law from the assumptions that all matter consisted of particles and that the forces between them were central forces, i.e., acting along the lines joining the particles.

The brilliance of von Helmholtz's work led to a cancellation of his remaining years of duty with the army and his appointment as Professor of Physiology at Königsberg (now Kaliningrad, Russia) in 1849, where he remained for 6 years. It was at this point that von Helmholtz gave up the practice of medicine and devoted most of his formidable talents to his laboratory work in physics and physiology. This concentration led to an explosion of researches on his part of almost bewildering variety: the determination of the velocity of nerve impulses (1850), the invention of the ophthalmoscope (1850), a theory of complex colors (1852) and other studies of the eye accommodation and acuity (1855), plus two papers on electric currents in conductors (1853).

This exhibition of talent brought him to Bonn in 1855 as Professor of Anatomy, but 3 years later he moved to Heidelberg, as Professor of Anatomy and Physiology. The high esteem in which his colleagues held von Helmholtz is reflected in a comment by Emil DuBois-Reymond (1818–1896), a celebrated physiologist (Fig. 3-1). The incident is described by von Helmholtz's biographer:

While Helmholtz believed that he had been successful in the anatomy lectures, du Bois writes to him on April 27, 1856, on Lehnert's [a Minister in the Prussian government] authority that it had been reported to the ministry that his lectures on anatomy were inadequate. Du Bois replied to Lehnert that, "while all things are possible, and stupidity probable, this was not only improbable but impossible". [9]

According to von Helmholtz, the basis of the complaint was apparently "the fact that I brought a good deal of physiology and chemistry into my anatomy, which restricted the amount of anatomy proper, and they made jokes at the introduction of a cosine into physiological optics [10]." Too much "higher" mathematics for the would-be physicians, no doubt!

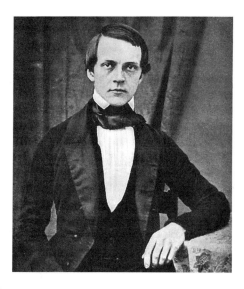

FIGURE 3-1. Herman von Helmholtz in 1848. (From Königsberger [6].)

Since this is a book on the history of acoustics, and not of physiology or anatomy, the reader may be wondering when acoustics will be mentioned again. In fact, von Helmholtz's first work in acoustics appears to have taken place in 1856 at Bonn with his important paper "On Combination Tones [11]," preceding his book on *On Sensations of Tone* (1862) by 6 years. All of von Helmholtz's publications on sound, with one exception, appeared in the period from 1856 to 1869. Thus, his "acoustics period" coincided with his presence at Bonn and Heidelberg.

von Helmholtz made his final move in 1877 to Berlin as Profesor of Physics, where he remained until his death in 1893. His later years were nearly as full of great discoveries as his earlier ones, but they were not in acoustics, and therefore we shall not concern ourselves with them.

3.2. *On Sensations of Tone*

While the book *On Sensations of Tone* is generally regarded as a treatise covering most of acoustics, it was written primarily to combine its author's knowledge of both physiology and physics into a detailed study of what one might call the physics, physiology, and psychology of music. In much of the text, he is reporting on, or summing up, his own researches in the field, and the same time trying to teach. It is certainly a treatise on the acoustics of his day. In reviewing the book here, we shall be largely reviewing von Helmholtz's researches in acoustics. One should note that, although the book first appeared in German in 1862, the standard English translation by

Ellis was made from the fourth German edition of 1877, after the conclusion (with the exception of one paper in 1878) of von Helmholtz's work in the field.

The text is divided into three parts. The first is entitled "On the Combination of Vibrations," the second, "On the Interruptions of Harmony," and the third "On the Relationship of Musical Tones."

Combinations of Vibrations

In Part I, von Helmholtz exhibits his teaching skill by leading the reader through the actual methods of experiment. He describes the way in which (in his day) one could demonstrate that pitch is due only to frequency, that the quality of a musical sound is due to the presence of higher harmonics (the "upper partials"), as well as the way in which the ear can separate out the various components of a complex tone. He worked with little more than a piano or stringed instrument, tuning forks, his siren, and his famous resonators, and clearly demonstrated how the reader, whether a physicist without a knowledge of music or physiology, or a physiologist without knowledge of physics, can perform the demonstrations himself or herself. The opening chapters of this text are well worth rereading today. In fact, one feels the urge to assemble the equipment and repeat the demonstrations. It was his further practice to describe virtually everything in non-mathematical, physical terms, putting all the necessary calculations in a collection of appendices [12].

In his work, one can see the gradual development of the understanding of a musical tone in this part of the nineteenth century. As we have noted in Chapter 2, Fourier began the chain of events with the idea that almost any periodic curve could be represented as the sum of sinusoidal curves of the frequencies nf_0, where f_0 is the base or fundamental frequency and $n = 1, 2, 3, \ldots$. Ohm then developed his law for acoustics, namely, the concept that the ear can recognize all these harmonics (the so-called upper partials) in addition to the fundamental (which latter defines the pitch of the musical tone), in effect carrying out a Fourier analysis of the note. That this was all the signal present in the musical tone (other than various scraping noises) was not yet fully accepted by the musical community at the time of von Helmholtz [13]. In addition, August Seebeck disputed Ohm's law for acoustics in 1843 [14]. In opposition to these criticisms, von Helmholtz observed

We have to assert, and we shall prove the assertion in the next chapter, that upper partial tones are, with the exceptions already named (i.e., such noises as the scraping of the violin bow), a general constituent of all musical tones, and that a certain stock of upper partials is an essential condition for a good musical quality of tone. [15]

von Helmholtz took it as a fundamental principle that the quality of a musical tone depended on the mixture of higher harmonics (the upper partials) that are present in the tone itself. For the voice, vibrating strings and vibrating air masses in musical instruments, these partials are the harmonics of the fundamental simple tone. Scraping, rubbing, or breathing noises, all which of which occur in the use of musical instruments, are indeed noise, and are not part of the music [16]. However, there is another class of sounds which, though not harmonic, are part of the musical scene. These are the overtones sounded in vibrating rods, shells, and bells. In these instances, the overtones are inharmonic, i.e., they are not integral multiples of the fundamental frequency.

At the beginning of von Helmholtz's researches, it was known that the ear could discern the pitch and the so-called quality of the tone. It was thought that this determination of quality rested on the ear's ability to detect the spectrum of the acoustic signal. Since von Helmholtz was demonstrating the validity of Ohm's acoustic law, that the ear analyzes the various harmonics or upper partials of the musical tone, he then set out to determine the effect of the phase of the individual components on the perception of the tone. He was able to develop apparatus to shift the various harmonics by any amount up to 90°, and obtained the result that the phase shift had no effect whatsoever. He therefore concluded that

the quality of the musical portion of a compound tone depends solely on the number and relative strength of its partial simple tones and in no respect on their differences of phase. [17]

von Helmholtz also applied himself to the study of the motions of a violin string. To follow its motions, he invented a vibration microscope (Fig. 3-2) [18], which was based on the idea of Lissajous figures. The lenses M and L form a microscope, with the lens L mounted on one prong of the tuning fork. If the observer is viewing a fixed spot, the motions of the fork move that point up and down in the vertical direction. If now that point is a mark on a violin string moving in the horizontal direction, then a typical Lissajous pattern is formed (Fig. 3-3) [19]. By viewing the patterns for the bowed violin string, von Helmholtz was able to arrive at the actual motions of that string. In addition, von Helmholtz worked out the mathematical behavior of the string [20] and thus arrive at an agreement between theory and experiment.

Next, von Helmholtz set himself to the study of speech sounds, especially vowels. Somewhat before von Helmholtz, a Dutch scientist Franz C. Donders (1818–1889) [20] had made a study of vowel sounds, identifying them with various tones on the musical scale. von Helmholtz made his own identification which differed from that of Donders in a number of instances, but, as von Helmholtz pointed out, Donders was working with Dutch pronunciation while von Helmholtz worked with North German, and, of course, some individuals' speech sounds will be higher on the frequency scale than others. This variability is shown in Table 1 [21].

FIGURE 3-2. von Helmholtz's vibration microscope [18].

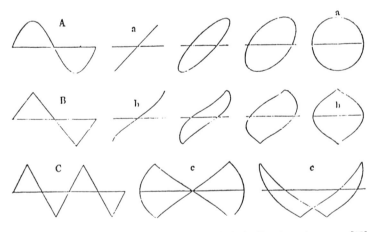

FIGURE 3-3. Lissajous' figures from von Helmholtz's vibration microscope [18].

TABLE 1. Pitch of varioius vowels in Dutch (Donders) and German. (von Helmholtz) From von Helmholtz [21]. In the notation of von Helmholtz, middle C, corresponding to 256 Hz, is labeled c'.

Vowel	Pitch according to Donders	Pitch according to Helmholtz
UH	f'	f
O	d'	$b'\flat$
A	$b'\flat$	$b''\flat$
Ö	g?	$c'''\sharp$
Ü	a''	g''' to $a'''\flat$
E	$c'''\sharp$	$b'''\flat$
I	f'''	d''''

The identification of the vowels with various fundamental pitches led von Helmholtz first to an attempt at improving the correspondence of vowel and pitch by addition of appropriate amounts of harmonics, then to attempts at artificial speech and, finally, to an study of both the speech and hearing organs. von Helmholtz worked with tuning forks and, since the sound made by such forks damps out very rapidly, he developed an ingenious system for sustained vibrations. This involved the two sets of apparatus shown in Figs. 3-4 and 3-5. In the first [22], an intermittent current (the generation of which will be described below) is applied to the coils of an

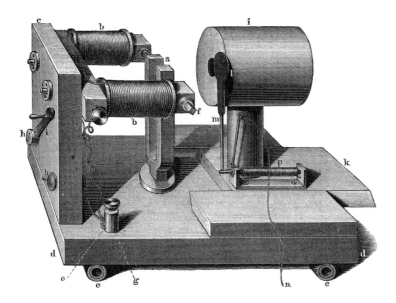

FIGURE 3-4. von Helmholtz's tuning fork [22].

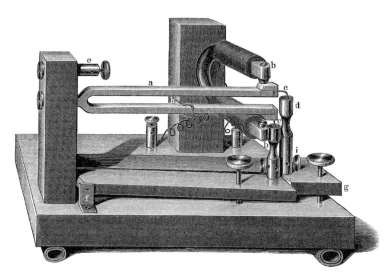

FIGURE 3-5. Source of intermittent current. (From von Helmholtz [23].)

electromagnet bb, the pole faces of which lie opposite the two prongs of a magnetized tuning fork. The frequency of the tuning fork was then identical with the frequency of the intermittency. Thus, each time the current surged through the coil, the prongs were pulled apart by the electromagnet.

Since the tuning fork did not sound loudly by itself, von Helmholtz added a resonator i, tuned to the pitch of the fork. The intensity of the resultant sound could be controlled by adjusting the lid l to cover the hole of the resonator by an appropriate amount.

The intermittent current was provided by the arrangement shown in Fig. 3-5, the idea for which von Helmholtz attributes to one Neeff [23]. A tuning fork and electromagnet are arranged in a fashion similar to that of Fig. 3-4, but this time the two are connected in series with one another and with a battery. A wire c connects the tuning fork with a cup partially filled with mercury. When the wire is in contact with the mercury, a direct current passes through the system. But, if the tuning fork is made to vibrate, the wire rises from the mercury in each oscillation, and then re-enters it. Hence the current in the circuit becomes intermittent. Each time the current is restored in the circuit, it energizes the electromagnet, causing its pole faces to attract the prongs of the fork and thus sustain the vibration. Thus, a form of alternating current is established, with a frequency identical to that of the fork.

von Helmholtz set up a series of these tuning forks, with frequencies = 2, 3, 4, . . . times that of the fundamental. von Helmholtz could thus approximate his vowel sounds by using different amounts of the higher harmonics. A near contemporary John G. McKendrick (1841–1926), wrote that it was von Helmholtz's opinion that "each vowel is characterized by a certain harmonic or partial tone, of constant pitch, whatever the pitch of the tone on which the vowel is sung or spoken. Attempts were then made, notably by von Helmholtz and Koenig, to fix the pitch of the characteristic partial tone or vocable and there appeared to be considerable differences in the results of the two distinguished observers, differences amounting to as much, in some cases, as three semitones [24]."

The Ear

Starting with research of this character, von Helmholtz soon passed to the study of the ear itself. If we compare the illustrations of the parts of the ear shown in von Helmholtz's book with those of Thomas Young 60 years before (Chapter 1), or if we read the detailed description of the various parts of the ear in von Helmholtz's text and compare it with those in Chladni or Young, we can see the enormous advance in the knowledge of the structure of the different parts of the ear over the 50-year interval. This is demonstrated by Fig. 3-6 [25], which shows a view of the cochlear nerve, and Fig. 3-7 [26], which shows the terminal cells of the nerves. von Helmholtz describes these illustrations as follows:

The construction of the cochlea is much more complex. The nerve fibres enter through the axis or modiolus of the cochlear into the bony part of the partition, and then come on to the membranous part. Where they reach this peculiar formations were discovered quite recently [1851] by the Marchese [Alfonso] Corti [1823–1888] [27], and have been named after him. On these the nerves terminate.

The expansion of the cochlean nerve is shewn in Fig. 48 [our Fig. 3-6]. It enters through the axis (2) and sends out its fibres in a radial direction from the axis

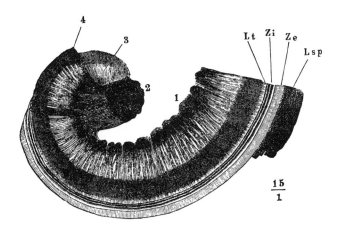

FIGURE 3-6. The cochlea. (From von Helmholtz [25].)

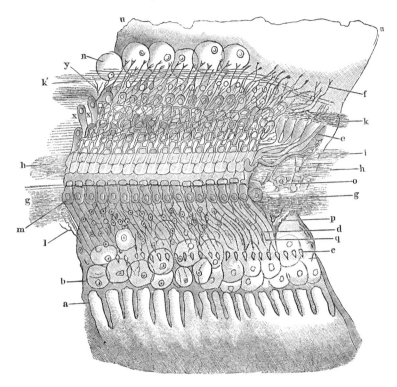

FIGURE 3-7. The hair cells. (From von Helmholtz [26].)

through the bony partition (1, 3, 4) as far as its margins. At this point the nerves pass under the commencement of the *membrana basilaris*, penetrate this in a series of openings, and thus reach the *ductus cochlearis* and those nervous elastic formations which lie on the inner zone (Zi) of the membrane. [28]

And, further,

These formations (the terminal cells of the nerves) are shewn in Fig. 51 [our Fig. 3-7], as seen from the vestibule gallery; *a* is the denticulated layer, *c* the openings of the nerves on the internal margin of the *membrana basilaris*, its external margin being visible at u–u; *d* is the inner series of Corti's rods, *e* the outer; over these, between *e* and *x* is seen the pierced membrane, against which lie the terminal cells of the nerves. [29]

The substantial and enduring nature of von Helmholtz's work on the ear can be confirmed by the many references to him in the book by Georg von Békésy [30] (1899–1972), published almost a century later. von Békésy began his chapter on the anatomy of the ear by remarking that "a good description of the gross anatomy of the ear is to be found in Helmholtz's *Sensations of Tone*, which first appeared in 1863 [31]."

von Helmholtz had developed a theory of resonance of the fluid in the cochlea, according to which the nerve fibers acted like vibrating strings, each resonating to a different pitch. He therefore assigned a specific region for every pitch (the so-called place theory). Noting that skilled musicians "can distinguish with certainty a difference in pitch arising from half a vibration in a second, in the doubly accented octave [32]," he concluded that some 1000 different pitches might be distinguished in the octave between 50 and 100 Hz. It was known that there were 4500 nerve fibers in the cochlea [33], which represented about one fiber for each two cents of musical interval. von Helmholtz was tempted to assign an individual hair cell to each component of the frequency, but could not demonstrate this experimentally, and was forced to observe:

we cannot precisely ascertain what parts of the ear actually vibrate sympathetically with individual tones. [34]

Further progress here also had to wait another century.

von Helmholtz was responsible for establishing the basic function of the bones of the middle ear. Wever and Lawrence point out in their 1954 book that the "middle ear apparatus is an acoustical transformer ... matching the external and internal impedances. ... The conception that the middle ear serves this physical purpose ... finally was given a clear formulation by von Helmholtz in 1863 [35]."

Interruptions of Harmony

In the second part of his book, von Helmholtz took up the phenomenon of beats and also that of combination tones. von Helmholtz's first paper in acoustics, "On Combination Tones," came out some years before his book [36], but we shall discuss this paper within the framework of the book itself.

In both the article and the book, von Helmholtz picked up the problem of the Tartini tones, discussed in Chapter 1. In the century that had passed since the work of Sorge and Tartini, no significant progress had been made in understanding the effect, the consensus remaining that the sounds were a form of beats. There were also reports that other tones might be heard. In particular, Hällström [37] theorized that the difference tone $f_1 - f_2$ could interact with the two fundamentals f_1, f_2, producing $2f_1 - f_2$, $2f_2 - f_1$, and so on.

von Helmholtz began a careful study of the phenomenon. As we have noted above, he was concerned with the structure of single musical tones and demonstrated that such tones, called compound tones by him, were rich in harmonics of the fundamental. This fundamental determined the pitch. It was therefore possible that these were "beating" with one another, or with the fundamental. von Helmholtz therefore carried out his experiments with tones from which all the harmonics had been removed, or with his siren, which had no harmonics to begin with. He called such tones simple tones, since they could be described by a single sine function, which in the field of mechanics we call simple harmonic. von Helmholtz also made the distinction between objective tones, which could exist in the air, outside the ear, and subjective tones, which existed only somewhere in the ear.

In addition to hearing the ordinary difference tones of Tartini, however, von Helmholtz reported a whole new class of combination tones—summation tones. These tones were fainter than the corresponding difference tones, and thus more powerful sources were necessary to hear them. Although he could hear the summation tones with organ pipes, he found the use of Dove sirens [38] most effective. Here also, the elimination of harmonics in the source was an important gain. And, with the siren, he could also detect combinations such as $f_1 + 2f_2$ and $2f_1 + f_2$ in addition to the simple summation $f_1 + f_2$.

von Helmholtz concluded that, if these summation tones had any reality, the beat theory of Young and others was dead, and a replacement was necessary. He reviewed the theory of superposition in the following way:

It has always been assumed that various wave trains that become excited simultaneously in the air or in another elastic medium are simply superposed on one another without any mutual interaction and one has believed that this assumption has been adequately proven through well-known experiments on the possibility of distinguishing from one another the simultaneously sounded tones of different instruments or human voices, each with its own tone, pitch and timbre. Conversely, it should be considered that the theoretical mechanism of such an undisturbed superposition holds only for the case of infinitely small oscillations while it could be shown at the same time from the equations of motion of the air that in wave trains of unbounded strength of amplitude there cannot be such an undistorted superposition. [39]

von Helmholtz suggested a quadratic dependence of the elastic force acting in the bones of the middle ear. Such a phenomenon would produce both sum and difference frequencies in the inner ear. Thus, as he put it, "the combination tones will not have merely a subjective existence but will also be objective even if they consist only of the oscillating particles of the ear itself [40]."

von Helmholtz's conclusion was in contradiction to that of Riemann (1826–1866) [41], who maintained that the mechanism of amplification of the middle ear had to transfer the signal from the outer ear with perfect, i.e., linear, accuracy. As we shall see later, Riemann was correct as to the linearity of the middle ear, and one had to look further (into the cochlea) to find an answer to the problem.

However, von Helmholtz was also alert to the possibility that such nonlinearities could exist outside the human ear and, in fact, by setting up sympathetic vibrations in thin membranes, he was able to detect such combination tones [42]. Although waves of finite amplitude were studied throughout this period, from Stokes to Rayleigh, no further progress was to be made on combination tones for the next 40 years.

von Helmholtz devoted the first chapter of Part II of his book to combination tones, and the remaining five chapters to beats themselves. It must be kept in mind that the two concepts—beats and combination tones—were very much intermingled in the minds of all acoustical scientists of the time (other than von Helmholtz), so that a thorough analysis of the phenomenon of beats was of great importance to him.

von Helmholtz began by analyzing the phenomenon of interference in acoustics, using both his siren and the tuning fork to provide easy demonstrations. Interference is the phenomenon of the linear superposition of two sounds of identical frequency. von Helmholtz then experimented with the sounds of not quite identical frequencies, and studied the resulting beats. But he was not satisfied with the physical conclusion of the existences of rises and falls of the amplitudes of oscillation that are the basis of the beat phenomenon. These had been amply displayed by the techniques developed by Koenig (see Chapter 6). von Helmholtz now looked for the physiological basis of what one heard, and attempted to answer the question "what becomes of the beats when they grow faster and faster [43]?" This involved the theory of musical consonance. It was known that, when beats are low in number (less than $10\,\mathrm{s}^{-1}$), they are perceived as a succession of pulses of the sound, but when they increase in number, the ear does not separate them, but hears the combination of sounds as an unpleasant one. This unpleasantness peaks at around $30\,\mathrm{s}^{-1}$ and then subsides.

What von Helmholtz was able to do was to produce intermittent sounds by means of his siren. He found that when the repetition rate of the intermittency was low, pulses of the musical tone were heard, but when the rate became higher, the same unpleasantness was experienced as was the case with beats. From this, he concluded that beats and intermittent sounds produced the same effect; they were in essence the same phenomenon.

The reason for the unpleasantness, he found in the physiological experience that

beats produce an intermittent excitement of certain auditory nerve fibres. The reason why such an intermittent excitement acts so much more unpleasantly than an equally strong or even a stronger continuous excitement, may be gathered from the analogous action of other human nerves. Any powerful excitement of a nerve deadens its excitability, and consequently renders it less sensitive to fresh irritants. [44]

von Helmholtz also worked on the problem of the lowest audible frequency that could be recognized as a musical tone. He was aware that "the strength of the vibrations of the air for very deep tones should be extremely greater than for high tones, if they are to make as strong an impression on the ear [45]," thus anticipating the shape of the low-frequency end of the audibility curve (see Chapter 8). He further noted the presence of upper partials in the deep tone, the perceived strength of which (due to the shape of the hearing curve, see Chapter 8) is very often greater than that of the primary tone:

To discover the limit of deepest tones, it is necessary not only to produce very violent agitations in the air but to give these the form of simple pendular vibrations. Until this last condition is fulfilled we cannot possibly say whether the deep tones we hear belong to the prime tone or to an upper partial tone of the motion of the air. [46]

One might note that musical tones all have their upper partials, and we now know that even a tone that is a simple one at the source, quickly developed summation tones with frequencies of two and three times the fundamental [47], and von Helmholtz knew that such summation frequencies were also developed somewhere inside the ear. Under such circumstances, one had to rule out many previous experiments in determining this threshold. von Helmholtz's conclusion was that the hearing threshold was located at the tone E more than two octaves below middle C (this E being located at 42.203 Hz in the tempered scale with standard A at 440 Hz). He added, "I think I may predict with certainty that all efforts of modern art applied to produce good musical tones of a lower pitch must fail . . . because the human ear cannot hear them [48]."

von Helmholtz's book is full of gems of insight from his brilliant mind and his habits of careful observation. One of the many examples of this was provided when, in discussing the low tones made by his siren, he made the following observation:

when I laid my head on it (the violently agitated bellows), my whole head was set into such powerful sympathetic vibration that the holes of the rotating disc which vanish to an eye at rest, became again separately visible through an optical action similar to that which takes place in stroboscopic discs. [49]

The image of von Helmholtz's rather solid head vibrating back and forth so as to provide a stroboscopic detector is too much to pass over without mention.

The Relationship of Musical Tones

In his preface to the third German edition (1870), von Helmholtz asked his readers "to regard this section as a mere compilation from secondary sources; I have neither time or preliminary knowledge sufficient for original studies in this extremely difficult field [50]." These caveats apply much more strongly to the author of the present volume, and therefore Section III of von Helmholtz's book will be given much briefer treatment than it deserves.

In discussing the relationship of musical tones, von Helmholtz first noted the effect of

historical and national differences of taste. Whether one combination is rougher or smoother than another depends solely on the anatomical structure of the ear, and has nothing to do with psychological motives. But what degree of roughness a hearer is inclined to endure as a means of musical expression depends on taste and habit; hence the boundary between consonances and dissonances has been frequently changed. [51]

Also,

The system of scales ... does not rest solely upon inalterable natural laws, but is also, at least partly, the result of estheticial principles, which have already changed, and will still further change with the progressive development of humanity. [52]

As the Romans put it, "de gustibus non disputandum est."

von Helmholtz divided music into three basic types, homophonic, polyphonic, and harmonic. These he regarded as a progression in history. Homophonic or one-part music was, according to von Helmholtz, "the original form of music for all people," still existing among various Asiatic nations of his time [53]. He pointed out that well-spoken prose is a form of polyphonic music, and cited a number of examples.

The second type was polyphonic or music of several parts, which was the backbone of church music in the Middle Ages. In polyphonic music, he wrote, "consonance was not the object in view ... but its opposite, dissonance, was to be avoided [54].

The third music form was harmonic music, in which "independent significance [was] attributed to the harmonies as such [55]." von Helmholtz argued that this was encouraged by the Protestant choral singing of the Reformation. "It was a principle of Protestantism that the congregation itself should undertake the singing." [56]

The bulk of this third part of von Helhmoltz's book discusses in great detail the tonality of music, consonances and discords, and the systems of musical keys. He noted that consonances are made more prominent by the use of dissonances, and in the some way, "the feeling for predominant tonality and satisfaction which arises from it is heightened by previous deviations into adjacent keys." [57] On the subject of pitch, he adds

There is nothing in the nature of music itself to determine the pitch of the tonic of any composition ... [rather,] the pitch of the tonic must be chosen so as to bring the compass of the tones of the piece within the compass of the executants, vocal or instrumental. [58]

In other words, the composer must write for the voice or the instrument, and not expect the instrument to adapt to the composer's composition.

In concluding the entire volume, von Helmholtz wrote:

I have carried (this work) as far as the physiological properties of the sensation of hearing exercise a direct influence on the construction of a musical system, that is, as far as the work especially belongs to natural philosophy. ... I prefer leaving to others to carry out such investigations (of the esthetics of music), in which I would feel myself too much of an amateur, while I myself remain on the safe ground of natural philosophy in which I am at home. [59]

And there we shall leave him.

3.3. John Tyndall (1820–1893) (Fig. 3.8)

While the contributions of Tyndall to research in acoustics were small in comparison with those of the more famous von Helmholtz and Rayleigh, his lectures and his book had a profound effect on the knowledge of the subject of sound among English-speaking people, and on the teaching of acoustics in the colleges during and after his lifetime. In the 1860s he was recognized as one of the foremost physicists in Great Britain [60], and was also very well known in the United States through his demonstration lecture tours there.

Tyndall's ancestors were English Protestants, and only moved to Ireland after the overthrow of James II in the latter part of the seventeenth century. John Tyndall was therefore born in County Carlow in 1820. His parents were poor and were not able to finance any advanced education for him. John therefore started out as a surveyor for the Ordnance Survey Office at less than one British pound a week (then equal to about five US dollars). He read everything he could lay his hands on, including an early edition of the *Encyclopaedia Britannica* [61], and looked everywhere for a better position. None was found, and soon Tyndall found himself in difficulties for his outspoken attitudes toward British government policy. Tyndall was a natural born controversialist, and was engaged in many debates about both government and religion, and soon lost his government employment.

Jobs were hard to come by, either in Ireland or in England in the 1840s, but Tyndall ultimately secured a position with an engineering firm in Manchester (England) at three guineas a week (£3.15). By 1847, even this job appeared to be insecure, but Tyndall was able to obtain a position with a new private school at Queenwood, near Stockbridge in Hampshire. The school emphasized science and engineering, and Tyndall was able to begin his teaching career.

FIGURE 3-8. John Tyndall. (From A.S. Eve and C.H. Creasey [60].)

Tyndall had his eye on higher things than teaching at Queenwood. After a year or so in that position, and armed with money saved from his salary as an engineer in Manchester, Tyndall traveled to Marburg in Germany, where the chemist Robert Bunsen was the university's most famous professor. In 2 years, Tyndall completed the work for the doctorate, obtaining his degree with a thesis entitled "On a screw surface with inclined Generetrix [sic] and on the Conditions of Equilibrium on such Surfaces." After extensive European travel, including visits to scientists in other universities, Tyndall returned to Queenwood in 1851.

During the next year, Tyndall made a number of attempts to obtain a professorship at various institutions, both in Britain and in her colonies, all without success, although he had recommendations from an amazing collection of distinguished scientists—Faraday, William Thomson (Lord Kelvin), Joule, Becquerel, and Bunsen, among others. His research in this period was mainly in the magnetism of crystals, and for his work he was elected to the Royal Society in 1852.

Success finally came in 1853 when he was appointed Professor of Natural Philosophy at the Royal Institution in London. The Royal Institution had been the base for demonstration lectures and research, first by Sir Humphrey Davy (1778–1829) and then by Michael Faraday. Faraday was approaching retirement, and Tyndall was destined to succeed him.

The Royal Institution gave Tyndall an opportunity to display his skills in popular lecturing, and he took full advantage of it. It was a time when the educated public flocked to such lectures and Tyndall had a superb talent for delivering lectures and developing brilliant demonstrations to accompany them [62]. His first lectures were on heat, and these lectures were subsequently published as a book [63]. In the course of writing the lectures, Tyndall wanted to learn more about Julius Robert Mayer, whose early work on the conservation of energy was just beginning to be recognized. Tyndall wrote to different German scientists, including von Helmholtz, for information about Mayer and thereby began a long-term friendship with von Helmholtz. At the same time, Tyndall became involved in a controversey with Joule [64], in which Tyndall supported the idea that precedence was to be given to Mayer, while Joule opposed that view. Kelvin and Peter Guthrie Tait (1831–1901) were brought in on Joule's side, and von Helmholtz on Mayer's. It was through Tyndall's efforts that Mayer became better known in Britain and that he was awarded the Copley medal by the Royal Society in 1871, but Tyndall's good relations with Kelvin and Tait were never restored.

Tyndall worked on problems of fluorescence in the 1860s and was awarded the Rumford Medal by the Royal Society for his work. He also undertook a lecture tour in the United States in 1872, which was a great success.

Throughout his life, Tyndall was an avid mountaineer, making mountain-climbing expeditions to the Alps, becoming a expert mountaineer. He put these vacations to professional use by becoming an acknowledged authority on glaciers, publishing a number of research papers on that subject.

On Sound

Tyndall's book, *On Sound*, first appeared in 1867, just 4 years after the publication of von Helmholtz's volume. In the preface to the first edition, Tyndall acknowledges his use of material from von Helmholtz's work: "copious references to it (von Helmholtz's book) will be found . . . but they fail to give an adequate idea of the thoroughness and excellence of the work [65]." It is interesting to note the respect which von Helmholtz and Tyndall had for each other. In fact, von Helmholtz undertook a translation of Tyndall's book into German the same year [66].

The first edition of Tyndall's book contained seven chapters. The third edition (the one currently in print) is little changed from the first, except that two additional chapters were added—on transmission of sound through the atmosphere and on combinations of musical tones. We shall first review the original edition and conclude with remarks on the two added chapters. All page references will be to the third edition.

This book is the writing down the of lectures and demonstration descriptions of materials covered by Tyndall, both in his lectures at the Royal Institution, and in his travels abroad. Unlike the work of von Helmholtz, which placed its major focus on music, Tyndall's lectures attempted to cover the whole existing span of acoustics. It provides a striking comparison with the book by Chladni, demonstrating the progress made in the field over a period of 60 years.

The lectures begin with the vibrations of sonorous bodies and the propagation of sound through gases, liquids, and solids.

Most of this subject has already been covered in this book, and we shall not spend time on it here, except to point out the work of Karl Friedrich Sondhauss (1815–1886) [64] on the focusing of sound by the use of spheres filled with a gas with a different sound velocity than that of air, and the careful measurements of Wertheim and Chevandier [65], on the velocity of sound in different directions in wood, which anticipated much later work in the propagation of shear and compressional waves in solids.

Chapter 2 begins with the relationship between noise and music and introduces the subject of the determination of the waveform of musical sounds. Tyndall attributes to Lissajous an experiment demonstrating the waveform of sound emitted by a tuning fork. Lissajous attached a small mirror to one prong of a tuning fork, balancing the weight of this mirror with a similar piece of metal on the other prong. Shining an intense but narrow beam of light onto the mirror, and directing this reflection to the second mirror, Lissajous was able to spread out the oscillations in time by gently rotating the second mirror (Fig. 3-9) [66]. One could therefore make out the waveform of the oscillations. This marked the beginning of a period

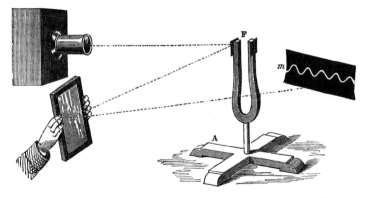

FIGURE 3-9. Visualization of waveform. (From Tyndall [69].)

of vigorous study of ways of visualization of sound waves, about much more will be said in Chapters 4 and 6. The rest of the chapter deals largely with sirens, paralleling the work in von Helmholtz's book.

Chapter 3 follows the vibrations of a single string, including detailed demonstrations of harmonics, traveling waves, and the "clang" of piano wires [70]. Chapter 4 reviews the vibrations of rods, plates, and bells, while Chapter 5, continuing the subject of rods, introduces the phenomenon of the conversion of longitudinal vibrations into transverse vibrations in a solid. Tyndall noted that the longitudinal vibrations, which he produced by passing a piece of cloth or leather treated with resin (or with resin on his fingers), along the wire, were more acute (i.e., of higher frequency) than the transverse vibrations [71]. Even to the present day, we use this technique of exciting the longitudinal vibrations in Kundt's tube experiment.

Another technique adopted by Tyndall [72], but taken this time from Biot, involved the use of polarized light (Fig. 3-10) [73]. The polarizer and analyzer are set in the position to allow no light through. If a piece of glass is introduced between the polarizer and the analyzer, and the glass is subjected to stresses, then colored bands are seen in the transmitted light. This has become a popular technique in modern stress analysis.

Tyndall was able to include in his Chapter 5 the work that had very recently been reported by August Kundt (1839–1894) in 1866 [74], but the description of Kundt's work will be deferred until our Chapter 6.

Chapter 6 of Tyndall's book begins with a rather poetic description of the "rhythm of friction," and the various quasimusical sounds that are produced from the motion of fluids through different openings. Without mentioning the name of Pieter Leonhard Rijke (1812–1899) [75] Tyndall recounts a number of demonstrations of the musical sounds produced in a glass tube placed over a flame, but his special emphasis was placed on work that he himself did, following up on an earlier experiment by John Le Conte (1818–1891) [76]. The middle of the nineteenth century was the heyday for

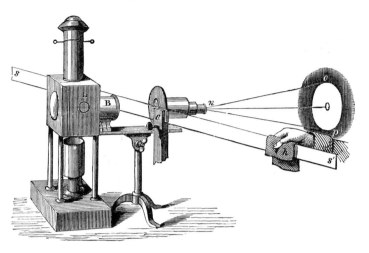

FIGURE 3-10. Use of polarized light to reveal strains in glass. (From Tyndall |73|.)

gas lighting, and, while attending an evening musical recital, Le Conte, a professor at South Carolina College (now the University of South Carolina) was observing the flame of a gas lamp, for which the gas pressure was such as to make the flame flare. When the music was played, the flame rose and fell in time with the music. As Le Conte expressed it, "a deaf man might have seen the harmony."

Many scientists followed Le Conte's lead and developed techniques involving gas flames [77]. We shall note here only those due to Tyndall. Tyndall quickly realized that the technique rested on getting the flame close to a point of instability. He wrote

Before you bur a bright candle-flame: I may shout, clap my hands, sound this whistle, strike this anvil. . . . Though sonorous waves pass in each case through the air, the candle is absolutely insensible to the sound. . . . I now urge from this small blow-pipe a narrow stream of air through the flame of the candle, and producing thereby an incipient flutter, and reducing at the same time the brightness of the flame. When I now sound a whistle the flame jumps visibly. [78]

An example of this "sensitive flame" is shown in Fig. 3-11 [79].

The eighth and ninth chapters of Tyndall's book repeat much of the material we discussed in reviewing von Helmholtz's book—the superposition of vibrations and sound waves, including interference, beats, and combination tones (Chapter 8), and the relation of various musical tones in consonance, dissonance, and harmony. The chapters contain descriptions of many demonstrations of the interaction of two beams of sound. Especially interesting is Lissajous' extension of the method of visualizing the vibrations of a single tuning fork to the interaction of two such forks of slightly different frequencies, so that beats could be observed (Fig. 3-12) [80].

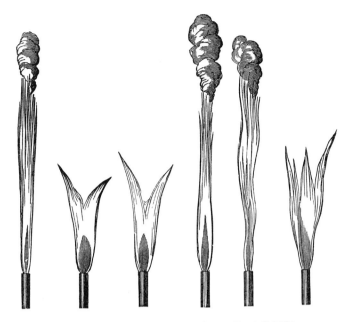

FIGURE 3-11. Sensitive flames. (From Tyndall [79].)

The most interesting new material added to the third edition of Tyndall's work was a summation of his work on sound transmission in the atmosphere, and, in particular, on the effects of fog on such transmissions. Tyndall had succeeded Faraday as Scientific Advisor to the "Elder Brethren of Trinity House." This was a guild, established in the reign of Henry VIII, and sometimes called the Corporation of Trinity House, which super-

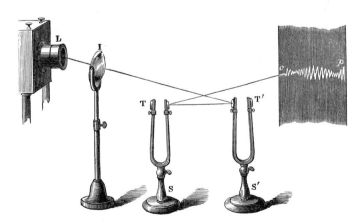

FIGURE 3-12. Visualization of beats. (From Tyndall [80].)

vised lighthouses and pilots in England (later, Rayleigh was to succeed Tyndall), and the problem of sound transmission in the atmosphere was thus of great practical interest.

The problem of fog for ships at sea at that time needs no elaboration here. Lights from the various lighthouses disappeared quickly in the fog, and the community turned to sound, with whistles, bells, sirens, and even gunfire used to penetrate the white blankets. In his work on fog, Tyndall paralleled some of the work done by Joseph Henry in America, which we shall recount in Chapter 5. Tyndall noted that popular opinion essentially followed that of Derham in 1708, who concluded (erroneously) that fog was a major obstacle to the passage of sound (see Chapter 1). Using all sound sources available, including a great siren lent by the US Lighthouse Board (see Fig. 3-13) [81], and the yachts available to Trinity House, Tyndall began in 1873 a systematic study of sound propagation over water under various weather conditions in the Straits of Dover. What he found was great inconsistencies in the propagation. Sometimes the whistles could be heard further than guns or sirens; sometimes the reverse was true. Sometimes the distance at which the sounds were heard was greater when the wind was against the sound than when it was in the same direction. And sometimes, even in clear air and relatively calm seas, there was very poor transmission. In this latter connection, Tyndall cited a letter he received from a former Confederate Army officer, R.G.H. Kean, who had watched the battle of Gaines Mills (one of the Seven Days Battles before Richmond, June 28, 1862, in the American Civil War), along with General G.W. Randolph, C.S.A. (1815–1867), Kean's father-in-law. In overlooking the battlefield from an adjacent hill (the width of the valley was about a mile and a half), Kean wrote

I distinctly saw the musket-fire of both lines, the smoke, individual discharges, the flash of the guns. I saw batteries of artillery on both sides come into action plainly in sight. . . .
    Yet looking for nearly two hours, from about 5 to 7 p.m. on a midsummer afternoon, at a battle in which at least 50,000 men were actually engaged, and doubtless at least 100 pieces of field-artillery, through an atmosphere optically as limpid as possible, not a single sound of the battle was audible to General Randolph and myself. [82]

In some of his experiments, Tyndall was able to observe the same degree of silence. He concluded that the air contained invisible "acoustic clouds," (another way of describing inhomogeneities) which reflected the sound back to the observer. To test this, he set up an observation post at the base of the cliff on which the sound source was located, thus somewhat shielding himself from the direct sounds. He wrote

the somewhat enfeebled diffracted sound reached me and I was able to hear with great distinctness, about a second after the starting of the siren-blast, the echoes striking in and reinforcing the direct sound. [83]

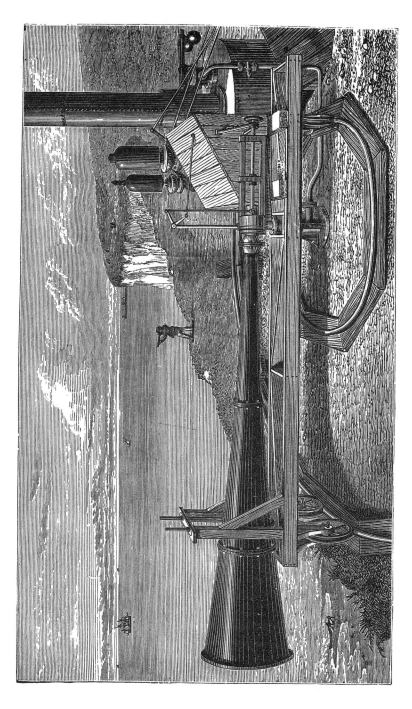

FIGURE 3-13. Tyndall's fog-horn [81].

It was characteristic of Tyndall that, when he set up a hypothesis for large scale behavior (such as sound propagation in the atmosphere and light scattering in the sky), he attempted to demonstrate the phenomenon by the use of table-top apparatus. In this case, he prepared the ingenious structure shown in Fig. 3-14 [84]. The central part of the structure $t - t'$ formed a glass-enclosed pipe, while the top and bottom of the unit contained tubes that allowed carbonic acid gas [carbon dioxide] to flow down from the upper vents and coal gas to rise from the lower ones. Thus a structure of some 24 alternate layers of gases of different density (and sound velocity) lay in front of the sound emanating from the bell P in the box at the left. Tyndall's sound detector at the right was one of his sensitive flames.

When the two gases were flowing, no sound from the bell reached the detector. However, if the sources of the two gases were shut off and the gas in the pipe allowed to mix uniformly, the sound of the bell was again detected. Tyndall also found that he could accomplish the same result with only a few alternating layers, and that the sound was being reflected from his simulated acoustic cloud, thus fully confirming his large-scale experiments.

Tyndall also made sets of observations in the Straits of Dover in the presence of fog and also of heavy rain, and found that the sound transmission was often better than that obtained in clear air. The primary reason for the attenuation of the signal was the inhomogeneity of the atmosphere. Of course, as we now know, the propagation of sound through the atmosphere

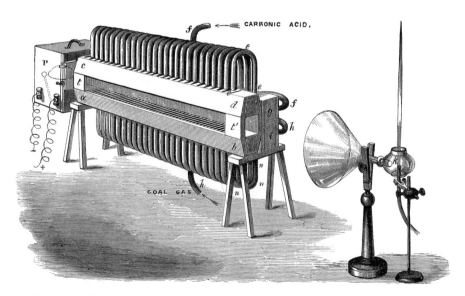

FIGURE 3-14. Tyndall's apparatus for measurement of sound scattering from inhomogeneous gas layers [84].

is affected by the temperature structure of the atmosphere, the amount of water vapor in the air, the surface state of the ocean, and is a very complex problem that is still being worked on in the twentieth century, but Tyndall found the right track (see Chapter 7).

Dayton Miller well summed up Tyndall's career by remarking that "his contributions to science are more due to his very genial and inspiring personality, and to his exceptional ability as a popular lecturer, rather than to his researches [85]." While his methods of sound measurement by means of flame have long since been replaced by more reliable electronic devices, the conclusions he reached both in light scattering and in atmospheric sound propagation were correct and enduring.

Notes and References

[1] Herman von Helmholtz, *Die Lehre von den Tonempfindungen* (1862). Translated as *On Sensations of Tone*, by Alexander J. Ellis from the 4th German edition (1877). Reprinted by Dover, New York, NY, 1954, p. 1.

[2] John Tyndall, *On Sound*, D. Appleton, New York, NY, 1867. Third edition, reprinted under the title, *The Science of Sound*, Citadel Press, New York, NY, 1964. Preface to the First Edition.

[3] Herman von Helmholtz, *Die Lehre von den Tonempfindungen*. This volume includes extensive commentary by Alexander Ellis, as well as an appendix by him on musical pitch.

[4] John Tyndall, *On Sound*.

[5] J.W. Strutt, 3rd Baron Rayleigh, *The Theory of Sound*, 1877. American reprinting of the second edition (1894), Dover, New York, NY, 1945.

[6] Much of the following material is taken from the biography *Herman von Helmholtz*, by Leo Königsberger, 1902, English translation by Francis A. Welby, Clarendon Press, Oxford, UK, 1906. Reference should also be made to the Preface to the reprinting of Ref. [1], written by Henry Margenau.

[7] This fact, plus the rheumatic fever of Lord Rayleigh during his young manhood, and the not-so-well-known hearing and speech handicaps of the French acoustician Sauveur, as well as the deafness of Beethoven and Edison, all serve as an encouragement for those with physical handicaps. They have served as such particularly for the author, who, in the words of his wife, has "enjoyed bad health" for more than 60 years.

[8] For details of Mayer's career and work, see R.B. Lindsay, *Men of Physics: Julius Robert Mayer*. Pergamon Press, Oxford, UK, 1973.

[9] See biography by Königsberger, Ref. [6, p. 148].

[10] Königsberger, loc. cit.

[11] Herman von Helmholtz, *Ann Phys. Chemie.* **99**, 497–540 (1856).

[12] von Helmholtz's towering contributions to mathematical physics (the von Helmholtz equation, the von Helmholtz free energy) are assurances that he was superbly prepared mathematically but, for pedagogical reasons, declined to write his book in such mathematical detail.

[13] Herman von Helmholtz, *On Sensations of Tone*, p. 58.

[14] A. Seebeck, *Ann Phys. Chemie* **60**, 449–481 (1843); **63**, 353–380 (1844).

[15] Herman von Helmholtz, loc. cit.

[16] Of course, twentieth-century music can take virtually any noise in rhythmic repetition and make "music" out of it, as anyone who has attended a performance of the stage production "Stomp" can testify.

[17] Herman von Helmholtz, *On Sensations of Tone*, p. 126. Koenig later made some corrections to this conclusion, see Chapter 6.

[18] Herman von Helmholtz, *On Sensations of Tone*, pp. 80–83.

[19] Herman von Helmholtz, *On Sensations of Tone*, pp. 384–390.

[20] F.C. Donders and W. Berlin, *Arch Holl. Beiträge Natur- Heilkunde*, **1**, 157–162 (1858). See also Donders book, *The Physiology of Speech Sounds, and Especially Those of the Dutch Language* (in Dutch) van der Post, Utrecht, 1870, p. 24ff.

[21] Herman von Helmholtz, *On Sensations of Tone*, p. 109.

[22] Harman von Helmholtz, *On Sensations of Tone*, pp. 120–121. Curiously, these two figures (3-4 and 3-5) are reproduced and exhibited in the Alexander Graham Bell Museum at Baddeck, Nova Scotia, in the same frame as models of two early telegraphs, and as an introduction to apparatus developed by Bell, but without mention of the name of Helmholtz.

[23] Herman von Helmholtz, *On Sensations of Tone*, pp. 121–122. Neeff published his article in *Ann Physik* **46**, 104–109 (1839), giving himself no first name or initial, but locating himself at Frankfurt-am-Main in Germany.

[24] John G. McKendrick, *Nature* **65**, 182–189 (1901–1902). Professor McKendrick was a Fellow of the Royal Society, and the article in *Nature* was based on a paper on experimental phonetics, given at a meeting of the British Association. The particular quotation is from p. 187.

[25] Herman von Helmholtz, *On Sensations of Tone*, p. 139.

[26] Herman von Helmholtz, *On Sensations of Tone*, p. 141.

[27] A. Corti, *Z. Wiss. Zool.* **3**, 109–169 (1851).

[28] Herman von Helmholtz, *On Sensations of Tone*, p. 139.

[29] Herman von Helmholtz, *On Sensations of Tone*, p. 140.

[30] G. von Békésy, *Experiments in Hearing*, McGraw-Hill, New York, NY, 1960. Reprinted by the Acoustical Society of America, Woodbury, NY, 1989. We shall have much more to say about von Békésy in Chapter 8.

[31] G. von Békésy, *Experiments in Hearing*, p. 11.

[32] Herman von Helmholtz, *On Sensations of Tone*, p. 147. The sensitivity of the musician's ear was described by W. Preyer, *Ueber die Grenzen der Tonwahrnehmung*, June 1876. An English adaptation appeared in *Proc. Musical Association* for 1876–1877, pp. 1–32.

[33] Herman von Helmholtz, *On Sensations of Tone*, p. 147. von Helmholtz attributes this enumeration of the fibers to one Waldeyer.

[34] Herman von Helmholtz, *On Sensations of Tone*, p. 145.

[35] E.G. Wever and M. Lawrence, *Physiological Acoustics*, Princeton University Press, Princeton, NJ, 1954, p. 75. Some earlier ideas of this function were given by Charles Bell, *The Anatomy of the Human Body* 1803, III, pp. 373–454 and W.F. Weber, *Berichte Gesellsch. Wiss. Leipzig, Math.-Phys. Cl*, **3**, 29–21 (1851).

[36] Herman von Helmholtz, *Niederrh. Sitzungsber.* 1856; *Berl. Monatsber.* 1856; *Ann. Physik Chemie* **99**, 497–540 (1856). A partial English translation of the latter paper is given in *Benchmark/19*, pp. 229–238.

[37] G.G. Hällström, *Ann. Physik Chemie* **24**, 438–466 (1831).

[38] H.W. Dove, *Ann Physik Chemie* **82**, 596–598 (1851).

[39] Herman von Helmholtz, *On Sensations of Tone*, p. 232.

[40] Herman von Helmholtz, *On Sensations of Tone*, p. 237.

[41] B. Riemann, *Z. Ration. Medicin.* **29**, 129–143 (1867).

[42] Herman von Helmholtz, *On Sensations of Tone*, p. 238.

[43] Herman von Helmholtz, *On Sensations of Tone*, p. 166.

[44] Herman von Helmholtz, *On Sensations of Tone*, p. 169.

[45] Herman von Helmholtz, *On Sensations of Tone*, p. 174.

[46] Herman von Helmholtz, *On Sensations of Tone*, p. 175.

[47] A.L. Thuras, R.T. Jenkins, and H.T. O'Neil, *J. Acoust. Soc. Am.* **6**, 173–180 (1935). See also R.T. Beyer, *Nonlinear Acoustics*, Navy Sea Systems Command, Washington, DC, 1976.

[48] Herman von Helmholtz, *On Sensations of Tone*, p. 175.

[49] Herman von Helmholtz, loc. cit.

[50] Herman von Helmholtz, *On Sensations of Tone*, p. vii.

[51] Herman von Helmholtz, *On Sensations of Tone*, p. 234.

[52] Herman von Helmholtz, *On Sensations of Tone*, p. 235. And von Helmholtz never heard any twentieth-century music!

[53] Herman von Helmholtz, *On Sensations of Tone*, p. 237.

[54] Herman von Helmholtz, *On Sensations of Tone*, p. 244.

[55] Herman von Helmholtz, *On Sensations of Tone*, pp. 236–237.

[56] Herman von Helmholtz, *On Sensations of Tone*, p. 246. The situation remains the same today. Protestant congregations sing out most vigorously, while, speaking as a Catholic, the author must admit that Catholic congregations remain peculiarly inept at group singing.

[57] Herman von Helmholtz, *On Sensations of Tone*, p. 250.

[58] Herman von Helmholtz, *On Sensations of Tone*, p. 310.

[59] Herman von Helmholtz, *On Sensations of Tone*, p. 371.

[60] Much of the following material is adapted from A.S. Eve and C.H. Creasey, *Life and Work of John Tyndall*, Macmillan, London, UK, 1945.

[61] A feat repeated in his youth by the late Julian Schwinger, on a much larger *Britannica*.

[62] At the International Congress in Acoustics in London, 1974, the author was privileged to attend a demonstration lecture on acoustics at the Royal Institution, in the very room in which Davy, Faraday, and Tyndall had lectured. The lecture was given by Professor Charles Taylor from Bristol, and the demonstrations were acoustical demonstrations of Tyndall, modified to make use of modern electronic equipment. The effect was overwhelming.

[63] J. Tyndall, *Heat*. This was a popular way of presenting one's research in this period, first by lecturing, and then by publishing the lectures. As we shall see, his book *On Sound* developed in this manner.

[64] Ref. [41, p. 56ff.].

[65] John Tyndall, *On Sound*, p. x.

[66] von Helmholtz also translated Thomson and Tait's *Natural Philosophy*. It is heartening to this latter-day translator to find such a distinguished figure as Helmholtz engaging in what is generally regarded as a humdrum pursuit.

[67] Carl Sondhauss, *Ann. Physik Chemie*, **85**, 378–384 (1852); *Phil. Mag.* **5**, 73–77 (1852).

[68] G. Chevandier and E. Wertheim, *Comp. Rend. Paris* **20**, 1637–1640 (1845).

[69] John Tyndall, *On Sound*, pp. 94–95.

[70] In his Introduction to the 1954 reprinting of von Helmholtz's book (Ref. [1, first page of Introduction]), Henry Margenau pointed out the debt of the scientific community to the translator, Alexander Ellis, for his rendering of the German "Klangfarbe" as "quality," rather than Tyndall's phrase of "clangtint," and of "upper partial" for "Obertone," rather than Tyndall's use of "overtone." The use of "clang" here should be replaced by "tone."

[71] John Tyndall, *On Sound*, p. 200ff.

[72] John Tyndall, *On Sound*, p. 209ff.

[73] John Tyndall, *On Sound*, p. 210.

[74] A. Kundt, *Ann Physik Chemie* **127**, 497–523 (1866).

[75] Rijke's work [*Phil. Mag.* **17**, 419–422 (1859)] will be discussed in Chapter 6.

[76] John Le Conte, *Amer. J. Sci.* **25**, 2 (1858); *Phil. Mag.* **15**, *235 (1858)*.

[77] A substantial list of these, and a discussion of some of them, may be found in A.T. Jones, *Sound*, Van Nostrand, New York, NY, 1937, pp. 232–238 and 438–439.

[78] John Tyndall, *On Sound*, p. 276.

[79] John Tyndall, *On Sound*, p. 280.

[80] John Tyndall, *On Sound*, p. 390.

[81] John Tyndall, *On Sound*, Frontispiece.

[82] John Tyndall, *On Sound*, p. 325.

[83] John Tyndall, *On Sound*, p. 331.

[84] John Tyndall, *On Sound*, p. 334.

[85] D.C. Miller, *Anecdotal History of the Science of Sound*, Macmillan, New York, NY, 1935, p. 81.

4
Lord Rayleigh and His Book

That child will either be very clever or be an idiot.
John Holden Strutt
on seeing his first grandson [1].

Lord Rayleigh's influence on acoustics (and, indeed, on physics in general) has been so great as to be overwhelming and daunting to anyone attempting to say something briefly that can do justice to him and not be repetitious of what has already been said many times. Nevertheless, we must make the attempt. In this chapter we shall summarize Rayleigh's life, and follow his work in acoustics through the appearance of the second edition of his famous book (1894). His subsequent work in acoustics will be discussed in the framework of other developments in acoustics.

John William Strutt, the future 3rd Baron Rayleigh (*our* Lord Rayleigh), was born on the family estate in Terling, Chelmsford, Essex, in 1842. His grandfather, Colonel John Holden Strutt, had been offered a baronetcy by George III, but declined it, suggesting that it be assigned to his wife [2]. Thus, Lady Strutt (née Fitzgerald, daughter of the Duke of Leinster and sister of an Irish revolutionist who died in the rising of 1798, Lord Edward Fitzgerald) became the first Baroness Rayleigh, while her husband remained Colonel Strutt.

4.1. Rayleigh's Early Life

John Howard, who maintains the Rayleigh Archives at the Hanscom Field Research Center in Bedford, MA, has liked to say of Lord Rayleigh that he had three things going for him: he was bright, he was born rich, and he married well (in the latter case, his wife was the niece of the Marquess of Salisbury, who was the Prime Minister of England several times in the period 1885–1902, while her brother, Arthur Balfour, was also Prime Minister, 1902–1905) [3]. However, there was one thing that he did lack—robust health. He was beset by numerous illnesses throughout his youth and early manhood. He started a term at Eton, spent much of it in a sanitorium recovering from "a mild case of smallpox" [4], and then caught diphtheria on holiday and that was the end of his attendance at Eton. After some private tutoring, he entered Harrow, where he had various chest problems, and was soon taken away from the school. After 4 years in a small private academy in Torquay, where he developed an interest in photography and in mathematics, he went on to Trinity College, Cambridge. From that point on, he moved forward rapidly. He had already attended a meeting of the British Association for the Advancement of Science (the B.A.) when he was still in preparatory school. While at Cambridge, he not only went to the summer meetings of the B.A. but also met with such scientific leaders of the day as Sir Charles Lyell and Sir John Herschel [5].

At Cambridge, he studied mathematics under Routh and physics under Stokes, and achieved the academic honor of Senior Wrangler in 1865, undergoing some criticism along the lines of what we would today call reverse discrimination (e.g., why waste such an honor on a dabbling Peer?). In 1866, he became a Fellow of Trinity College.

In 1868, he traveled to America. Here also, doors were open to him; he met with Secretary of State Seward and President Andrew Johnson at the White House, and traveled through the newly reconstructed South.

During this early period, Rayleigh continued his work in photography, bought his first apparatus, and began experiments in magnetization, reading a paper on the subject at a meeting of the British Association. In 1870 (Fig. 4-1) he began a correspondence with James Clerk Maxwell (1831–1879) that lasted until Maxwell's death. It is of interest to note that Rayleigh began his work in science as a private project, buying his own apparatus and

FIGURE 4-1. John William Strutt, age 28. (From the Hon. Guy Strutt.)

setting up his laboratory in his home. As he was to say on later occasions, he did the work because he enjoyed it, and because it gave him personal satisfaction. Upon being awarded the Order of Merit, he remarked "the only merit of which he personally was conscious was that of having pleased himself by his studies, and any results that may have been due to his researches were owing to the fact that it has been a pleasure to him to become a physicist [6]."

4.2. His First Researches

Rayleigh's interest in acoustics began in his Cambridge days [7]. Professor William Fishburn Donkin (1814–1869), a professor of astronomy (and the author of an early text on sound [8]), told Rayleigh that he ought to learn to read German. Whether it was Donkin's suggestion or Rayleigh's own choice, he began this effort by making his way through von Helmholtz's volume *On Sensations of Tone*. He was particularly taken by von Helmholtz's work on resonators and began his own study, which led to a lengthy paper in the *Philosophical Transactions of the Royal Society* in 1870 [9]. This was his fifth publication and it, along with a paper commenting on the work of Dr. Sondhauss on resonators that appeared somewhat earlier [10], marked the beginning of his acoustical research, a work that encompassed 128 articles and extended up to the year of his death.

In these two early papers, Rayleigh exhibited a way of undertaking scientific research that would become quite characteristic of him. He was apparently a consistent peruser of the scientific literature. He would read a paper that caught his interest, and decide that he could extend it, simplify it, or generalize it. And, it so doing, he usually improved the subject considerably. In the paper on resonances, Rayleigh introduced the concept of acoustic conductivity, the ratio of the volume velocity of flow of fluid through some acoustical element to the difference in the acoustic velocity potential between its ends. This was in exact analogy with Ohm's law for electricity and marked the beginning of a long history of such analogies in acoustics [11].

In 1850, Sondhauss published a paper [12] giving the resonant frequency of a "flask or bottle-shaped vessel" with a long cylindrical neck, and also one with a neck that was very short compared with the diameter (today we would could them both von Helmholtz resonators. Early in 1870 [13], Sondhauss wrote another paper giving a single expression that would reduce to each of the previous formulas under the limiting conditions. Rayleigh (then, the Hon. J.W. Strutt) compared the experimental results of Wertheim [14] with values calculated from Sondhauss's formula, and those calculated by a formula due to von Helmholtz [15].

Rayleigh's work in this paper was modest, and he softened his criticism of Sondhauss' work by some concluding felicitous comment:

FIGURE 4-2. Rayleigh, age 33. (From the Hon. Guy Strutt.)

In the foregoing remarks I have naturally dwelt most on my differences with Dr. Sondhauss, but I should be sorry to have it supposed that I write in a hostile spirit, or do not recognize the claims of one to whom the science of acoustics is so largely indebted. [16]

Rayleigh married Eleanor Balfour in the summer of 1871, and 6 months later, came down with rheumatic fever. He nearly died, and his lungs were badly affected. After his recovery, he had put on weight: "he got up with a middle-aged figure, and remained always easily put out of breath [17]."

Because of his weakened physical condition, it was decided that Rayleigh should avoid the oncoming British winter [18], and he and his wife set off in November, 1872, for a cruise up the Nile river on a sailing houseboat. And it was on this occasion that he began the writing of his famous book, spending every morning working on it in his cabin.

Shortly after their return to England, the second Baron Rayleigh died, and his son succeeded to the title, and to the estate at Terling in Essex, where he began setting up the laboratory he would use for the rest of his life.

Rayleigh's early fame as a physicist was established by his work on the scattering of light in the atmosphere. In 1871, when James Clerk Maxwell became the Cavendish Professor of Physics at Cambridge, Rayleigh was apparently the next possibility. Sir William Thomson (later, Lord Kelvin) wrote to von Helmholtz that year "Did you meet Strutt when you visited his family in England? I hear he would have been the new professor in Cambridge if Maxwell had not accepted [19]."

When Maxwell died in 1879, the professorship was indeed offered to Rayleigh (Fig. 4-3). He accepted it, in large measure because there was an agricultural depression at the time and his farm tenants were having diffi-

FIGURE 4-3. Rayleigh, age 37. (From the Hon. Guy Strutt.)

culties in making rent payments [20]. In other words, Lord Rayleigh was generous toward his tenants and consequently short of cash. Four years later, when conditions had improved, Rayleigh gave up the post and returned permanently to his home and laboratory at Terling.

Rayleigh's taking on of the Cambridge Professorship raised a few eyebrows. *Punch* devoted a paragraph to him, noting that "Lord in the Professor's Chair . . . Cambridge till now has never seen, nor Oxford neither," but added "The name is a happy augury. Such blended rays of rank and science blend in this Lord-High Professor's aureole, that he would be more than mortal did not his very gait proclaim his race," "Verus et incessu patuit Strutt [21]!"

4.3. His Interaction with Tyndall

In his early career, Rayleigh interacted strongly with Tyndall on at least two occasions. Tyndall had done a great deal of work on the scattering of light in passing through various vapors and through fluids with particles suspended in them (what we now call the Tyndall effect). In particular, he was able to simulate the blueness of the sky by such suspensions. These experiments stimulated Rayleigh into developing the mathematical theory of scattering of light by the molecules of the air, thus deriving his famous fourth-power scattering law in 1871 [22].

The second interaction involved Rayleigh's work on the reciprocity law for acoustics. In 1860 [15], von Helmholtz enunciated his reciprocity law, namely, that, as Rayleigh later put it, "if a uniform frictionless gaseous medium be thrown into vibration by a simple source of sound of given period and intensity, the variation of pressure is the same at any point B when the source of sound is at A as it would have been at A had the source of sound been situated at B [23]." In two papers [23], [24] Rayleigh generalized this law to apply to all acoustical systems, and included the effect of damping. This theorem has found widespread use in modern times in the calibration of microphones and other transducers [25], and has been extended to an inhomogeneous medium by Leonid M. Lyamshev [26].

In 1875, Tyndall produced a paper involving the following experiment. A high-pitched reed was mounted in a short tube and air blown through the tube [27]. Tyndall used his sensitive flame [28] to detect pressure variations at a second point. Now, if a piece of cardboard or glass used as a screen was placed between the reed and the flame, it was found that the effect on the flame was different, depending on whether the screen was placed close to the flame or close to the reed. As Rayleigh pointed out, "the motion of the screen is plainly equivalent to an interchange of the reed and flame, there is to all appearances a failure in the law of reciprocity [29]."

Rayleigh was deeply concerned with this difficulty, and paid a visit to the Royal Institution, where Tyndall demonstrated the experiment to him. On the same occasion, Tyndall's assistant demonstrated the directivity of the sound source to Rayleigh. Since the principle of reciprocity required that the source be a simple one, i.e., one that emits sound equally in all directions, it became clear that the experiment did not overturn the principle. Rayleigh took this opportunity, however, to generalize the theorem to include "double sources," and what we today would call dipole sources. This was one of many occasions in which Rayleigh took a difficulty proposed by someone else, and turned it into a positive scientific advance.

When Tyndall's health began to fail in the late 1880s, he retired from his lecturing at the Royal Institution, and was replaced by Rayleigh. The economical bent of nineteenth-century scientists is made clear from the following excerpts from letters between Rayleigh and Kelvin, and Rayleigh and Tyndall:

Rayleigh to Kelvin (16 Feb. 1888)

I am now established in the Royal I[nstitution]. The apparatus has been allowed to fall behind altogether, of which I may give you an idea when I say that there is not an ohm in the place! [30]

Tyndall to Rayleigh (16 Dec. 1887)

You are quite right, but our poverty as to apparatus was self-imposed. We did not buy, but we borrowed, and paid for the loan. This was Faraday's plan and mine. It answered. [31]

This was the day of experimental science by means of "string and sealing wax," when new assistant professors did not require at least a quarter of a million dollars just to get their research started.

4.4. Summary of his Later Career

Rayleigh's return to Terling did not sever him from active relations with science in either London or Cambridge. He served terms as Secretary (1885–1896) and President (1905–1908) of the Royal Society, and was successively awarded its Royal medal (1882), the Copley medal (1899), and the Rumford medal (1914). He was also Chancellor of the University of Cambridge from 1908 until his death in 1919.

Enormous though the work of Rayleigh in acoustics was, it formed only a part of his total research, which involved standardization of the basic quantities in electricity, optics, the nature of gases, and the discovery of argon (for which he was awarded the Nobel prize in 1904). Rayleigh was known to remark that he liked to conduct three or four researches at one time, so that when he experienced difficulties with one of them he could turn to another, and then return, fresh, to the original work. While many of us may follow a similar program in our own research work, it was Rayleigh's talent that he could do it in several totally different branches of physics, and make noteworthy contributions to all of them.

4.5. Other Early Work in Acoustics

In the nineteenth century, it was frequently observed, usually in some country setting, that the echo of a sound was often heard as a higher harmonic of the source [32]. Rayleigh actually observed such a phenomenon and quickly realized that it could be explained by his fourth-power law of light scattering. Applying that law to problems of sound waves for situations in which the dimensions of the scatterer are small in comparison with the wavelength, he noted that "on a composite note . . . a group of small obstacles will return the first harmonic, or octave, sixteen times more powerfully than the fundamental [33]." The branches of trees in a wood could provide such scattering.

While working on his book, Rayleigh published a paper on "Some General Theorems Relating to Vibrations. [34]" This paper included some work on the reciprocity theorem mentioned above, a proof that the addition of a mass to a vibrating system could never lower the period of its motion, and the introduction of the dissipation function. This sweeping together of a collection of notes on specific problems of acoustics, optics, or another field became a characteristics of Rayleigh. The "Acoustical Observations" were later numbered, are reached a total of eight, and there were several similar publications that could have been added to the series.

In 1876, Rayleigh published a lengthy paper on waves [35], in which he considered a special case of waves in a uniform canal of rectangular cross section in which the wavelength is large compared with the depth of the canal, but the maximum height of the wave is small in comparison with the latter quantity. Aspects of this problem had been explored earlier by Airy [36]. Rayleigh was interested in the case in which the cross section undergoes a gradual alteration because of a change in the vertical dimensions of the tube. In attacking this problem, he considered the water in the canal to be moving with a velocity equal and opposite to that of the wave, so that the problem becomes one of steady motion. In this way he anticipated an approach commonly used in the analysis of shock waves.

In the same paper, he reviewed the work of J. Scott Russell (1808–1882) on a solitary wave [37], and pointed out the difference in subsequent behavior between a positive wave (elevation of the medium) or a negative one.

We have mentioned earlier the observations of Chladni and others on binaural hearing and its use in determining the direction of a sound source. Rayleigh picked up this problem in 1876 [38] and considered the possibility that the effect was due to a difference in the sound intensity between the two ears. On the basis of experiment, however, he concluded that this was not the basic cause of the phenomenon. In its way, this was a pioneering paper in psychological acoustics, and Rayleigh would return to the subject again and again [39], [40], [41], ultimately coming to the (correct) conclusion that the phase difference of the two sounds was the determining factor.

This first paper on the perception of the direction of a sound source was part of Acoustical Observations I. To get a glimpse of the breadth of the work in such a collection, we might note that this same paper contains remarks on the acoustical shadows cast by various objects, the audibility of different consonants at a distance, the interference pattern produced by two tuning forks, the presence of octave sounds in tuning forks, the influence of a flange on the correction for the open end of a pipe, and some comments on the length corrections for the pitch of organ pipes.

In 1877, Rayleigh's publications in acoustics were in transition. He was publishing papers and at the same time finishing off his book, and some of the material was discussed in both places. Thus, in a paper on progressive waves [42] in 1877, he refers to the velocity of a group of water waves, notes that the explanation of the slower velocity of the group had been given by Stokes [43], and also mentions work by Froude [44]. Rayleigh then notes that he has considered the problem more generally in his book, arriving at the result that the group velocity c_{gr} would be given by the derivative $d(kc_{ph})/dk$ [$= d\omega/dk$], where c_{ph} is the phase velocity and $\omega = kc$; more succinctly, $c_{ph} = \omega/k$, $c_{gr} = d\omega/dk$.

4.6. The Theory of Sound

As remarked above, Rayleigh began work on his book while cruising up the Nile in 1872. The first volume appeared in 1877, and Volume Two followed in 1878 [45]. The entire text was revised in 1894. In the revision, Rayleigh maintained the structure of the first edition, identifying the new sections by adding a letter to the number of the preceding section. Thus, there are 397 numbered sections in the first edition, with more than 60 sections added in the second edition.

The coverage of the field is so comprehensive that there might not be any sensible way of describing it here except perhaps to recommend its reading by the interested observer [46]. Fortunately, however, there is another way. Reviews of the volumes in *Nature* were written by von Helmholtz [47], [48]. This was a rare occasion for any field, in which the two most significant figures in the history of the science were put in the position of one commenting on the work of the other, and we could hardly do better than by reprinting von Helmholtz's remarks in their entirety. This is done in the Appendix.

von Helmholtz, who was 20 years senior to Rayleigh, observed that Rayleigh, who had "acquired a respected name in the domain of science," has undertaken to give a complete and coherent theory of the phenomena of sound . . . and does this with the application of all the resources furnished by mathematics, since without the latter a really complete insight into the causal connection of the phenomena of acoustics is altogether impossible [47]."

It is in fact, the extensive and brilliant use of mathematics that most distinguishes the Rayleigh book from its predecessors. In the works that we have discussed of Chladni, Young, and Tyndall, there was only a minimal use of mathematics, and von Helmholtz, while he filled in mathematical detail in many of his appendices (and certainly knew the subject as well or better than anyone), deliberately kept the main text on a nonmathematical level. Rayleigh not only used mathematics, he created it, and many of his derivations and developments found later use in other branches of physics.

In his review of Volume II, von Helmholtz made the following remark:

At the end of the volume Lord Rayleigh has placed the words; "The End." We hope that this may be only the provisional, not the definite end. [48]

After listing a number of topics, mainly in music, that needed further elucidation, von Helmholtz concluded

I believe I am speaking in the name of all [physicists and students of physics] if I express the hope, that the difficulties of that which yet remains will incite him to crown his work by completing it. [48]

Rayleigh responded to some of these problems in the second edition, but there was to be no Volume Three.

4.7. The Period 1877–1894—Between Editions

Threshold of Audibility

In 1870, Ludwig Boltzmann (1844–1906) and August Toepler (1836–1912), working at the University of Graz in Austria, attempted a measurement of the faintest sound that the human ear could detect [49]. In that paper, the authors used von Helmholtz's theory of the open organ pipe, combined with measurements of the maximum condensation within the pipe. They arrived at a value for the maximum displacement velocity of 0.022 cm s^{-1} at a frequency of 181 Hz. This translates into a peak pressure of 0.97 dyn cm^{-2} (rms value, 0.69 dyn cm^{-2}), which is considerably above the accepted value today.

In 1877, Rayleigh published a note on this same problem, without being aware of the work of Toepler and Boltzmann [50]. By measuring the longest distance at which he could still hear the sound of a whistle, whose frequency was known (2730 Hz), and by careful measurement of the amount of air blown through the whistle, he was able to arrive at a value for the displacement amplitude of 8×10^{-8} cm, i.e., slightly less than one ten millionth of a centimeter. Rayleigh remarks, "I am inclined to think that on a still night a sound of this pitch whose amplitude is only one hundred millionth of a centimetre, would still be audible." He was quite right. The velocity amplitude that he measured was equal to 1.4×10^{-3} cm s^{-1}. Modern measurements indicate that the actual value of the audibility threshold at this frequency is nearly one hundred times less than Rayleigh's value. It is of interest to note that Rayleigh chose a frequency that was very close to that of the maximum sensitivity of the ear.

In a later publication (1894, Ref. [51]), Rayleigh returned to the problem, this time using tuning forks as his sound source, and calculating the energy involved by using an observing microscope to determine the actual amplitude of motion of the tines of the fork. In these experiments, Rayleigh obtained a value of velocity amplitude of the sound in air that was about one order of magnitude smaller than that achieved in his early work.

The Rayleigh Disk

In his measurements on the threshold of audibility, Rayleigh had to follow a complicated path to determine the intensity of his sound source. In 1881, in the course of his work on the determination of the absolute value of the electrical ohm [52], Rayleigh noted that a light disk, suspended in an air flow, or in a sound beam, tended to line itself up with its plane perpendicular to the direction of the fluid motion. He therefore proposed a new instrument to measure the absolute value of the sound intensity [53]. If a thin disk is suspended in an air flow or sound beam, as shown in Fig. 4-4 [54], with its plane at an angle of 45° to the flow direction, then the particle velocity at P_1 and P_2 will be a minimum, and the pressure a maximum, so that a torque is exerted on the disk. With proper arrangement of light source and mirrors, even very small deflections of the disk can be determined [55], [56].

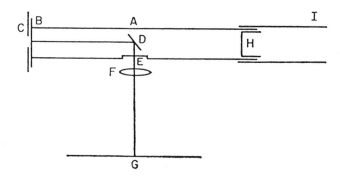

FIGURE 4-4. The Rayleigh disk [54]. A mirror D is suspended by a silk thread. Light is emitted from the slit C, and its reflection from the mirror is focused at G. If a standing acoustic wave is introduced in the tube, with a node at the mirror, the mirror will be rotated slightly by the beam and the light at G will be deflected.

Wilhelm König later determined the torque on the disk [57] to be given by

$$L = \tfrac{4}{3} a^3 \rho u^2 \sin 2\theta. \tag{4.1}$$

Here a is the radius of the disk, ρ the mean density of the fluid, u the displacement velocity, and Θ the angle of deflection of the disk from its position (see Fig. 4-4). Rayleigh made use of this relation in the second edition of his book, noting that "if the stream be alternating instead of steady, we have merely to employ the mean value of u^2 [58]." Since the sound intensity is proportional to the mean value of u^2, the torque is seen to be directly proportional to the sound intensity. This device soon became a standard one for measurement of the intensity of airborne sound.

Vibration Theory

Another subject to which Rayleigh made important contributions at this time was vibration theory. In his paper at a symposium on the accomplishments of Lord Rayleigh in acoustics, at the Seattle meeting of the Acoustical Society of America in 1988, Stephen Crandall remarked that "the first ten chapters of *The Theory of Sound* constitute an unprecedented exposition of vibration theory [59]." In writing his text, Rayleigh in a sense was the organizer of vibration theory, and in fact, later texts by S.P. Timoshenko (1878–1872) in 1928 [60] and J.P. den Hartog (1901–   ) in 1940) [61] followed the Rayleigh pattern, which began with harmonic motion, systems with one degree of freedom, and vibrating systems in general, proceeding, in order of presentation (and perhaps, difficulty) with transverse vibrations of strings, vibrations of bars, membranes and plates, and finally (in the second edition), the study of curved plates and shells. Crandall noted that Rayleigh gave the preliminary work on nonlinear stiffness and nonlinear damping, subjects that were later to be associated with the names of Duffing [62] and J.B. Van der Pol [63], respectively.

Rayleigh continued his interest in vibrations after the appearance of his first edition. In 1888–1889, he added four more papers, three on the vibrations of different shells [64–66] and one on the free vibrations of an infinite plate [67]. These further advanced Rayleigh's contribution to the field. As Crandall summed it up, "[in] the underlying framework [in the study of vibrations], laid down by Rayleigh the sequence is: one-degree-of-freedom systems, systems with a finite number of degrees of freedom, and finally continuous systems. The primary tool is the concept of natural models with their associated natural frequencies, modal damping parameters, and normal coordinates [68]. And we owe that to Rayleigh [59]."

Musical Acoustics

The portions of Rayleigh's book that involve vibration theory laid the foundations for much of research on musical instruments in the modern era. As Thomas Rossing (1929–   ) put it at the Seattle symposium, "I do not know of a musical acoustician who does not keep a well-thumbed copy of [Rayleigh's book] in his/her personal library [69]."

Rayleigh made great efforts to establish the precise frequency of his tuning forks. In principle, the method remained that of Sauveur from the previous century. As Rayleigh put it,

the absolute frequencies of vibration of two musical notes can be deduced from the interval between them, i.e., the ratio of their frequencies, and the number of beats which they occasion in a given time when sounded together. [70]

The degree of Rayleigh's concentration and patience can be appreciated when we recall that, in a case when the number of beats was about four per second, he would count the beats for 10 minutes. Every time he reached the number 10 he made a mark on a paper. On making measurements on four successive days, he arrived at frequencies of 67.09, 67.04, 67.29, and 67.19, a variation of only one part in six hundred.

In terms of specific research on musical instruments, one can mention his detailed investigations of the modes of vibrations of bells. Whether the bell was a church bell, such as the one in the church tower at Terling, a specially cast one for measurement in his laboratory, or the simple bells of a thin wine glass, Rayleigh was interested in determining their modes of vibration and comparing them with those calculated theoretically, although he noted, somewhat wryly,

A complete theoretical investigation [of air-pump receivers finger-bowls, claret glasses as well as church bells] is indeed scarcely to be hoped for; but one of the principal objects of the present paper is to report the results of a new experimental examination of several church bells, in the course of which some curious facts have disclosed themselves. [71], [72]

Aeroacoustics

The study of the intertwining of fluid dynamics and acoustics that forms the basis of aeroacoustics was just getting under way in the second half of the nineteenth century. This intertwining involved the ancient phenomenon of Aeolian tones, edge tones, and turbulence.

Aeolian tones were reported by Athanasius Kircher (1602–1680) in 1650 [73], and frequently commented upon after that date. The tones are generated when wind sets a string into vibration. Rayleigh's first remarks on the subject occurred in his "Acoustical Observations" of 1879 [74], [75]. In that paper, he first expresses his doubt in the conventional idea that the string moves in the direction of the wind, and, using the draft created by the wire in his chimney (according to Alan Powell 1928–   ), "perhaps the first wind tunnel" [76]) he attached an illuminated bead to his string and noted that the motion of the string was at right angles to the direction of the wind.

Another comment on the aeolian tones was made by Rayleigh after a very simple observation, made upon "taking the waters" at Bath. In 1884, Rayleigh wrote in his journal:

I find in the baths here that if the spread fingers be drawn pretty quickly through the water (palm foremost) they are thrown into transverse vibration and strike one another. This seems like the aeolian string. . . . It is pretty certain that with proper apparatus these vibrations might be developed and observed. [77]

In citing the above passage in his biography of his father, the 4th Baron Rayleigh observed that it was another 30 years before this idea was pursued [78]. In that later paper, Rayleigh pointed out that sound radiation could occur in the flow of fluid past a wire when the wire was not permitted to vibrate, a point that was to be taken up by Yudin many years later [79].

Edge tones were first remarked upon by Sondhauss in 1854 [80], and later, by A. Masson in 1855 [81]. If air, emerging from a narrow slit, falls on a sharp wedge of wood or metal (Fig. 4-5) [82], musical tones can be produced. Series of swirling air masses (later called vortex trails or streets) pass out from the slit one either side, and a second stream of vortices is produced at the wedge tip. This second set of vortices apparently must keep pace with the first, so that the distance d between slit and edge makes a kind of resonance distance. This alternating nature of the detachment of the vortices (from the wedge) causes a vibration of the wedge piece just as was the case for the wire. Here again, the forces produced by these turbulent vortices produce sound.

In 1878, V. Strouhal (1850–1922) [83] determined an empirical relation between the diameter of string or wire d and the frequency of the emitted sound f, when a wind of speed v blows past it:

$$f = 0.185\,v/d. \tag{4.2}$$

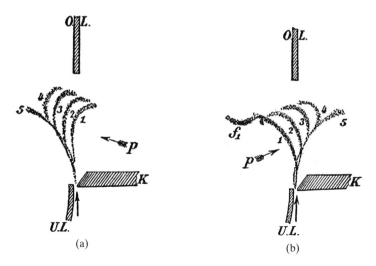

FIGURE 4-5. Edge tones [82]. Motion of the air sheet in the mouth of an organ pipe [Van Schaik]. *O.L.* and *U.L.* represent, respectively, the upper and lower lips of the pipe, and *K* the languet. Each diagram shows several positions of the air sheet, and the numbers give the order in which the positions were seen. In (a) the sheet was moving outward, and in (b) it was moving inward.

In his book [84], Lord Rayleigh noted that the Strouhal number could depend on the shear viscosity η of the medium only through the combination $U\rho d/\eta$, where ρ is the fluid density and u the volume velocity. This combination was later used by Osborne Reynolds (1842–1912) in describing the transition of flow from laminar to turbulent (Fig. 4-5), and is now known as the Reynolds' number [85].

Another way of producing sound was observed by Rayleigh as early as 1877. He considered the effect of a rigid sphere undergoing periodic oscillations in a fluid, and showed that it would act in a fashion similar to that of an acoustics dipole, which was defined by him as "the limit of two equal opposite simple sources whose distance is diminished and intensity increased without limit in such a manner that the product of the intensity and distance is the same as for two unit sources placed a unit distance apart [86]."

The next step was to regard turbulent motion of a fluid as a kind of inhomogeneity. Now the mathematical theory of the scattering of sound from small-scale inhomogeneities had been considered by Rayleigh [87]. He wrote out an equation of motion in which the d'Alembertian of the pressure

$$\left[\nabla^2 - \left(\frac{1}{c^2} \right) \frac{\partial^2}{\partial t^2} \right] p, \tag{4.3}$$

where

$$\nabla^2 = \frac{\partial^2}{\partial x^2} + \frac{\partial^2}{\partial y^2} + \frac{\partial^2}{\partial z^2}$$

was placed on the left side, and all other terms, including those involving the scattered signal, were placed on the right. In reviewing this approach, Alan Powell later remarked

We arrive at the standard wave equation by assuming small perturbations of a uniform stationary acoustic medium, dropping all the second-order terms. If we re-derive the wave equation, but assume the presence of a given plane wave and small patches of variations in the density and compressibility of the medium, and keep the largest of the new terms, then we get the inhomogeneous wave equation. Its right-hand side, that is, the source strength, depends on the product of the incident wave amplitude and the deviations in density and sound speed. Kirchhoff's solution to the equation gives the scattered waves—secondary waves to Rayleigh— immediately. One term is of monopole character, the other is a space derivative, and Rayleigh shows how to use Green's theorem to reveal its basic dipole nature. [88]

This analysis, which served Rayleigh well, was much later to provide the starting point for Lighthill's work on aerodynamic noise and Westervelt's scattering of sound by sound. (See Chapter 10.)

Surface Waves

In 1885, Lord Rayleigh began another field of acoustics in a short paper in which

it is proposed to investigate the behaviour of waves upon the plane free surface of an infinite homogeneous isotropic elastic solid, their character being such that the disturbance is confined to a superficial region, of thickness comparable with the wavelength. [89]

He found that these waves propagate along the surface of a solid with a velocity somewhat smaller than the velocity of shear waves in the solid, and that the amplitude of the disturbance falls off rapidly as one penetrates the solid. These waves are today called Rayleigh waves, and we shall encounter them again in Chapter 8. With his usual prescience, Rayleigh commented in the paper,

It is not improbable that the surface waves here investigated play an important part in earthquakes and in the collision of elastic solids. [90]

We shall encounter the fulfillment of this statement in Chapter 10.

Nonlinear Acoustics

Many of Rayleigh's contributions to nonlinear acoustics have already been described (Aeolian tones, acoustic streaming, the Rayleigh disk). The best known of his works in this field is on the radiation pressure produced by a sound wave in a fluid [91]. In this analysis he began with the case of a vibrating string and progressed to the conclusion that the pressure was equal to $p_{rad} = (\gamma +1)E/2$ for gases, where E is the energy density of the incident wave. This quantity became known as the Rayleigh radiation pressure, and differs from the value of $p_{rad} = E$ obtained by Langevin [92].

Near the end of his long career, Rayleigh investigated the problem of cavitation, assuming an incompressible fluid, with the motion being determined by that of the inner boundary of the bubble [93]. He then calculated the high pressures that are produced in a liquid as a result of the collapse of the cavitation bubble, a conclusion well substantiated in later work on cavitation effects on propellers and in the field of sonoluminescence.

FIGURE 4-6. The Rayleigh family in 1896. (From Fourth Baron Rayleigh, *Rayleigh* [11].)

4.8. The Second Edition of *The Theory of Sound* (1894)

The second edition of *The Theory of Sound* appeared in 1894. As Rayleigh noted in his Preface, two new chapters were added to Vol. I, one of which, on curved plates or shells, was a logical sequel to his chapter on the vibrations of plates. The second, on electrical vibrations, is an exposition of alternating current theory as of 1894, with some remarks on transmission of electric current through cables, and on the theory of the telephone. The field of electroacoustics was in its infancy, and Rayleigh was preparing his readers for understanding the connection been alternating currents and sound. Volume 2 contained four new chapters, entitled "Capillarity," "Vortex Motion and Sensitive Jets," "Vibrations of Solid Bodies," and "Facts and Theories of Audition," covering many of the topics discussed in previous portions of the current text.

The work of Rayleigh after 1894 will be covered in the appropriate places in the later portions of this book. We are not yet finished with the contributions of this most formidable figure in the history of acoustics!

Notes and References

(*Note*: Papers published before Rayleigh acceded to the title are listed under the name of J.W. Strutt, while later ones are under the name Rayleigh.)

[1] Rayleigh (4th Baron), *Life of John William Strutt, 3rd Baron Rayleigh*, Augmented edition, University of Wisconsin Press, Madison, WI, 1968, p. 9. Considerable use will be made of this text in following the life of Lord Rayleigh.

[2] Rayleigh (4th Baron), *Life of John William Strutt*, p. 4.

[3] John Howard, personal communication.

[4] Rayleigh (4th Baron), *Life of John William Strutt*, p. 13.

[5] The author has long maintained that scientific meetings were indeed, meetings, where one met people, rather than merely hearings, where one heard papers. Lord Rayleigh's experiences well confirm that judgment.

[6] Quoted in *Lord Rayleigh, The Man and His Work* (R.B. Lindsay, ed.), Pergamon Press, Oxford, UK, 1970, p. 12.

[7] R.B. Lindsay, Historical Introduction to the reprinting of Rayleigh's *The Theory of Sound*, Dover, New York, NY, 1945, p. xxv. This article has been very helpful in the preparation of this text.

[8] W.F. Donkin, *Acoustics*, Clarendon Press, Oxford, UK, 1870. Of this text, Lord Rayleigh wrote, in the Preface to volume 1 of his *Theory of Sound:* "[this book] though little more than a fragment [because of the early death of its author] is sufficient to shew that my labours would have been unnecessary had Prof. Donkin lived to complete his work."

[9] J.W. Strutt, *Phil. Trans. Roy. Soc. (London)* **161**, 77–118 (1870). This paper was apparently submitted after Ref. [6], but it was published sooner.

[10] J.W. Strutt, *Phil. Mag.* **40**, 211–217 (1870).

[11] For subsequent uses of this analogy, see G.W. Stewart and R.B. Lindsay, *Acoustics*, Van Nostrand, New York, NY, 1930, pp. 51–55, and Allan D. Pierce, *Acoustics*, McGraw-Hill, New York, NY, 1981. Reprinted by the Acoustical Society of America, Woodbury, NY, 1989, pp. 320–322.

[12] C. Sondhauss, *Ann. Physik* **81**, 235–257 (1850).

[13] C. Sondhauss, *Ann. Physik* **140**, 53–76, 219–241 (1870).

[14] W.M.G. Wertheim, *Ann. Physik* **82**, 463–464 (1851).

[15] Hermon von Helmholtz, *J. Reine Angew. Math.* **57**, 1–72 (1860).

[16] Ref. [6, p. 217].

[17] Rayleigh (4th Baron), *Life of John William Strutt*, p. 59.

[18] Rayleigh (4th Baron), *Life of John William Strutt*, 1, p. 60. The writer has great empathy with Lord Rayleigh at this point; he himself contracted rheumatic fever in March, 1933, nearly died, and also put on much weight during his recovery. It was further recommended that he avoid the ensuing New York winter, but his family resources were not equal to those of the Rayleigh family, and he had to "tough it out" in New York, in what turned out to be that city's coldest winter.

[19] Rayleigh (4th Baron), *Life of John William Strutt*, p. 50.

[20] Rayleigh (4th Baron), *Life of John William Strutt*, p. 100. During Rayleigh's years at Terling, the estate was gradually changed from wheat farming to dairying. At his death, there were 800 cows on the farm, and the sign "Lord Rayleigh Farms" had became a familiar one in southeastern England. It remains so today.

[21] *Punch*, Dec. 13, 1879. The Latin quotation ("The true Strutt is revealed by his walk,") is a modification of a line from Virgil's *Aeneid*, book 1, 1. 405: "Et vera incessu patuit dea [Venus]." Rayleigh had a lively sense of humor, and would tell the story of the woman to whom he had been introduced (before he succeeded to the title). She observed, "I always associated the name Strutt with peacocks."

[22] J.W. Strutt, *Phil. Mag.* **61**, 107–120, 274–279 (1871). See also Ref. [1, pp. 52–54].

[23] Rayleigh, *Proc. Roy, Soc. (London)* **25**, 118–122 (1876).

[24] Rayleigh, *London Math. Soc. Proc.* **4**, 357–368 (1873).

[25] Victor Nedzelnitsky, Chapter 5, in *AIP Handbook of Condenser Microphones* (George S.K. Wong and Tony F.W. Embleton, eds.), AIP Press, Woodbury, NY, 1995, pp. 103–119.

[26] L.M. Lyamshev, *Soviet Phys. Dokl.* **4**, 405–409 (1959).

[27] J. Tyndall, *The Science of Sound*, *Proc. Roy. Inst.* (1875). Reprinted by Citadel Press, New York, NY, 1964.

[28] See Chapter 3.

[29] Rayleigh, Ref. [22, p. 120].

[30] Ref. [1, p. 231].

[31] Ref. [1, p. 232].

[32] Ebenezer Cobham Brewer, *Sound and its Phenomena*, Longmans, Green, London, U.K., 1854, p. 305.

[33] J.W. Strutt, *Nature* **8**, 319–320 (1873). We would today call this the second harmonic.

[34] J.W. Strutt, *London Math. Soc. Proc.* **4**, 357–368 (1873).

[35] Rayleigh, *Phil. Mag.* **1**, 257–279 (1876).

[36] Airy, *Tides and Waves*, Art. 401.

[37] J.S. Russell, Report 14th Meeting, *Brit. Assoc. Adv. Sci.*, p. 311 (1844).

[38] Rayleigh, *Nature* **24**, 32–33 (1876).

[39] Rayleigh, *Phil. Mag.* **13**, 214–232 (1907). This idea that a phase difference could be detected bothered Rayleigh, because, as he put it, "when we admit that phase differences at the two ears of tones in unison are easily recognized, we may be inclined to go further and find less difficulty in supposing that phase relations between a tone and its harmonics, presented to the same ear, are also recognized." For further discussion, see David M. Green, *An Introduction to Hearing*, Lawrence Erlbaum, Hillsdale, NJ, 1976, Chapters 7 and 8.

[40] Rayleigh, *Phil. Mag.* **16**, 235–246 (1908).

[41] Rayleigh, *Proc. Roy. Soc. (London)* **A83**, 61–64 (1909).

[42] Rayleigh, *London Math. Soc. Proc.* **9**, 21–26 (1877).

[43] G.G. Stokes, *Mathematical and Physical Papers* (5 vols.) Cambridge University Press, Cambridge, UK 1880–1895.

[44] W. Froude, *Papers*, Institute of Naval Architects, London, 1955.

[45] Rayleigh, *The Theory of Sound*, Vol. 1, Macmillan, London, UK 1877; Vol. 2, Macmillan, London, UK 1878. Reprint of second edition, Dover, New York, NY, 1945.

[46] One could, in fact, spend most of one's career working through the book and the six volumes of published papers. One can rarely pick up either the book or the collected papers without finding something of real interest that was previously missed.

[47] Herman von Helmholtz, *Nature*, **17**, 237–239 (1878).

[48] Herman von Helmholtz, *Nature*, **19**, 117–118 (1878).

[49] A. Toepler and L. Boltzmann, *Ann. Physik* **141**, 321–352 (1870).

[50] Rayleigh, *Proc. Roy. Soc. (London)* **26**, 248–249 (1877). In his *Collected Works*, Rayleigh mentions his lack of knowledge of the work of Toepler and Boltzmann, and acknowledges the precedence of the Austrian work at the beginning of the reprinted article.

[51] Rayleigh, *Phil. Mag.* **38**, 365–370 (1894).

[52] See also, Rayleigh, *The Theory of Sound*, 2nd ed., vol. II, p. 433ff.

[53] Rayleigh and Arthur Schuster, *Proc. Roy. Soc. (London)* **32**, 104–141 (1881). The experimental setup is discussed on p. 110.

[54] Rayleigh, *Phil. Mag.* **14**, 186–187 (1882).

[55] R.W. Stephens and A.E. Bate, *Acoustics and Vibrational Physics*, 2nd ed., Arnold, London, UK, 1966, pp. 296–299.

[56] Rayleigh, *Theory of Sound*, 2nd ed., vol. II, pp. 43–45.

[57] W. König, *Ann. Physik* **53**, 51 (1891). He is not to be confused with Rudolph König, whom we shall encounter in Chapter 6.

[58] Rayleigh, *The Theory of Sound*, 2nd ed., vol. II, p. 44.

[59] Stephen Crandall, *J. Acoust. Soc. Am.* **83**, S43–S44 (1988). Only an abstract of the paper has yet appeared, and the author is grateful to Stephen Crandall for making a draft of the paper available to him.

[60] S.P. Timoshenko, *Vibration Problems in Engineering*, Van Nostrand, New York, NY, 1928.

[61] J.P. den Hartog, *Mechanical Vibrations*, 2nd ed., McGraw-Hill, New York, NY, 1940.

[62] G. Duffing, *Erzwungne Scwhimgungen bei veränderlicher Eigenfrequenz*, Braunschweig, 1918.

[63] Balthasar Van der Pol, *Phil Mag.* **2**, 978–992 (1926); *Z. Hochfreq. Tech.* **28**, 178 (1926); **29**, 114 (1927). A modern treatment of Duffing's and van der Pol's equations may be found in John Guckenheimer and Philip Holmes, *Nonlinear Oscillations, Dynamical Systems and Bifurcations of Vector Fields*, Springer-Verlag, New York, NY, 1983.

[64] Rayleigh, *Proc. Roy. Soc. (London)* **45**, 105–123 (1988).

[65] Rayleigh, *Proc. Roy. Soc. (London)* **45**, 443–448 (1889).

[66] Rayleigh, *London Math. Soc. Proc.* **20**, 372–381 (1889).

[67] Rayleigh, *London Math. Soc. Proc.* **20**, 225–244 (1889).

[68] These researches are recounted in Chapters Xa and Xb of volume I of the second edition of Rayleigh's *The Theory of Sound*.

[69] Thomas Rossing, *J. Acoust. Soc. Am.* **83**, S43 (1988). The author is indebted to Professor Rossing for supplying him with a draft of his paper.

[70] Rayleigh, *Nature*, **19**, 275–276 (1879).

[71] Rayleigh, *Phil. Mag.* **29**, 1–17 (1890).

[72] The bells at Terling are discussed in Rayleigh, *The Theory of Sound*, 2nd ed., pp. 389–393.

[73] A. Kircher, *Musurgia Universalis*, Rome, 1650.

[74] Rayleigh, *Phil. Mag.* **7**, 149–162 (1879).

[75] See also Rayleigh, *The Theory of Sound*, 2nd ed., II, pp. 412–414.

[76] Alan Powell, *Rayleigh Symposium*, Seattle, WA, 1988. *J. Acoust. Soc. Am.* **83**, S44 (1988). The author is grateful to Professor Powell for making a copy of his talk (from which this quotation is taken) available.

[77] Cited in Rayleigh (4th Baron), *The Life of the Third Baron Rayleigh*. Arnold, London, UK, 1924, p. 136.

[78] The pursuit was finally carried out by Rayleigh in *Phil. Mag.* **29**, 433–444 (1915).

[79] E.Y. Yudin, *Zh. Tekhn. Fiz.* **14**, 501 (1945).

[80] C. Sondhauss, *Ann. Physik* **91**, 214–240 (1854).

[81] A. Masson, *C.R. Acad. Sci. Paris* **36**, 257–260, 1004–1005 (1855).

[82] W.C.L. van Schaik, *Arch. Nerland.* **25**, 281 (1892). The drawings are reproduced in A.T. Jones, *Sound* Van Nostrand, New York, NY, 1937, p. 336.

[83] V. Strouhal, *Ann. Physik* **5**, 216–251 (1878).

[84] Rayleigh, *The Theory of Sound*, vol. II, p. 413.

[85] O. Reynolds, *Phil. Trans. Roy. Soc. (London)* **174**, 925 (1883).

[86] Rayleigh, *The Theory of Sound*, 2nd ed., vol. II, p. 147.

[87] Rayleigh, *The Theory of Sound*, vol. II, p. 150.

[88] A. Powell, loc. cit.

[89] Rayleigh, *Proc. London Math. Soc.* **17**, 4–11 (1885).

[90] Ibid.

[91] Rayleigh, *Phil. Mag.* **10**, 364–374 (1905).

[92] For a comparison of Rayleigh and Langevin pressures, see R.T. Beyer, *Nonlinear Acoustics*, Navy Sea Systems Command, 1974, Chapter 6; R.T. Beyer, *J. Acoust. Soc. Am.* **98**, 3032 (1995).

[93] Rayleigh, *Phil. Mag.* **34**, 94–98 (1917).

5
Inventors to the Fore!

It must be confessed that, among the civilized peoples of our age, there
are few in which the highest sciences have made so little progress as in
the United States.

De Tocqueville, 1850

de Tocqueville's remark was quite accurate for the early part of the nineteenth century. It was difficult for Americans to point with pride to any
native-born scientist during the nation's early period, always excep-
ting Benjamin Franklin. Audubon was born in France and Agassiz in
Switzerland. The bustling life of the American frontier was not conducive to
much thoughtful reflection. On the other hand, inventive genius paid off,
and there was much of that. Eli Whitney invented the cotton gin in 1793,
and began the first assembly line for the manufacture of rifles in 1798;
Robert Fulton launched his steamboat in 1815 and, for the rest of the
century, from Morse to Elias Howe to Colt to John Deere to McCormick to
Bell (born in Scotland, but an American citizen after 1884) to Edison,
invention was the order of the day, and most of the inventors were American. A number of such inventions, riding on the crest of electrical advances,
were acoustic in character, and we shall track them and their inventors.

5.1. Electromagnetism

In order to follow the work of our acoustical inventors, it is first desirable to
recount the almost explosive development of the knowledge of electricity in
the early part of the century. These discoveries, made largely by Europeans,
paved the way for the work of Henry, Bell, Edison, and others.

We begin our story with the Italian Alessandro Volta (1745–1827) [2], a
professor first at the University of Pavia and then at Padua, who invented
the electrical battery in 1800. Then in 1819, Hans Christian Oersted (1777–
1851) [3], a Professor of Physics at the University of Copenhagen reported
that there was a magnetic field surrounding a current-bearing wire [Hunt

points out in his book on electroacousics [4] that the effect had actually been observed much earlier, in 1802, by an Italian, Gian Domenico Romagnosi (1761–1835) in "Articolo sul Galvanismo," *Gazzetta di Trentino*, 3 Aug 1802]. In 1820, the French physicist, D.F.J. Arago (1786–1833) [5] found that a current in a copper wire could induce magnetism in a piece of soft iron.

In the same year, André Marie Ampère (1775–1836) [6], Professor of Physics at the École Polytechnique in Paris, discovered that two parallel wires, each bearing a current, would attract or repel one another, depending on whether the currents were in the same or opposite directions. From this, Ampère inferred that a bar magnet consisted of electrical current loops, all lying in planes perpendicular to the axis of the magnet. In the same volume of the journal in which Ampère's work appeared, Jean Baptiste Biot (1774–1862), Professor of Physics at the Collège de France, and Félix Savart (1791–1851), a physician from Strasburg, who later worked at the Collège de France, established the famous law bearing their names, a law connecting the strength of an electric current, the intensity of the magnetic field, and the distance from the wire [7].

The construction of the first galvanometer [accomplished independently by two Germans, first by Hans Christian Schweigger (1779–1857) [8], the inventor of the ammeter, and then by Johann Christian Poggendorf (1796–1877) [9] (for many years the editor of the *Annalen der Physik*), occurred in 1820. In 1825, an Englishman, William Sturgeon (1783–1850), bent a piece of iron into the shape of a horseshoe, varnished it for insulation, and then wound a loose coil of wire around it. Apparently, the wire was bare but the turns were far enough apart to prevent shorting contacts. He had, in fact, made the world's first electromagnet [10].

5.2. Joseph Henry

The stage was now set for the first world-class American physicist—Joseph Henry (1799–1878) [11]. Born in Albany, New York, of Scottish ancestry (his immigrating grandfather, who had embarked for America on the very day of the battle of Bunker Hill, had changed the family name from Hendrie [12]). At an early age, because his father was suffering from a mortal illness, Henry was sent to a tiny village away from Albany, in the care of relatives, and he grew up in that location. He was largely self-educated. An episode is recounted in which he, at the age of ten, pursued a pet rabbit under the village church (the church apparently had no basement). In the pursuit, he noticed loose boards in the church's flooring, and, being an inquisitive boy, he removed them. He thus found himself in a small room that housed the village library. And, in picking out books at random, he suddenly discovered the joy of reading unassigned material—reading just for the fun of it. He returned again and again to this library by the unconventional route, and continued his reading. Ultimately, his clandestine visits were discovered but, fortunately, it led not to his punishment, but to his admission to the library as a regular user of the collection.

At fourteen he returned to Albany. At first apprenticed to a silversmith, he studied evenings at Albany Academy, taking courses in geometry and mechanics, as well as studying English grammar on the side. He soon left the apprenticeship and began a life of teaching in local schools and of private tutoring.

Somewhat like von Helmholtz, Henry found that the only professional field in science that might be open to him was medicine, and he began such study in the Albany Academy, working as an assistant to a lecturer in chemistry and, later as librarian at an organization of townspeople interested in science, the Albany Institute. He was conducting various scientific experiments by this time, and reported on them at the institute.

After a variety of other part-time employments, including that of surveyor, Henry finally began a regular academic life as Professor of Mathematics and Natural Philosophy at the Albany Academy—age 26 (Fig. 5-1 [13]).

We remarked above that the first electromagnet was constructed by Sturgeon in 1825. The following year Henry saw it in New York City and was fascinated by it and its possibilities. He set to work building a better and more powerful magnet, i.e., one that would lift a greater weight per weight of the magnet itself. He was the first to use insulated wire in winding his coils, and this enabled him to wind them closely at an angle perpendicular to its core. By 1831, he had an electromagnet that could lift almost 31 times its own weight—a total lift of 750 lbs.

At this time, Joseph Henry in America and Michael Faraday in England were working on parallel research tracks, but the only communication

Figure 5-1. Joseph Henry, age 32. (From *Papers of Joseph Henry* [13].)

between them was through publications. During this period, Henry moved to Princeton as a professor (1832). This disruption of his research no doubt slowed his scientific progress. In addition, as his biographer Michael Coulson pointed out, "Again we are compelled to lament the fact that Henry did not bequeath to us a diary comparable to that of Faraday, to which he confided the reasons for undertaking his projects [45]." One can only guess as to how close Henry was to making the discovery of mutual induction. Faraday discovered this effect in his laboratory on September 24, 1831, and reported it to the Royal Society on November 24 of the same year.

Coulson estimated that Henry could not have learned of this publication until March 1832. Henry quickly wrote an article reviewing his own researches in the matter, an article that was published in July 1832 [15]. While this article notes Henry's earlier, but not yet published work, the credit for the discovery of mutual induction remains with Faraday. On the other hand, this same article by Henry reported his discovery of a second phenomenon—self-induction—and credit for that belongs to Henry.

After his years at Princeton, Henry moved on to Washington, DC (1846) as the Secretary of the newly founded Smithsonian Institution, and there he remained for the rest of his career. He frequently served as an adviser to US presidents, including Pierce and Lincoln, and in 1862, was one of the founders, and first president, of the National Academy of Sciences.

Henry played several acoustical roles during his lifetime. In 1856, he had been engaged in measuring the acoustical properties of the lecture hall in the Smithsonian, when he was asked by President Pierce to look into the plans for the addition of rooms to the US Capitol. The result was a detailed study of the existing state of knowledge of room acoustics [16]. As his biographer put it,

Instead of the microphone and electronic measuring devices, Henry had to employ a pair of clapped hands, a tuning fork, or his own voice. His ear was his measuring instrument. Yet the conclusions he reached are consistent with our modern conception of the satisfactory acoustical construction of a large room in which speeches are to be made and heard. [17]

In Chapter 3, we discussed the work of Tyndall relative to sound propagation in the atmosphere. Henry also became involved in the problem of the erratic nature of reception of signals from fog horns and participated in studies in the waters off Pt. Judith and Block Island in Rhode Island [18]. Henry noted that sound echoes could be produced under all weather conditions, even foggy ones, and therefore concluded that Tyndall's idea of large, "acoustic clouds" of invisible water vapor that reflected the sound on clear days would not be tenable. Henry proposed that the reflections were coming from the waves on the water surface, when the sound waves were refracted down onto it. The issues were not yet settled, and discussions about this phenomenon would continue into the next century.

5.3. Henry and the Telegraph

Henry's work with electromagnetic induction was not acoustical, but it turned out to have acoustical consequences. Henry had noted that an electromagnet could be operated at some distance from its operating battery, provided that the wires were in a series connection. This was a form of signaling at a distance—a "telegraph." As described by Hunt in his book *Electroacoustics*,

Henry exhibited to his classes at the Albany Academy in 1831 a telegraph using an "intensity" magnet operated by an "intensity" (that is, a series-connected) battery at the end of a mile of wire. . . . This demonstration telegraph was the first to make use of the linear attraction of an armature by an intermittent electromagnet . . . and the first to use electromagnetism to activate an electroacoustic transducer. Henry's receiving trandsucer operated as a polarized "sounder"; the position of an armature would reverse when the current was reversed and make evident the transmitted signal by the sound produced by the impact of the armature against its stops. [19]

The clicking of the telegraph key reminds us that this electromagnetic device had an acoustical component. While the credit for putting everything together in the "printing telegraph" goes to Samuel F.B. Morse (1791– 1872), who patented his device in 1837, the role of Henry should not be forgotten. As Hunt writes,

it appears that Morse was actually anticipated by others in regard to almost every individual feature of his system, as might be illustrated by citing the Henry "intensity" magnet and relay, the Gauss–Schilling [Karl Friedrich Gauss, 1777–1855, Baron Paul Schilling, 1786–1837 [20]] code, the earth return of Steinheil [Karl August von Steinheil, 1801–1870] . . . [n]evertheless Morse's imaginative combination of these features into a workable system suitable for commercial telegraphy did represent invention of the highest type. [21]

5.4. Cooke and Wheatstone

The idea of telegraphy—the writing at a distance—has a long history. An article by an unnamed (except for the initials C.M.) author in 1753 [22] described in detail a method of using electrostatic charge to transfer signals over 26 separate wires (one for each letter of the alphabet). The book by Marland lists a dozen or so "telegraphs", all prior to or contemporary with the work of Henry and Morse [23]. The most significant of these were those developed by Cooke and Wheatstone, who, like some of the "duplex" telegraphs that were later to be developed, worked both in cooperation with, and in opposition to, one another.

William F. Cooke (later, Sir William, 1806–1870) was a physician who studied at Heidelberg and saw some of the early attempts at telegraphy there. In 1836–1837, he developed an instrument in which the telegraphic key closed a dc circuit that allowed a clock mechanism to turn a dial on instruments at each end of the system. The dial turned until the appropriate letter was reached, at which point the sender would break the circuit. The sender would then begin a new letter. It was a slow and laborious process, and worked only over short distances. Cooke tried to sell this system to the British railroads, but was unsuccessful.

In 1837, Cooke met up with Charles Wheatstone, who had also begun work on telegraphy. Wheatstone contributed his knowledge of electrical science, while Cooke had great skills as an organizer and commercializer. They set up a partnership in 1837 and produced instruments in which an electromagnet was placed in the field of a permanent magnet and then pivoted so as to turn when a current passed through it. Shortly after a meeting with Joseph Henry in London in April, 1837, they switched to using a magnetic needle suspended in the field of an electromagnet (a setup that Cooke had employed earlier), and on July 10, 1837, they were granted a patent on their device [24]. This and later refined models (employing a number of separate wires, see Fig. 5-2) were used in England for some years as railroad signaling devices.

5.5. The Telephone: First Stirrings

Like the telegraph, the telephone had many fathers, although history assigns the credit only to one—Bell. We begin our story with Robert Hooke, who wrote in 1664:

I have, by the help of a distended wire, propagated the sound to a very considerable distance in an instant, or with as seemingly quick a motion as that of light. [25]

While he did not give details, it is thought that Hooke's idea may have been the forerunner of the "string telephone" that children use in connecting two paper cups or metal cans with a stretched string between them [26]. This ability of a wire was also known to Chladni, who describes an experiment in which the ends of a wire are held between the teeth of two observers, who speak softly to one another [27].

The next advance was more of a wild idea, somewhat in the spirit of Jules Verne. A French electrician, Bourseul (1829–1912), wrote in 1854

Suppose that a man speaks near a movable disk, sufficiently flexible to lose none of the vibrations of the voice; that this disk alternately makes and breaks the

FIGURE 5-2. Wheatstone's telegraph. (a) Cooke and Wheatstone's five-needle telegraph, 1837; (b) Wheatstone's "Hatchment Dial"; and (c) Alphabet of the single-needle telegraph. (From Hubbard [24].)

(a)

(b)

(c)

FIGURE 5-3. Philip Reis, inventor of the telephone. (From Brockhaus Enzykl [29].)

connection with a battery; you may have at a distance another disk which will simultaneously execute the same vibrations. [28]

The first to make an instrument that actually provided some of the characteristics of a telephone—and the first to use the word "telephone" in this connection—was Philipp Reis (1834–1874) (Fig. 5-3 [29]), a science teacher in southern Germany [30], whose name and models of his instrument were prominent features in the lengthy courtroom fight over Bell's telephone patent, Reis died early, before the time of Bell's patent, and was therefore unable to present his views in the ensuing controversy.

In Reis' telephone [31] (Fig. 5-4), the transmitter consisted of a stretched membrane bc (which was often the skin of a sausage) was used as the diaphragm, and a contact finger cd rested lightly against a small metal electrode g. In principle, the membrane would vibrate and the connection between the contact finger and the electrode would vary with the vibrations. Exactly how this variation took place, and how faithful it could be to the vibrations of the air was a matter of conjecture during the 1870s and 1880s. As we shall see later, the US Supreme Court decided that it was not an operative instrument.

Hunt noted in his book that Reis' receiver, shown in the lower portion of Fig. 5-4, was even more interesting than his transmitter. It consisted of a coil of wire surrounding an iron rod, m, and mounted in a resonant box, w. The rod was attached to a wooden diaphragm, i. Since the instrument was relying on length changes in the iron rod (i.e., magnetostriction, see Chapter 6), Hunt observed that "it can be accorded the distinction of standing as the first magnetostriction loudspeaker—perhaps the first loudspeaker of any type [32].

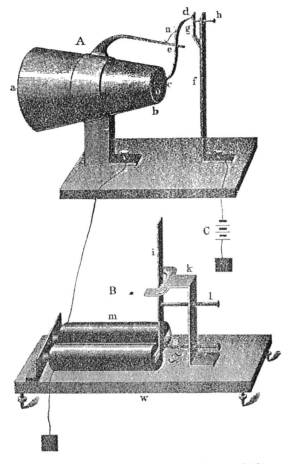

FIGURE 5-4. Reis's telephone. (From Prescott [31].)

5.6. Alexander Graham Bell. The Early Years

It is a remarkable thread that connects the teaching of the deaf with the invention of the telephone, and it is one well worth following. In Chapters 1 and 2, we commented on the two schools of thought in the teaching of the deaf, namely, the oralists and the signers, and their potential financial supporters. In his efforts at establishing the future Gallaudet College, which was in the signers' camp, Edward Gallaudet had the backing of Amos Kendall (1789–1869), a successful financier who was associated with Samuel Morse in the development of the telegraph service [33]. A prominent figure in the oralist community was Gardiner Hubbard, a Boston lawyer, and, later, chairman of the Board of Trustees of the Clarke Institution for the Deaf. Hubbard was also interested in multiplexing the telegraph, i.e., in having a single line carry a number of messages simultaneously. And he was concerned about the education of his deaf daughter. At all three levels, he would come to interact with a young Scot named Alexander Graham Bell.

Bell was born in Scotland in 1847 to a family that had in the two preceding generations been strongly interested in speech and public speaking [34], [35]. His father, Alexander Melville Bell (1819–1905), taught diction and elocution, first in Scotland, then in London. He developed a list of what he maintained were the fundamental speech sounds of all languages, which he associated with different shapes of the tongue and lips, invented symbols for them (based on tongue and throat positions involved in their uttering) and published a book on "visible speech," which he attempted to sell to speech teachers, ultimately including teachers of the deaf. While it was never especially successful, his "visible speech" served as a forerunner of the universal phonetic alphabet currently in widespread use.

Melville Bell's three sons (of whom Alexander was the middle one) were associated with their father's work almost from childhood, and when the other two sons died in early manhood, Alexander [36] was left to carry out work with and for his father. He began his teaching career at the age of 16, as a teacher of music and elocution, at a private school in Elgin, Scotland.

Even in this early period, Alexander Graham Bell began experimenting with speech and tuning forks, using the latter to establish the resonant frequencies of his own voice. Once, in early childhood, Bell and his father paid a visit to the home of Sir Charles Wheatstone, primarily to see his improved model of von Kempelen's talking machine that Wheatstone had constructed some quarter of a century earlier (see Chapter 2). Wheatstone also lent the elder Bell his copy of von Kempelen's book explaining the apparatus. Thus, the boy was early stimulated into trying to build models that could reproduce the human voice [37].

About this time, Bell also became acquainted with an associate of his father, Alexander Ellis, a famous phonetician of his day [38], who introduced young Bell to the work of von Helmholtz. Bell was particularly impressed by von Helmholtz's tuning-fork sounder (Fig. 3-4). He misinterpreted what the instrument could do, thinking that it was a transmitter of sound, and this set him into thinking about the possibilities of transmitting sound by wire. A quotation by a Boston magazine of the period, with reference to work of Bell's father, is even more appropriate to the son: "he found it [here, we might say, the telephone] because he happened to take the way to the place where it was, while the learned men were misled by their learning to seek for it where it was not [39]."

Meanwhile, Alexander Bell had taken up teaching of the deaf in England. Upon the death of Alexander's two brothers, Melville Bell, who had already visited America, and given a set of Lowell lectures in Boston in

1868, moved with his wife and son Alexander to Brantford, Ontario, Canada, in 1870.

After the senior Bell had again lectured in Boston (1871), he was approached by two schools for the deaf—the Boston School (later the Horace Mann School), and Clarke School for the Deaf in Northampton, Massachusetts, to teach his visible speech method in those schools. Since Melville Bell felt he was too busy in Ontario, he recommended his son. In June of 1871, Alexander Bell began his career in the United States as a teacher of the deaf, at the same time taking on students for private instruction.

As we mentioned above, Gardiner Hubbard had a daughter (Mabel) who lost her hearing at the age of five through scarlet fever. Hubbard would not rest until he found the best means for instructing her in her deafness. He was not satisfied with the signing method of the Gallaudets, and, when he heard about Samuel Gridley Howe's work with the oral method, he contacted him and the two joined ranks to start an oralist school in Northampton, helped materially by a considerable gift from a businessman, John Clarke. In the summer of 1872, Hubbard and Bell met for the first time.

In 1873, Bell was offered a professorship at the newly created Boston University. With his work in that position and with his private tutoring (Mabel Hubbard was now one of his students), he was left with little time for his researches on the telegraph (Bell was attempting to devise a scheme that would allow two or more electrical signals to pass simultaneously along a wire). He was also impressed with the manometric flame of Koenig [40], in which the pronounciation of different vowels produced different flame patterns. Bell wrote "If we can find the definite shape due to each sound, what an assistance in teaching the deaf and dumb!!"

In the same period, Bell met Elisha Gray (1835–1901), who was working on the transmission of tones by wire. Gray's device employed a circuit-interrupter and a polarized relay, so that, when the key of a telegraph closed the primary circuit, the relay vibrated with the frequency of the circuit interrupter. He referred to this device as a telephone [41]. Gray was also involved in the creation of the Western Electric Company, and therefore had substantial backing. Bell now had a formidable rival.

Whenever he could, Bell continued to work on his so-called harmonic telegraph. The basis of the Morse system of telegraphy was the interruption of a continuous current by the use of the telegraph key, i.e., the Morse code. In the period around 1870, there were many individuals trying to improve the telegraph. Up to that time, messages could only be sent one at a time, in one direction, over a single telegraph wire. Bell came up with the idea that different interrupters could be used, each one tuned to the frequency of a separate tuning fork. The tuning fork sounder of von Helmholtz showed how an interrupter could be established at the frequency of a tuning fork. Bell thought in terms of having a number of such interrupters, each at a different frequency of interruption. The currents from these could all be transmitted simultaneously on the same wire, and sorted out at the receiving end by the use of other tuning forks, thus achieving multiplexing [42]. This constituted the co-called "singing telegraph." Hubbard was very much interested in this idea and the two interacted on the problem. Here again there was competition, since Thomas Edison (1847–1935) had invented his own multiplexer and had sold it to the Western Union Co. shortly before.

5.7. Bell's Telephone

But Bell's mind was drawn away from the telegraph. In 1916, Bell described his thought process during the autumn of 1874:

I was at work on a problem of transmitting musical sounds by a telegraphic instrument by an intermittent current of electricity, and I had dreams that we might transmit the quality of a sound if we could find in the electric current any undulations of form like these undulations we observe in the air.

I had gradually come to the conclusion that it would be possible to transmit sounds of any sort if we could only occasion a variation in the intensity of the current exactly like that occurring in the density of the air while a given sound is made.

If you were to take a phonautograph tracing of, we will say, the vowel /ah/, you would have a vibration of a certain shape. If you could get an electrical current in which that would cause variations, you would get a current that would transmit the whole /ah/.

I had reached this idea and had gone one step further. I had obtained the idea that theoretically you might, by magneto electricity, create such a current. If you could only take a piece of steel, a good chuck of magnetized steel, and vibrate it in front of the pole of an electromagnet, you would get the kind of current we wanted . . . it struck me that the bones of the human ear were very massive, indeed, as compared with the delicate thin membrane that operated them, and the thought occurred that if a membrane so delicate could move bones relatively so massive, why should not a thicker and stouter piece of membrane move my piece of steel. And the telephone was conceived. [43]

These ideas were developed at his family home in Brantford, Ontario. It was at this point that he ceased to follow the technique of the use of an interrupted current to produce voice signals; he now worked with fluctuations of a continuous current, fluctuations that faithfully reflected those of the density of the air in the presence of a sound wave.

This change in strategy led to a disagreement with Gardiner Hubbard, who wanted the research emphasis placed on the multiple telegraph. The situation was complicated by the fact that Bell was on the verge of proposing marriage to Hubbard's daughter. But, *amor omnia vincit*, and Bell was able to marry Mabel Hubbard and pursue his telephone research to a successful conclusion, both in the same year [44] (Fig. 5-5 [45]).

In March, 1875 [46], Joseph Henry, who had participated in the development of the telegraph 40 years before, now made a contribution to the

FIGURE 5-5. Alexander Graham Bell, age 34. (From Mackenzie [45].)

telephone. Bell came to visit the old scientist at the Smithsonian. Henry was enthusiastic about Bell's work, and urged him not to publish what he had done thus far, but to perfect the invention himself. To Bell's statement that he didn't have much knowledge of electricity, Henry responded with a laconic and commanding "Get it!"

Bell took Henry's advice, and worked steadily on the invention. By the summer of 1875, he had found that the vibration of a permanent magnet over an electromagnetic coil would lead to a sound from the receiver of his instrument. He then concentrated on the use of diaphragms. The sequence of events describing Bell's invention, recorded in his own handwriting), is now located in the Bell Museum at Brantford (Fig. 5-6) [47].

On 14 Feb. 1876, Bell filed for a patent on his telephone—just in time, for on the afternoon of the same day, Elisha Gray filed a caveat (i.e., his intention to file for a patent in the very near future). While Bell had won this round, the patent being granted on March 7, 1876 (Fig. 5-7), lawsuits would abound for the next decade, until the US Supreme Court ruled in 1887 that Bell had indeed invented the telephone, and that his patent was a valid one [48].

The basic form of the Bell telephone is shown in Fig. 5-8 [49]. A person would speak into the tube T, which would act on the membrane stretched across the mouth of the tube. A light (permanent) bar magnet ns is attached to the diaphragm. Just beyond the magnet was another electromagnet through which a constant current is passing. When the bar magnet moves under the action of the voice, an additional current is induced in the circuit. This current in turn activates the receiving electromagnet R, thus moving the armature r in regular response to the original voice.

From the table of events given by Bell, it is clear that the telephone he had at the time of the patent filing was not very useful. Complete sentences

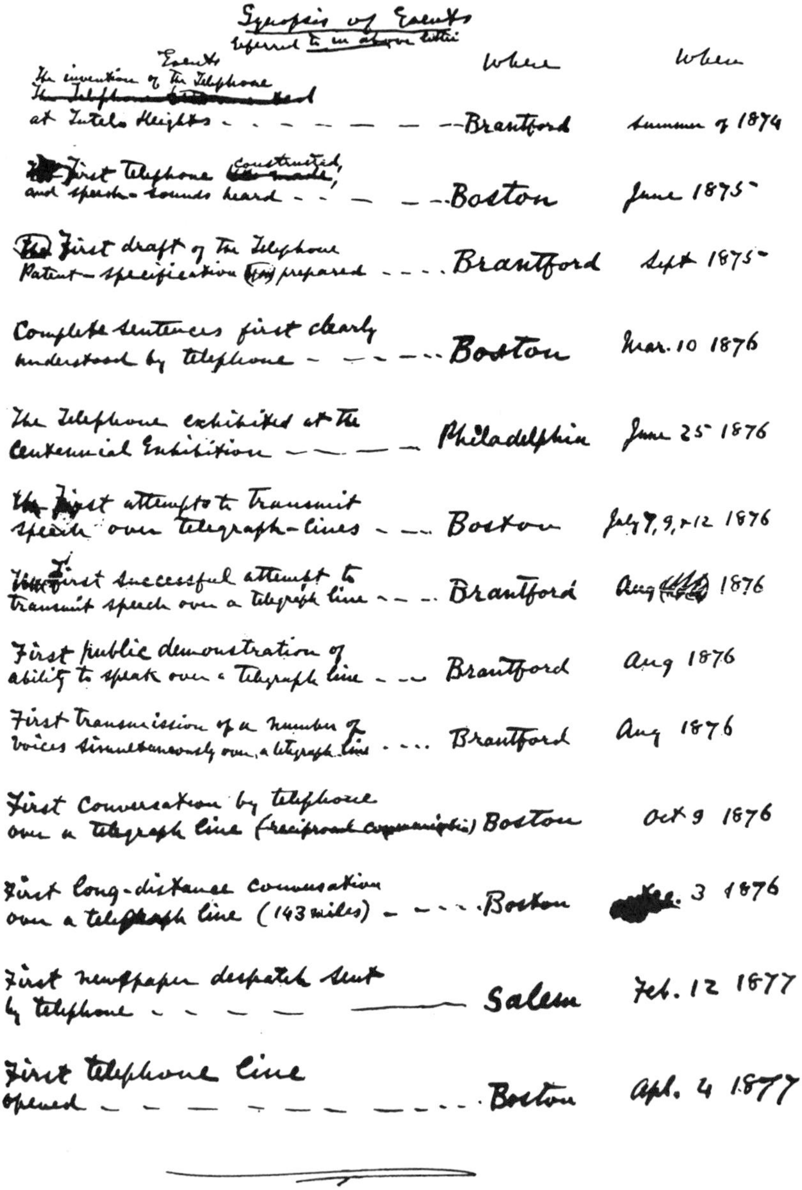

Synopsis of Events
referred to in above letter

Events	Where	When
The invention of the Telephone at Tutela Heights	Brantford	Summer of 1874
First Telephone constructed, and speech-sounds heard	Boston	June 1875
First draft of the Telephone Patent-specification prepared	Brantford	Sept 1875
Complete sentences first clearly understood by telephone	Boston	Mar. 10 1876
The Telephone exhibited at the Centennial Exhibition	Philadelphia	June 25 1876
First attempt to transmit speech over telegraph-lines	Boston	July 7, 9, & 12 1876
First successful attempt to transmit speech over a telegraph line	Brantford	Aug 1876
First public demonstration of ability to speak over a telegraph line	Brantford	Aug 1876
First transmission of a number of voices simultaneously over a telegraph line	Brantford	Aug 1876
First Conversation by telephone over a telegraph line (reciprocal communication)	Boston	Oct 9 1876
First long-distance conversation over a telegraph line (143 miles)	Boston	Dec. 3 1876
First newspaper despatch sent by telephone	Salem	Feb. 12 1877
First telephone line opened	Boston	Apl. 4 1877

SYNOPSIS OF TELEPHONE CHRONOLOGY IN MR. BELL'S HANDWRITING, WRITTEN AT THE TIME OF THE ERECTION OF THE BELL MEMORIAL AT BRANTFORD CANADA

FIGURE 5-6. Bell's log. Synopsis of telephone chronology in Mr. Bell's handwriting, written at the time of the erection of the Bell memorial at Brantford, Canada. (From Bruce [47].)

FIGURE 5-7. The telephone patent, March 7, 1876. (From Prescott [49].)

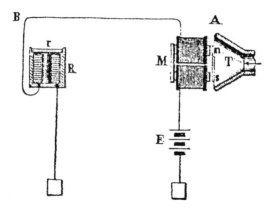

FIGURE 5-8. Bell's original telephone. (From Prescott [49].)

were not heard until 10 March 1876, when Bell uttered the famous call "Mr Watson, come here, I want to see you." [50a] But improvements came, and some 3 months later, Bell exhibited his telephone at the Centennial Exposition in Philadelphia. Bell had not intended to demonstrate his instrument at the exhibition, but was persuaded by his wife to do so, although it was now at a time when it was too late to have it exhibited in the proper place in the exhibit hall; hence it was assigned a small out-of-the-way location. And here, Joseph Henry made one more contribution. The Emperor of Brazil, Dom Pedro II (1825–1891), paid a visit to the exhibition on 25 June, 1876—the same day as the defeat of General George Custer on the Little Big Horn. On the Emperor's tour, he was accompanied by Sir William Thomson (later, Lord Kelvin) and Joseph Henry, as Director of the Smithsonian, as well as by other well-known acoustical researchers, including Joseph Le Conte, the brother of John Le Conte (see Chapter 3), Rudolf Toepler (see Chapter 6), and Elisha Gray. In the science exhibit, all eyes were on the enormous Corliss steam engine, and few paid any attention to the small out-of-the-way exhibit of Mr. Bell. But Henry observed the exhibit, and steered his famous guests over to it. The Emperor listened to the earpiece as Bell recited Hamlet's soliloquy, "To be or not to be," from the next room, and exclaimed, "My God! It talks!" [50b] And after that, the whole world listened.

Bell's success at Philadelphia gave work on the telephone great impetus. Technical problems were still facing him, involving the best type of unit for both the transmitter and the receiver. But even more immediate were the various lawsuits that had been initiated in attempts to deny the validity of Bell's patent and to establish subsequent patent applications for various improvements. An excellent account of this legal struggle is given by Hunt in Ref. [1, pp. 26–39]. We shall only try to pick out the scientific highlights of this period. Before doing that, however, we shall turn our attention to Edison, who had also became heavily involved in the telephone controversey.

5.8. Thomas Alva Edison (Fig. 5.9)

Thomas Edison was born in Milan, Ohio, in 1847, into a family that had, in previous generations, twice fled across the Canadian–American border, first, from New Jersey to Nova Scotia, as loyalists at the end of the American Revolution, and second, back to the United States from Ontario, where Edison's father had been a soldier in the failed rebellion of William Lyon Mackenzie in 1837 [51]. From these ancestors, he inherited the characteristic of holding steadfast to his own opinions, and also the genes to live a long and vigorous life.

Edison's early days, as a newspaper and candy salesman on the Grand Trunk Railway, along with his chemical laboratory that he operated on the

FIGURE 5-9. Thomas Alva Edison, age 36. In April, 1878, when he was thirty-one, Edison went to Washington to demonstrate the phonograph to the American Academy of Science, to members of Congress, and to President Rutherford Hayes. This picture was taken at the studio of the noted Civil War photographer Mathew Brady. (From Josephson [54].)

train, and as a telegraph operator in different cities of the Midwest, are well known from various books and motion pictures. At age 20, he came east to Boston, to work as a telegrapher for Western Union. His mind was already full of ideas on improvements in the telegraph and new uses for it. He applied for his first patent in 1868, curiously, for a device that would record voting in legislatures. It found no market, for politicians were not yet ready for such a clean-cut operation. He then made a resolution not to invent things for which there was not already a demanding market. He thus turned his back on the kind of research that Joseph Henry, Tyndall, and Rayleigh conducted, i.e., basic research, in favor of the practical applications of science and technology. He soon developed a "ticker," to record stock market sales. At the same time, he had begun work on a duplex telegraph system that would allow messages to be sent in opposite directions simultaneously on a single telegraph wire. As we noted earlier, this was a problem on which Alexander Graham Bell and others were also working, but with no immediate success.

It was at this time also that Edison first encountered the writings of Michael Faraday. He had earlier tried to read Newton, but the mathematics were beyond him (giving him a life-long prejudice against the subject). He later remarked "It (Newton's *Principia*) gave me a distaste for mathematics from which I have never recovered [52]." Faraday's mode of thought and nonmathematical style of writing were much more to Edison's liking.

After some failed enterprises in Boston, Edison moved on to New York in 1869, and began a period where he obtained patents on telegraphic devices, sometimes working by himself, sometimes for Western Union, and sometimes for companies run by the financier, Jay Gould (1836–1892). In 1874, Edison finally perfected his multiplexer, which enabled two telegraph signals to be sent in each direction on the same wire, thus quadrupling the usefulness of the telegraph instrument.

5.9. Further Work on the Telephone

The Bell telephone transmitter employed a magnetic unit in which the vibrations of a metal plate induced currents in an electromagnet, but the result was generally poor reception. Meanwhile, in July 1877, Edison developed his "carbon lampblack button microphone." As described by Edison's biographer, carbon black "was molded in the form of a button and introduced into the transmitter, placed between two metal plates adjacent to the diaphragm, the metal plates being connected to the battery circuit. That carbon button proved to be wonderfully elastic in varying its resistance in accordance with the pressure exerted upon it by the diaphragm [53]."

Western Union, which was emerging as the principal competitor to the newly formed American Bell Telephone Co. for the telephone market, acquired the right to make use of Edison's patent on the carbon microphone, and formed the American Speaking Telephone Co. It had also acquired the rights to the works of Elisha Gray, and of Amos Dolbear (1837–1910), a Tufts University professor, who had developed a telephone system of his own. Because of the superiority of the Edison microphone, Bell might have been forced out of the market, but his company was able to secure the rights to patents by Francis Blake (1850–1913) and Emile Berliner (1851–1929), both of whom had also invented carbon microphones. In addition, Berliner had filed a caveat on a carbon microphone prior to Edison's patent application and another set of lawsuits ensued, not ending until 1891, when the US Supreme Court declared that "Edison preceded Berliner in the transmission of speech. . . . The use of carbon in a transmitter is, beyond controversy, the invention of Edison [55]." Before that time, however, the Bell company had reached an agreement with Western Union, which, for a substantial financial settlement, withdrew from the telephone business and sold its existing lines, as well as its patent rights, to the Bell company. This settlement thus gave the American Bell Telephone Co. access to the Edison transmitter patent, and assured the Bell forces of a telephone monopoly in the United States.

During this same period, Edison and his associates had established an Edison Telephone Company in Great Britain, and Edison promised to develop a wholly new microphone. In early 1879, he invented the so-called chalk microphone [56]. Edison wrote of it

There was no magnet, simply a diaphragm and a cylinder of compressed chalk about the size of a thimble. A thin spring, connected to the center of the diaphragm extended outwardly and rested on the chalk cylinder, and was pressed against it. The chalk was rotated by hand. The volume of sound was very great. The voice, instead of furnishing the power, merely controlled the power. [57]

There was much publicity in Britain about this instrument, but its fame was short-lived, and it served mainly as a weapon in the patent fight with the Bell interests. Once the settlement described above was concluded, the chalk microphone faded from the scene. It did, however, provide one other point of historic interest. For a brief period, the Edison Company employed a red-bearded young Irishman to demonstrate the instrument in London. This the young man did with great flair and showmanship. His name was George Bernard Shaw (1856–1950) [58].

Other controversies now arose regarding improvements in the telephone system. A group at Brown University [E.W. Blake, Jr. (1836–1895), John Peirce (1836–1897), and W.F. Channing] introduced the truncated cone on the mouthpiece of the transmitter [59] (Fig. 5-10) (which style remained in telephone use into the 1930s), and also developed the hand-held "butterstamp" receiver [60] (Fig. 5-11), which also continued in use well into the

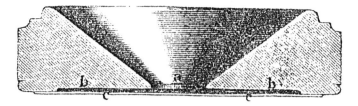

FIGURE 5-10. Peirce's mouthpiece. (From Prescott [59].)

twentieth century. At the same time, they also operated with a straight bar magnet, instead of Bell's horseshoe arrangement [61]. Upon writing to Bell and sharing some of these developments with him, Blake was informed by Bell that he had just invented the same thing. None of this group was interested in the commercial aspects of the subject, however, so that no patent suits resulted [62].

During the years immediately following Bell's first invention, he made many demonstrations of his instrument. Mostly, they were successful and created much enthusiasm. But one, at the meeting of the British Association for the Advancement of Science, in Plymouth, England (Bell had married Mable Hubbard shortly before and was on an extended

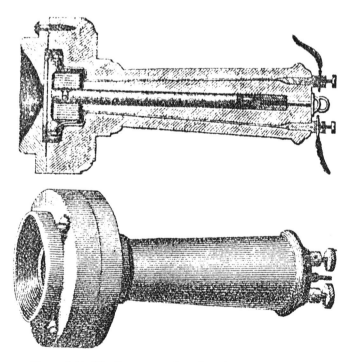

FIGURE 5-11. The "butterstamp" receiver. (From Prescott [59].)

honeymoon in Britain) was not so successful. The story is told by Bell himself:

I had one telephone in the hall and another about a mile away with a man in charge. When I had delivered my address I invited anyone to come up to the platform and talk (on the telephone) to this man.

There was a call for Lord Kelvin (Sir William Thomson), and when he mounted the platform, there was dead silence. Everyone was anxious to hear the words of wisdom that would fall from the lips of the great scientist.

He was silent for a minute, and then he cried "Hi diddle, diddle, the cat and the fiddle; follow that up!"

Then he placed the telephone to his ear. Next, addressing the audience, he remarked, "There he goes, he says the cow jumped over the moon."

There was much merriment over this, and after the meeting, Bell asked his helper whether he had heard Sir William clearly, "I did not hear him at all," was the reply. "Then what did you say in reply to him, Bell asked. "I said, 'Please repeat'." [63]

So much for the faithful recording of scientific experiments!

Bell's contributions to telephony and to the deaf are summed up in a most engaging tribute by Hunt:

Alexander Graham Bell had devoted his youth to teaching the deaf to speak and the mute to understand visible speech, but his greatest triumph of all was to endow the stuttering telegraph with the power of articulate speech. [64]

5.10. The Phonograph

Having assisted in the transmission of speech over wires through his invention of the carbon microphone, Edison turned his attention to the preservation of such speech. In the spring of 1876, he began his famous laboratory in Menlo Park, New Jersey [65], where he and his employees devoted themselves, full time, to the invention of new devices, and it was there that Edison invented the carbon microphone. But, even before that, he developed a device to allow the recording and repeating of telegraph messages. The instrument consisted of a disk of paper placed on a rotatable platen. An embossing point was attached to the armature of a telegraph so that each dot and dash of an incoming message would be recorded on the rotating paper. If the paper were placed on a second such disk, which had a contact point, the indentations of the paper could lift a marker up and down, thus converting them into signals to transmit down another telegraph wire.

This system was designed to send the coded messages at relatively high speeds—hundreds of words per minute. But Edison found that when the disk rotated a high speeds, the rattling of the contact marker up and down began to sound like a musical note. He therefore reasoned that if he used one diaphragm to mark the paper and have a second marker follow the indentations and drive another diaphragm, he might be able to record and play back human speech. He therefore prepared a paper tape with paraffin

FIGURE 5-12. The Edison phonograph. (From Josephson [67].)

wax, and pulled it past a diaphragm which had a small blunt pin attached to its center. He shouted the word "Halloo" into it, and then ran the paper through a second unit, activating the second diaphragm. He wrote later, "we heard a distinct sound, which a strong imagination might have translated into the original "Halloo [66]." It was July 18, 1877. The recording industry had been born.

The original "phonograph" (it was given that name by Edison) is shown in Fig. 5-12 [67]. Edison used a metallic cylinder with grooves in it. A piece of tinfoil was placed over the cylinder and a stylus, attached to the diaphragm of the receiving horn, made impressions on the foil in response to the spoken words. Then a second stylus rode up and down on the impressions on the foil, and thus drove and second diaphragm of the primitive "loudspeaker."

Edison's phonograph was an instant sensation. Sarah Bernhardt came to Menlo Park to view it (and, probably to be recorded), Edison brought it to the White House to demonstrate it to President Hayes [67]. But interest soon faded, and it remained a little-used novelty for nearly 10 years. Then the threat of other competition [particularly from Alexander Graham Bell's cousin, Chichester Bell (b. 1848)] stimulated Edison into making improvements on his device. He replaced the tinfoil by paraffin-coated paper. Much later (1903), because of the fragility of the wax, a gold coating was vaporized onto it, thus making a negative of the indentations. This negative could then used as a master to imprint other cylinders and therefore open up a market for large-scale sales [69].

One of the difficulties of the early phonograph was that it was powered only by the human voice, and the person or instrument, had to be as close as possible to the mouthpiece (Fig. 5-13) [70]. Elimination of this difficulty had to await the electronics of the twentieth century.

Madame Marie Roze "warbling a *scena* from an opera" into the phonograph, as pictured on the cover of *Frank Leslie's Magazine*, April 22, 1878.

FIGURE 5-13. An early recording session. (From Josephson [70].)

In all of these early instruments, the stylus cut a path into the wax that moved up and down, perpendicular to the surface of the cylinder. Such a method of recording has come to be known as "hill-and-dale recording." In 1891, Emile Berliner, whom we encountered in the early development of the telephone, introduced two major changes in the recording process. He switched to a flat disk, instead of the cylinder, and he began the practice of having the stylus move back and forth in the plane of the disk, both of which practices remained in vogue until the arrival of tape recording.

Overview

The fantastic developments of telegraph, telephone, and phonograph relied on parallel developments in the field of electricity. In the days of Henry, one made one's own electromagnets and batteries, and wire was rather hard to come by. The development of the dry cell and, later, the storage battery, gave reliable direct current sources of electricity, while the perfection of the dynamo (in the 1850s) and the invention of the incandescent light (also by Edison, in 1879) gave the researcher almost unlimited sources of electrical power in the laboratory and lighting to work anywhere and anytime [71].

Notes and References

[1] We are indebted to the introductory historical essay in *Electroacoustics*, by F.V. Hunt, 1954, reprinted by the Acoustical Society of America, Woodbury, NY, 1982, for many interesting details of the history of this subject. A much briefer account is given in *Memorial of Joseph Henry*, Government Printing Office, Washington, DC, 1880, pp. 58–59.

[2] A. Volta, *Phil. Trans. Roy. Soc. (London)* **90**, 403–431 (1800).

[3] H.C. Oersted, *Ann. Phys.* **16**, 273 (1819).

[4] F.V. Hunt, *Electroacoustics*, p. 12.

[5] D.F.J. Arago, *Ann. Chim. Phys.* [2] **15**, 92–102 (1820).

[6] A.M. Ampère, *Ann. Chim. Phys.* [2] **15**, 59–76, 170–218 (1820)

[7] J.B. Biot and F. Savart, *Ann. Chim. Phys.* [2], 222–223 (1820).

[8] J.S.C. Schweigger, *J. Chem. Phys.* **31**, 1–17 (1821).

[9] J.C. Poggendorf (as described by David Brewster, ed.), *Edinburgh Phil. J.* **12**, 105–114 (1825). In the nineteenth century, many of the leading journals were commonly referred to under the author's name; thus, *Annalen der Physik* was popularly known as *Poggendorfs Annalen*, *J. Reine. Angew. Math.* became *Crelle's Journal*, etc.

[10] W. Sturgeon, *Trans. Soc. Encourage. Arts, Manufactures, and Commerce (London)* **43**, 38–52 (1825).

[11] That is, one who worked fully or mostly in physics. Benjamin Franklin was first class in his physics experiments, but he was also first class in many other things, and physics was only a small part of his amazing career.

[12] Thomas Coulson, *Joseph Henry: His Life and Work*, Princeton University Press, Princeton, NJ, 1950. The author is indebted to this book for many details of Henry's career.

[13] *The Papers of Joseph Henry*, Smithsonian Institution Press, Washington, DC, 1972, vol. 1, Frontispiece.

[14] Ref. [12, p. 75].

[15] J. Henry, *Am. J. Sci.* **22**, 403–408 (1832).

[16] Ref. [12, p. 249].

[17] Ibid.

[18] An account of this is given by A.T. Jones, *Sound*, Van Nostrand, New York, NY, 1937, pp. 387–390. Henry's reports are given in the reports of the US Lighthouse Board for 1873–1877.

[19] F.V. Hunt, *Electroacoustics*, p. 17.

[20] Gauss was, of course, the famous mathemtical physicist; Baron Schilling came from a family of German origin but was in fact a diplomat in the service of the Tsar.

[21] F.V. Hunt, *Electroacoustics*, p. 18. The lengthy fight over patents, and the break between Henry and Morse are described in Hunt's book *Electroacoustics* (Ref. [1]).

[22] C.M., *Scot's Magazine* **15**, 73 (1753).

[23] E.A. Marland, *Early Electrical Communication*, Abelard-Schuman, London, UK, 1964, Chapters 1 and 2. The book contains an excellent survey of early telegraph systems, with many detailed drawings and photographs.

[24] Geoffrey Hubbard, *Cooke and Wheatstone and the Invention of the Electric Telegraph*, Routledge and Kegan Paul, London, UK, 1965, p. 58. Hubbard notes that in those days patents were signed by the monarch, and that King William IV signed the application of Cooke and Wheatstone just ten days before his death so that the oncoming of the telegraph almost coincided with the birth of the Victorian era.

[25] Quoted by George O. Squier, *Telling the World*, Williams and Wilkins, Baltimore, MD, 1933, p. 60.

[26] And used not only children. Natan Sharansky describes one such telephone system in the Soviet Gulag as late as 1980: "We also communicated through the radiators. You would press your [metallic] mug against the heating pipe and speak into it; to listen, you'd turn your mug upside down." Natan Sharansky, *Fear No Evil*, Random House, New York, NY, 1988, p. 248.

[27] E.F.F. Chladni, *Traité d'Acoustique*, Chez Courcier, Paris, 1809, p. 315.

[28] Quoted by F.V. Hunt, *Electroacoustics*, p. 25.

[29] *Brockhaus Enzyklopaedia*, F.A. Brockhaus, Wiesbaden, 1972, vol. 15, p. 612.

[30] F.V. Hunt, *Electroacoustics*, pp. 26–30.

[31] Figure 5-4 is taken from George B. Prescott, *The Speaking Telephone, Electric Light and Other Recent Inventions*, Appleton, New York, NY, 1878, p. 10.

[32] F.V. Hunt, *Electroacoustics*, p. 29.

[33] Kendall was a Dartmouth College graduate who early served as the tutor of the children of Henry Clay, and was later a member of the cabinet of Andrew Jackson.

[34] Ruth E. Bender, *The Conquest of Deafness*, The Press of Western Reserve University, Cleveland, OH, 1937, especially, Chapter XIV.

[35] Robert V. Bruce, *Bell: Alexander Graham Bell and the Conquest of Solitude*, Little Brown, Boston, MA, 1973. A goldmine of information about the Bell family and the early experiments of Bell.

[36] As in all families where fathers and sons share given names, the Bells used various arrangements. The father was usually referred to as Melville (and the older son as Melly), while Alexander Graham Bell was called Aleck or even Al. In adult life, he usually signed himself as A. Graham Bell, until his wife persuaded him to use all three names, a practice that we follow today.

[37] Robert V. Bruce, *Bell*, p. 35.

[38] R.V. Bruce, *Bell*, p. 40. It is of interest to note that Ellis achieved later fame as the translator of Helmholtz' famous book on acoustics.

[39] R.V. Bruce, *Bell*, p. 50.

[40] R.V. Bruce, *Bell*, p. 111. The work of Koenig will be discussed in Chapter 6. The manometric flame was similar to that developed by Tyndall (see Chapter 3).

[41] F.V. Hunt, *Electroacoustics*, pp. 26, 27.

[42] R.V. Bruce, *Bell*, Chapter 11.

[43] Quoted by Catherine Mackenzie, *Alexander Graham Bell*, Houghton Mifflin, Boston, MA, 1928, pp. 72–73.

[44] Catherine Mackenzie, *Alexander Graham Bell*, Chapters 8 and 9.

[45] Catherine Mackenzie, *Alexander Graham Bell*, opposite p. 102.

[46] The Coulson biography of Henry (Ref. [12]) gives the date as March 1873, but both Bender (Ref. [34]) and Bruce (Ref. [35]) give it as March, 1875. The latter date is consistent with the developments made by Bell during the period. In 1873, he was not yet thinking in terms of a telephone, let alone preparing a working model.

[47] Robert V. Bruce, *Bell*, between pp. 183 and 184.

[48] A good coverage of the sequence of events during this period can be found in the book by F.V. Hunt, *Electroacoustics* pp. 23–44. The Supreme Court decided thus, between Reis and Bell: "To follow Reis is to fail; to follow Bell is to succeed."

[49] G.G. Prescott, *The Speaking Telephone*, D. Appleton, New York, NY, 1879, p. 17.

[50a] Robert V. Bruce, *Bell*, p. 181.

[50b] Ref. [11, p. 315]. In his book (*Bell*, p. 197), Bruce states that the Emperor cried "I hear, I hear!"

[51] Matthew Josephson, *Edison*, McGraw-Hill, New York, NY, 1959, pp. 8–9. The author is indebted to this text for many details of Edison's life and career.

[52] M. Josephson, *Edison*, p. 33.

[53] M. Josephson, *Edison*, p. 145.

[54] M. Josephson, *Edison*, following p. 290.

[55] M. Josephson, *Edison*, pp. 145–148.

[56] M. Josephson, *Edison*, p. 150.

[57] *Edison's Notes for Meadowcroft*, Book II, pp. 13–14.

[58] M. Josephson, *Edison*, pp. 152–154.

[59] G.G. Prescott, *The Speaking Telephone*, p. 275.

[60] Ibid., p. 276.

[61] R.V. Bruce (Ref. [35, pp. 225–226]) acknowledges the work of Peirce on the mouthpiece and of Channing's use of the single bar magnet.

[62] A coverage from the Brown point of view is given in an article by Waltar Lee Munroe (or the classs of 1879) in the *Brown Alumni Monthly*, March 1939, pp. 279–283. As recently as Aug. 29, 1975, there appeared a letter on the subject in the *Christian Science Monitor* from a relative of Professor Blake, Susan Baker Keith, who wrote "In reply to your front-page question 'when was the telephone invented?' I am one of the few people alive who can accurately answer it. The telephone was invented by my great-uncle Eli Whitney Blake."

[63] C. Mackenzie, *Alexander Graham Bell*, pp. 185–186.

[64] F.V. Hunt, *Electroacoustics*, p. 22.

[65] This laboratory can still be seen at Henry Ford's Dearborn Village in Dearborn, MI.

[66] George P. Lathrop, "Talks with Edison," *Harper's Magazine*, Feb., 1890, p. 428.

[67] M. Josephson, *Edison*, following p. 290.

[68] M. Josephson, *Edison*, pp. 169, 244.

[69] Ibid., p. 320. Perhaps this was the forerunner of the "golden records" of the recording industry.

[70] Frank Leslie's Magazine, April 22, 1878. The picture is reproduced in M. Josephson, *Edison*, following p. 290.

[71] But even the greatest of our scientists and inventors still remain the people of their times. Rayleigh never introduced electric lighting either in his home or in his laboratory, and Edison detested the radio ("it's a lemon," he once wrote).

6
The Last Half of the Nineteenth Century

Full many a genius, in obscure, language bred,
Will end his days his praises yet unsung;
Full many a journal is born to lie unread
And waste its learning in an unknown tongue.
R.T. Beyer, *Physics Today*,
Jan. 1965, p. 44 (with apologies to Thomas Gray)

We have spent the last three chapters in following the work of such out-standing researchers in acoustics as von Helmholtz and Rayleigh, and the great inventions of Bell and Edison. Now it is time to return to the systematic examination of the subject as a whole.

6.1. Sound Velocity Measurements

In earlier times, as we noted in Chapters 1 and 2, all the measurements of sound velocity relied on human judgment. In a typical experiment, a flash at the sound source caused the observer to start the timing mechanism, and the hearing of the sound caused him to turn it off. In 1868, however, H.V. Regnault (1810–1878) used an electric circuit [1] (Fig. 6-1), connected with a tuning-fork chronometer, arranged so that the time of emission of the sound and the time of recording of it could both be made without human intervention. The circuit carrying a direct current was broken at W when a gun was fired. This caused the electromagnet M to release the stylus S, thus making a mark on paper on the drum D, which rotated with constant speed (controlled by a tuning fork chronometer). When the sound reached R, it moved the diaphragm, momentarily closing the circuit, and thus producing a magnetic field in the magnet M, again moving the stylus (this time in the opposite direction) on the recording paper.

Regnault carried out his measurements on pipes of various sizes, destined for the Parisian water system. By using multiple reflections, he was able to reduce his experimental errors, but found generally that the sound velocity in the pipe was larger than that measured in open air, and that the measured value decreased as the diameter of the pipe increased. When the diameter

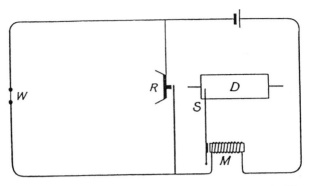

FIGURE 6-1. Regnault's method of measuring sound velocity [1].

of the pipe reached about one foot, the open-air value of the sound velocity was obtained.

At nearly the same time, the German physicist August Kundt (1839–1899) (Fig. 6-2) [2] came up with a "desk-top" method of measuring the sound velocity in a tube (Fig. 6-3) [3]. If a rod is clamped at its center, the stroking of it will set up sound waves in the metal whose wavelength is twice the length of the rod. These waves in turn generate sound waves in the gas surrounding the rod, having the same frequency as that of the sound in the metal. Kundt found that the air in the tube was in motion, and, by putting a fine powder in the tube, he found that the powder accumulated on the bottom at points such as B'. The distance between these accumulations of powder would therefore be one-half of the wavelength in air. Since the

FIGURE 6-2. August Kundt [2].

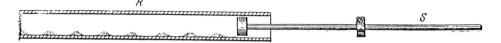

FIGURE 6-3. Kundt's tube [3].

sound frequency is the same in the metal and in the gas, we have

$$f = \frac{c_m}{\lambda_m} = \frac{c_g}{\lambda_g}, \tag{6.1}$$

where f is the sound frequency, and c, λ, the sound velocity and the sound wavelength. Since the sound velocity in the metal c_m could easily be calculated from elasticity measurements, this became a method of determining the velocity of sound in air (or in any other gas). Once again, the sound velocity in the air in the tube was larger than that in open air. von Helmholtz [4] proposed an empirical formula fitting the data of Regnault and Kundt,

$$c' = c\left(1 - \frac{k}{r}\right), \tag{6.2}$$

where r is the radius of the pipe, c' the measured sound velocity, and k is a constant, which could be evaluated by making measurements on two tubes of different radii. The proof was later provided by Rayleigh [4].

6.2. Sound Absorption in Fluids

We saw in Chapter 2 that Stokes had developed an expression for the sound absorption in fluids that depended on the shear viscosity of the fluid. In 1868, another advance was made by Gustav Kirchhoff (1824–1887) (Fig. 6-4 [5]). Kirchhoff, who is much more famous for his discovery of optical absorption spectra and for his work in electricity, derived an expression for the absorption of sound due to thermal conduction [6]. Like the formula deduced by Stokes for viscosity, the absorption coefficient was found to be proportional to the square of the sound frequency, and, for gases, equal in magnitude to about one-third to one-half of the viscosity contribution.

Direct measurement of sound absorption was still in the future. However, an ingenious experiment carried out near the end of the century by an American, Wilmer Duff [7] (Professor of Physics at Purdue University and later, at Worcestor Polytechnic Institute), gave the first evidence that the processes described by Stokes and Kirchhoff did not tell the whole story. Rayleigh described the experiment as follows

(Duff) compared the distances of audibility of sounds proceeding respectively from two and from eight similar whistles. On an average the eight whistles were audible only about one-fourth further than a pair of whistles; whereas, if the sphericity of the

FIGURE 6-4. Gustav Kirchhoff [5].

waves had been the only cause of attenuation, the distances would have been as 2 to 1. Mr. Duff considers that in the circumstances of his experiments there was little opportunity for atmospheric irregularities, and he attributes the greater part of the falling off to radiation. [8]

This paper by Duff stimulated Rayleigh into working on deriving an expression for the energy dissipated in sound propagation by radiation. He found, however, that the absorption due to radiation was much too small to account for the difference. Then, with a flash of insight, he wrote

it will I think be necessary to look for a cause not hitherto taken into account. We might imagine a delay in the equalization of the different sorts of energy in a gas undergoing compression, not wholly insensible in comparison with the time of vibration of sound. [9]

Thus Rayleigh anticipated the development of the relaxation theory of sound absorption in fluids, which was so widely explored in the twentieth century (Chapter 8).

6.3. Vibrating Systems

A number of specialized vibrations came under scrutiny in the second half of the nineteenth century. Franz Emil Melde (1832–1901), for many years Professor of Physics and Astronomy at the University of Marburg, Germany, developed a simple demonstration of waves on a string that became a lecture-room standby—Melde's Experiment. He stretched a string over a pulley (Fig. 6-5) by means of an attached weight, and con-

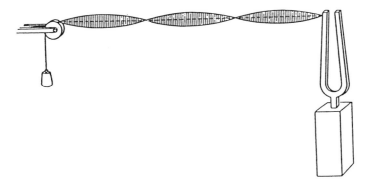

FIGURE 6-5. Melde's standing wave experiment. (From E.G. Richardson [10].)

nected the other end of the string to one prong of a tuning fork [10]. By driving the fork and attaching the proper weight, he was thus able to exhibit standing waves on the string, waves that involved one or more antinodes.

The passage of sound through tubes was investigated, first by Sir John Herschel (1792–1881) and then by Georg Hermann Quincke (1834–1924), a professor at Berlin, Würzberg, and Heidelberg. Herschel, in attempting to explain some optical spectra phenomena, suggested an experiment in which sound, passing through one tube, would be made to travel through two branches of unequal length, the tubes being reunited at the far end (Fig. 6-6) [12]. He thought that, with an appropriate difference in the lengths of the two branches, sound could be prevented from penetrating into the final tube. While Herschel never performed the experiment, Quincke did, producing the system that is today called the Herschel–Quincke interference tube [12]. In the time of Quincke, it was thought that the sound in the further tube would vanish only when the lengths of the two parallel tubes differed by an odd number of half wavelengths [13]. The Herschel–Quincke system was the forerunner of acoustic filtration devices, developed in the next century (see Chapter 8) [14].

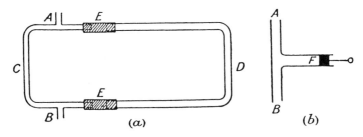

FIGURE 6-6. Quincke's interference tubes. (From E.G. Richardson [12].)

6.4. Structural Vibrations

The vibrations of solids provide us with an extremely complicated problem. Historically, the various scientists dealing with the problem have attempted to simplify it, concentrating their attention on beams of homogeneous isotropic material, plates of the same sort, and thin shells, all of which reduce the number of independent variables.

In his *Benchmark* collection of papers on vibrations, Artur Kalnins remarks "it is a commonplace that beams are one-dimensional plates, and that plates are two-dimensional generalizations of beams [15]." We have already remarked on Rayleigh's contributions to all three approximation areas (Chapter 4). Two higher-order effects in beam theory are the rotatory inertia of the beam and the transverse shear deformation. The first of these was developed by Lord Rayleigh [16] while the second had to await the twentieth century.

Rayleigh also contributed to the solution of the problem of thin shells. In the second edition of his book [17], Rayleigh hypothesized that, during vibrations, the two extensional strains and the shearing strain at the middle of the shell are all equal to zero. This was vigorously questioned by Love, especially the assumption on the middle surface. Eighty years were to elapse before the issue was settled (Chapter 8).

Other important analyses appeared at this time. These included Pochhammer's [18] discussion of the theory of waves in an infinitely long cylindrical rod, Rayleigh's [19] work on the problem of two-dimensional waves in a solid that was bounded by parallel planes [20], and the scattering of waves in an elastic solid, began by Alfred Clebsch, who treated the situation as a mathematical boundary-value problem [21].

6.5. Heat-Generated Sound

We have already referred twice to the use of flames in measuring the characteristics of sound, first by Tyndall (Chapter 3) and then by Koenig (above). A number of other connections between heat and sound were discussed by Rayleigh in the second edition of his book [22]. The best known of these is the one provided by Pieter Leonhard Rijke (1812–1899), a Professor of Natural Philosophy at the University of Leiden, who generated sound by placing a piece of metal gauze in the lower portion of a glass tube and heating the metal to red heat, and then removing the heater. Rijke found that the gauze had to be in the lower half of the tube and could not be within $\frac{1}{32}$ of the lower end of the tube [23]. It was Rijke's explanation that the wire gauze heats the surrounding walls of the tube, that an ascending column of (heated) gas pulls air through the gauze. This newly arrived air is itself heated and moves upward, repeating the process. Thus a succession of pulses of sound would be generated. The sound would last until the temperature of the air and the gauze became the same. If an electric current were run through the metallic gauze, a steady tone could be achieved.

Rayleigh also looked at this problem [24] and concluded that it was due to a series of node and antinode effects. As put by Richardson, Rayleigh argued as follows:

Under the action of the first cause, every upward movement of the air in vibration brings cold air onto the heated gauze, whereas the downward movements bring the already hot air back to the gauze. Thus the greatest temperature difference occurs in the "upstroke," consequently the greater heat transfer and consequent compression occur as the air goes up. [25]

6.6. The Birth of Seismology

The Chinese philosopher Heng Chang produced the first operational seismic wave detector or seismoscope in A.D. 132 [26], Chang's device "consisted of a column so suspended that it could move in one of eight directions. A ball was held lightly along each of these lines and, when thrown down by the rod, was caught in a cup below and so revealed the direction of the motion [27]." Its reinvention in the West did not occur until 1875, when it was done by Filippo Cecchi in Italy [28]. The use of a seismometer began in 1841 by David Milne (1805–1890) and developed rapidly thereafter. In 1889, accurate measurements were made in Potsdam, Germany, of an earthquake that had occurred 15 min earlier in Japan. The operations of vertical and horizontal seismometers are shown in Fig. 6-7 [29].

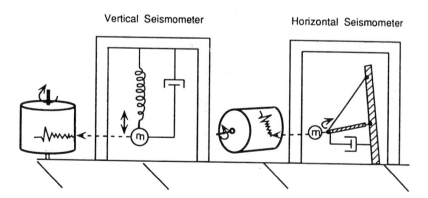

FIGURE 6-7. Seismometers. Schematics of inertial-pendulum vertical and horizontal seismographs. Actual ground motions displace the pendulums from their equilibrium positions, inducing relative motions of the pendulum masses. The dashpots represent a variety of possible damping mechanisms. Mechanical or optical recording systems with accurate clocks are used to produce the seismograms. (From Lay and Wallace [29].)

A compact seismometer was invented by an Englishman, John Milne (1850–1813), in 1892. Milne spent much of the first half of his career in Japan, returning to England in 1895. Very soon after his inventing the compact seismometer, there were large numbers of such instruments in use throughout the world, making possible a global study of earthquake phenomena. At the present date there are at least 3000 seismic stations worldwide, so that no earthquake goes unreported.

6.7. Devices for Making Sound Visible

Until the great developments of electroacoustics changed all the rules of acoustic experimentation at the beginning of the twentieth century, the greatest experimental efforts in acoustics were directed toward the visualization of sound by nonelectrical means. The ideas of Ohm and Fourier indicated that the sounds of speech and music could be reduced to combinations of simple tones. As Alfred Mayer (1836–1897) put it,

The ear has the sensation of a simple sound only when it receives a pendulum vibration, and it decomposes any other periodic motion of the air into a series of pendulum-vibrations, each of which corresponds to the sensation of a simple sound. [30]

(The term "pendulum vibration" was used by Mayer for what we would today call simple harmonic motion.)

We have already described the technique used by Tyndall (Chapter 3) in which he reflected a beam of light from a mirror attached to a tuning fork, and re-reflected the beam by a hand-held mirror. The vibrations of the tuning fork led to a line of light when the hand held the mirror steady, but the motions could be spread out by rotation of this mirror, so that the waveform became temporarily visible.

A second technique was the use of a soot-blackened glass plate. If a sharp pointer is attached to a tuning fork, and the plate is moved past the vibrating pointer at a constant rate, the scratchings in the soot produce a replication of the motions of the fork.

A master of acoustical techniques and instruments was Karl Rudolph Koenig (1832–1901) (Fig. 6-8 [31]). Born in Koenigsberg, Germany (now Kaliningrad, Russia), he studied at its university at the same time that von Helmholtz was a professor there, although there is no evidence that the two interacted during that time. Koenig moved to Paris as a young man, became a French citizen, and spent the rest of his life in Paris, operating a laboratory in his home near the Ile de la Cité, and developing the finest acoustical instruments of his day. He perfected the manometric flame, used for observing the waveform of various acoustical sounds, and developed a tuning fork chronometer (Fig. 6-9), Dayton Miller described a number of his exquisitely made tuning fork devices [32] designed as frequency standards,

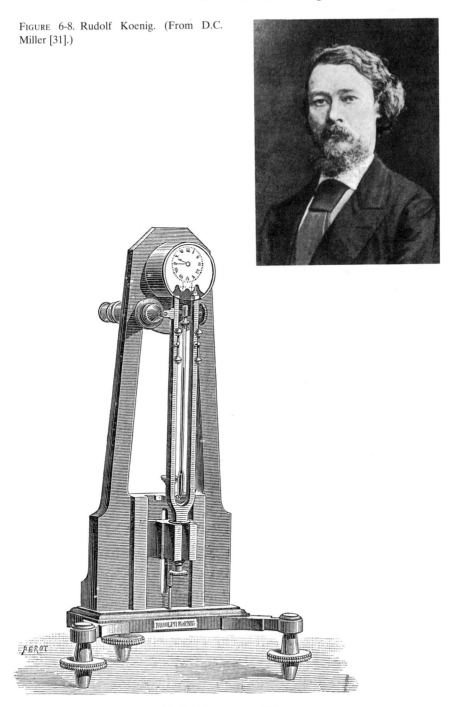

FIGURE 6-8. Rudolf Koenig. (From D.C. Miller [31].)

FIGURE 6-9. Koenig's fork chronometer [32].

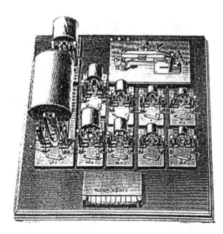

FIGURE 6-10. Koenig's apparatus for tone synthesis. (From D.C. Miller [33].)

for audibility measurements and for the synthesis of partial tones, used in the study of vowel sounds (Fig. 6-10). (One can recognize in this instrument miniatures of the fork-sounder developed by von Helmholtz and shown in Fig. 3-4.) Koenig was not so far back in time but Miller, a man known to many of this writer's contemporaries, visited him in his laboratory in the later 1890's and wrote of him

he probably did more to develop the science of sound than any other one man; he devoted his whole being and his whole life, for half a century to this one field of work, and he brought to it a mental and intellectual power and a manual dexterity not surpassed by any. [33]

In 1857, another French scientist, Edouard-Léon Scott, produced what he called a "phonoautograph," in which he used a hog-bristle, attached to a diaphragm, to mark the vibrations of sound on a piece of lamp-blackened paper [34], [35]. Koenig improved on this device (Fig. 6-11), adding a sound-collecting horn and a rotating cylinder on which to record the sounds. As Miller has pointed out, this was a direct antecedent of Edison's phonograph of 20 years later [36].

In 1878, Professor Eli Whitney Blake, Jr. (1836–1895), mentioned in Chapter 5 for his work on the telephone, made use of an optical system to amplify the very small vibrations of the mechanical systems involved in telephony [37]. Blake attached an iron plate to the back of a telephone mouthpiece. A stiff wire then connected this plate with a plane mirror (A in Fig. 6-12), hooking into the hole D as shown in the figure. He then let a beam of light from a heliostat (a device that reflects a beam of sunlight in a constant direction) be reflected from the mirror at an angle of approximately 45°. This latter beam then fell on a photographic plate that was moved across the field of view at a constant rate. This technique gave a substantial amplification to the minute vibrations of the telephone mouthpiece. Some results of Blake's work are shown in Fig. 6-13. We can easily

FIGURE 6-11. The Scott–König phonoautograph. (From D.C. Miller [35].)

recognize these as the forerunners of the pictures of speech sounds on oscilloscopes that have become so widespread in the present day.

The German scientist, August Toepler (1836–1912) (Fig. 6-14 [38]), is famous for two different techniques for rendering sound visible. In the first of these, he was interested in studying singing flames (Chapters 1 and 2) and followed up on the work of Stampfer and Plateau mentioned in Chapter 3 [39]. He focused a telescope on the flame but interposed a rotating disk with slits (Fig. 6-15). He was thereby enabled to view the flame in various portions of its vibration. This was an early stroboscope.

The second method developed by Toepler he called "Schlierenbeobachtung" ("striae observation"), because of the fact that he used it first

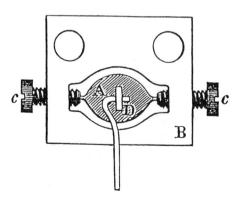

FIGURE 6-12. Blake's mirror, back view of mirror [37].

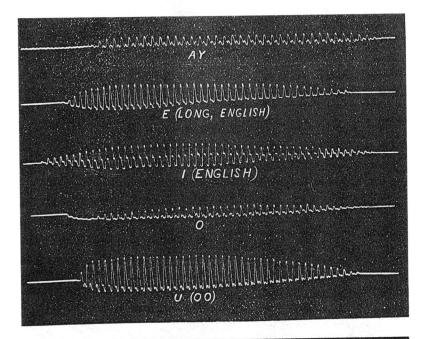

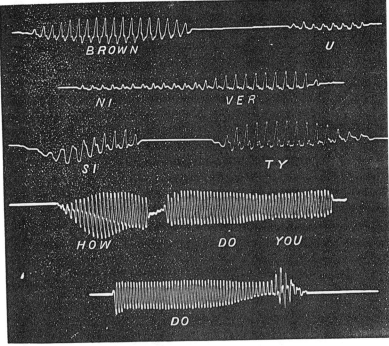

FIGURE 6-13. Blake recordings [37].

Figure 6-14. August Toepler. (From Ostwald [38].)

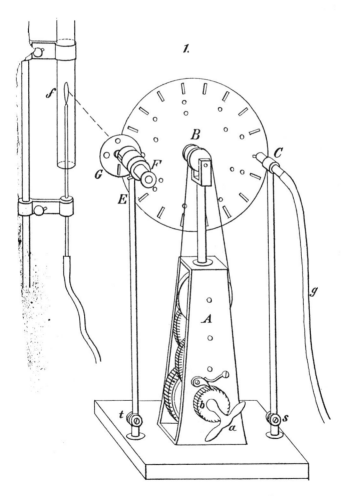

Figure 6-15. Toepler's stroboscope [39].

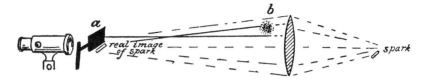

FIGURE 6-16. Schlieren apparatus. (From Wood [41].)

as a very reliable technique for observing inhomogeneous streakings or striae in glass [40]. Toepler used a long focal length, high-quality achromatic lens and focused the image of his spark source. He the then covered this image with an opaque screen. Figure 6-16 shows the arrangement as employed later by Robert W. Wood (1868–1955) [41]. If the rays of light were to encounter any region (such as B) where the index of refraction differs from that of the region as a whole, some of the light would be bent more or less than the rest of the rays, and the light would therefore be focused at a point outside the barrier. Thus, if a source gives rise to a spherical wave, a point in the medium experiencing the crest of the wave would have a different index of refraction than that of the medium at rest, light would therefore appear in the field of the viewing telescope. In his later paper, Wood used a stronger light beam and was able to get improved photographs of the result. Figure 6-17 [39] is due to Toepler and shows: (a) waves reflected from a plane barrier; (b) refraction and reflection from carbon dioxide in the bottom of the vessel, separated from the air in the rest of the chamber by a thin collodion film; and (c) reflection and refraction of a sound wave in air from hydrogen gas. The wavefronts are easily seen. In Wood's work, the sound source was a spark, which also provided the illumination. Figure 6-18 shows some of Wood's experimental results [41]. The method was so successful that the word "schlieren" almost immediately entered the English language.

6.8. Shock Waves

In 1742, a Belgian named Robens [42] had observed that projectiles traveling faster than sound seemed to experience a surprisingly high resistance in the air. This remained an isolated observation, but new information came forward during and after the Franco–Prussian war of 1870–1871. Early measurements of the speed of sound in air were made by measuring the time lag of the sound of gunfire behind the flash of the powder, but experiments with more powerful guns indicated that the speed of sound became greater for these intense signals. As a result of studies in this area, Felix Albert Journée published a note in 1888 on the speed of sound produced by gunfire. He wrote

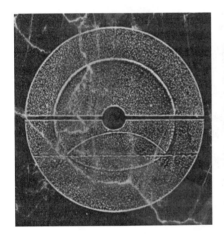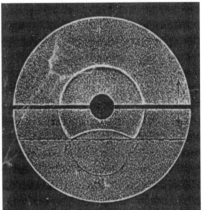

FIGURE 6-17. Toepler's schlieren pictures [40].

If one fires, on an iron plate, a gun whose bullet has a speed greater than the normal speed of sound in air, an observer located behind the target hears simultaneously the detonation and the shock noise ("bruit du choc") of the bullet against the target, as long as the distance of this target and the gun does not exceed a certain limit.

Beyond this limit, the noise of the detonation precedes that of the shock, and the interval of time which separates these two noises increases with distance from the target to the weapon. It can be stated, furthermore, that the distance, beginning at which the two noises become separate is that at which the speed of the bullet, gradually reduced by the resistance of the air, becomes equal to the speed of sound. [43]

While the term "shock wave" did not come into use until the early 1900s, it may be that Journée's "bruit du choc" was its ancestor.

During this same period of time, the Austrian physicist, Ernest Mach (1836–1916), who was first a Professor at Graz, then at Prague and, finally

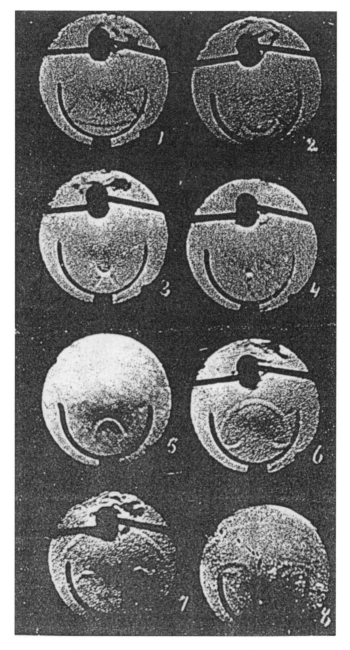

FIGURE 6-18. Wood's schlieren pictures [41].

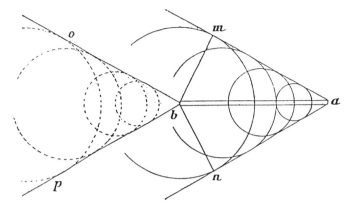

FIGURE 6-19. Mach's drawing of the wavefronts of sonic booms [44].

at Vienna, began a long study of the sounds produced by explosions. He soon transferred his interest to the behavior of bullets, and adapted Toepler's schlieren method to visualize the motion of the air in the vicinity of these high-speed objects. In Fig. 6-19 he portrayed the existence of the shock front formed by the nose of the bullet, and the second front created at its rear tip [44]. Figure 6-20, from a later paper [45], shows spark photographs of such fronts: (a) from a bullet traveling at $520\,\mathrm{m\,s^{-1}}$; and (b) from one traveling at $700\,\mathrm{m\,s^{-1}}$. The decrease in what is now called the Mach angle with increasing speed is clearly shown. Also apparent is the turbulent trail (the von Karman trail) behind the bullet.

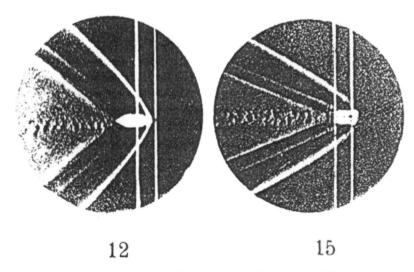

FIGURE 6-20. Mach's photographs of sonic booms [45].

6.9. Nonlinear Acoustics

Following up on the work of Poisson, Stokes, and Airy, discussed in Chapter 2, a forward step on the problem of nonlinear sound propagation was made by the Rev. Samuel Earnshaw in 1858 [46]. In what today looks to us to be rather crabbed mathematics [47], Earnshaw found an implicit expression for the particle velocity of the form

$$u(x,t) = u_0 \sin w\left[t - \left(\frac{x}{c_0}\right)\right.$$
$$\left.\times \left\{1 + \left(\frac{B}{2A}\right)\left(\frac{u}{c_0}\right)^{-2A/B-1}\right\}\right], \qquad (6.3)$$

where the pressure p and the density ρ are related by the expression

$$p = p_0 + A(\rho - \rho_0)/\rho_0 + \left(\frac{B}{2}\right)\left\{\frac{\rho - \rho_0}{\rho_0}\right\}^2 + \cdots \qquad (6.4)$$

with higher-order terms neglected. The explicit solution of Eq. (6.1) had to wait a century for its recognition in acoustics, although it had in fact been obtained by F.W. Bessel (1784–1846) in the astronomical problem for whose solution he invented Bessel functions (see Chapter 8) [48]. And, while neither Earnshaw's work (in English) nor Bessel's [49] (in German) was in "an unknown tongue," no one did put the two together for a long, long time (see Chapter 8) [50].

A contemporary of Earnshaw was the German mathematician, G.F. Bernhard Riemann (1826–1866), who published a lengthy paper on nonlinear wave propagation in 1860 [51]. Riemann considered the equations of motion of a fluid and found, by defining two relations P and Q:

$$P = u + \int_{\rho_0}^{\rho}\left(\frac{c}{\rho}\right)d\rho; \qquad Q = u - \int_{\rho_0}^{\rho}\left(\frac{c}{\rho}\right)d\rho, \qquad (6.5)$$

where u is the velocity of sound, that the quantities P and Q would be invariant along the lines $dx/dt = u + c$ and $dx/dt = u - c$, respectively. The quantities P and Q have become known as the Riemann invariants. If P is originally given as a function of x at some time t_0, then the particular values of P will remain the same as the initial values along the curve defined by $dx/dt = u + c$.

Since P is a velocity, we might consider the value at $t = 0$ to be $u_0 \sin kx$. Then, with the passage of time, the sinusoidal wave shape would be distorted, as indicated in Fig. 6-21) [52].

A curious inability to understand the directions toward which these results were pointing seems to have settled on researchers in the nonlinear field after this time. Following on the work of Stokes, Earnshaw and

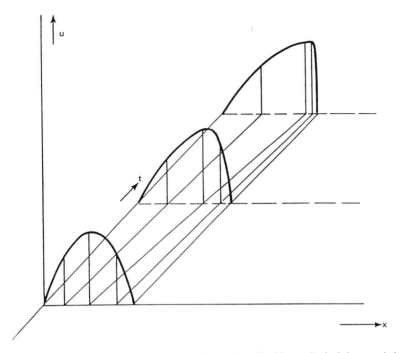

FIGURE 6-21. Riemann's invariants Wave distortion as described by method of characteristics. (From Beyer [52].)

Riemann had proven that a sinusoidal wave of sufficient amplitude would gradually become distorted as it propagated, thus leading to the production of higher harmonics of the original (fundamental) frequency. These harmonics are a particular form of the combination tones of von Helmholtz. In the meantime, von Helmholtz had also claimed the production of combination tones outside the ear (Chapter 3). It would seem in retrospect that the opponents of von Helmholtz would have been subdued by his theoretical support of the nonlinear process as well as by Riemann's analysis, but that did not happen. W. Preyer [53], R.H.M. Bosanquet [54], and Koenig [55] all continued to disagree with von Helmholtz [56], and no one seemed to connect Riemann's work with the subject at all.

Another attempt to establish the objective nature of Tartini tones was made by Sir Arthur W. Rücker and E. Edser [57] in 1895. In their experiment, a siren capable of producing two tones whose sum or difference frequency amounted to 64 Hz, sounded into a horn (a Koenig resonator, Fig. 6-22) and served to activate a tuning fork (also of Koenig's manufacture) designed to vibrate at 64 Hz. A mirror, attached to one prong of the fork, served as one of the mirrors in a Michelson inteferometer. Rücker and Edser went to rather great lengths to achieve isolation of the system. The tuning fork was attached to a large stone which was in turn "suspended on

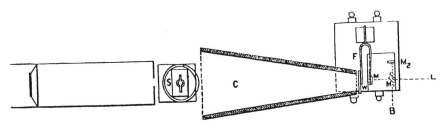

FIGURE 6-22. Rücker and Edser apparatus [57].

wires and india-rubber door-fasteners from a heavily weighted beam, which itself rested on india-rubber balls [58]." In the absence of vibrations of the fork, the interference bands from a sodium vapor light source could be seen clearly. The slightest vibration of the fork (by as little as one-quarter of a sodium wavelength (5.89×10^{-7} m) could be detected.

When the experiment was performed late an night, the interference bands were "absolutely clear and steady." In many separate experiments, Rücker and Edser detected the presence of both the sum and difference frequency when the siren was excited in the way described above. Apparently, however, not too many people read the *Philosophical Magazine*, and this work of Rücker and Edser was largely forgotten. Barton, in his 1932 text [59], pays more attention to critics of Rücker and Edser than to their work, and neither Jones (1937) [60] nor Richardson (1940) [61] mention the experiment at all in their texts.

6.10. Hearing

There was a considerable burst of activity in hearing research just after the mid-century. Marchese Alfonso Corti (1822–1876), who took a medical degree from the University of Pavia in 1847, and who had visited a number of laboratories in western Europe and learned a great deal about tissue preparation, discovered the organ in the cochlea that bears his name [62]. His drawing, a flattened-out version, is shown in Fig. 6-23.

While Corti gave a description of the cochlea, he did not think of the hair cells as being the auditory receptors, but rather believed that the nerve fibers were stimulated directly [63]. Albert von Kölliker (1817–1905), who received his medical degree at Zürich, and was later Professor of Anatomy at the University of Würzburg, was able to trace the nerve fibers up to the tympanic side of the basilar membrane [64].

In 1851, Ernst Reissner, a Professor at the University of Dorpat in what is now Estonia, discovered the membrane that has been named for him. This membrane further subdivides the cochlea [65]. A remarkably accurate drawing of the cross section of the cochlea by Adam Politzer (1835–1920) in

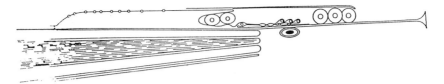

FIGURE 6-23. The organ of Corti. (After Corti [62].

1873 is shown in Fig. 6-24 [66]. This figure clearly locates both Reissner's membrane and the basilar membrane. The organ of Corti is located just above the basilar membrane, in its full size, not compressed as in the earlier figure of Corti (see also Fig. 6-25) [67].

The minute dimensions of the parts of the inner ear were established during this period. (See Fig. 2-10 for the location of the small connecting bones or ossicles—the hammer, anvil, and stirrup.) von Helmholtz [68] gives the diameter of the ear drum as 9–10 mm, while that of the oval window (Fig. 2-10) as 1.5–3 mm. The cochlea was even smaller, with its estimated volume being about 7 mm^3, while the basilar membrane has a length of 30 mm and a width varying from 0.1–0.5 mm [69]. von Helmholtz also measured the swing of the foot of the stirrup bone (the stapes) to be no greater than 0.1 mm and observed that

The mechanical problem which the apparatus within the drum of the ear had to solve was to transform a motion of great amplitude and little force, such as impinges

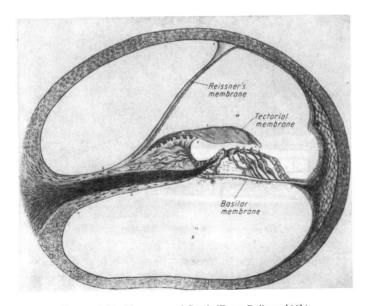

FIGURE 6-24. The organ of Corti. (From Politzer [66].)

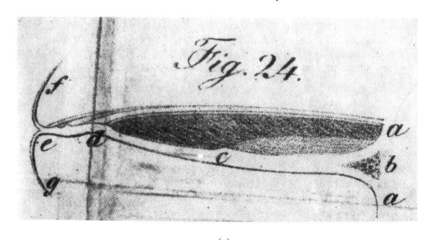

(a)

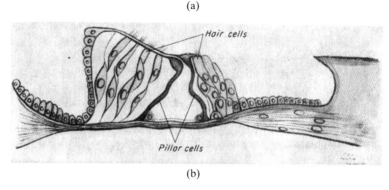

(b)

FIGURE 6-25. Other views of the organ of Corti. (a) The cochlear partition, from Lincke (1837). (b) The organ of Corti, from a drawing used at Harvard University around 1865. (From von Békésy [67].)

on the drumskin, into a motion of small amplitude and great force, such as had to be communicated to the fluid in the labyrinth. [70]

This transformer-like action of the ossicles and drum formed the basis for a detailed study by von Helmholtz, but his ideas on the shape and mobility of the drum membrane had to wait until well into the twentieth century for their ultimate testing [71]. The structure and dimensions of the drum and ossicles indicated that there is a ratio of about 21 between the drum membrane and the stirrup footplate. Thus the middle ear gives a considerable amplification to the force of the original sound.

Sound can also reach the cochlea through the bones of the skull. It was earlier thought that this bone conduction excited the ossicles into vibration in the same manner as in ordinary hearing (A. Lucae [72]) but work by Mach [73] and later by Petzold [74] suggested that relative motion between the sensory cells and the surrounding bone was the necessary feature, although this was not clearly established until much later [75].

The problem still confronted the scientists: What was happening in the cochlea and what gave rise to the hearing? Two rival theories developed. The first of these was the resonance theory, which was first advanced by Eduard Weber in 1841 [76], and developed by von Helmholtz (Chapter 3) [77]. In that theory, and in a frequency theory developed by W. Rutherford [78] (the latter theory is sometimes called the telephone theory of hearing), the signal from the oval window is presented immediately to the various portions of the cochlea. As we had noted earlier, von Helmholtz favored the idea that each hair cell in the cochlea reacted to a different frequency.

Opposed to these theories were the traveling and standing wave theories. In the first traveling wave theory, due to Hurst [79], a displacement wave proceeds along the basilar membrane and produces stimulation at the appropriate points along its path. As described by Wever and Lawrence, according to the Hurst theory, a displacement wave is formed in the basal portion of the basilar membrane,

a narrow bulge that moves along the membrane toward the apex. When the wave reaches the apex, it runs around the terminal wall to Reissner's membrane and returns along this membrane to the basal end of the cochlea. [80]

Several other ideas on traveling waves appeared before the end of the century. Pierre Bonnier hypothesized that the wave was a back-and-forth motion along the basilar membrane and spread slowly along that organ [81]. In ter Kuile's theory, it is a back-and-forth motion that is communicated to the fluid on the membrane. This wave also was presumed to travel at low speed ($0.5\,\mathrm{ms}^{-1}$ along the membrane) [82]. All these theories had in common some sort of traveling wave that was dampened out during its passage along the basilar membrane, and differed mainly in the assumptions on the mechanical characteristics of the membrane, including the amount of damping. There was at the time no way of testing their assumptions.

The standing wave theory was advanced by J.R. Ewald [83]. This theory assumes reflections of the waves on the basilar membrane, so that standing waves are set up in it (Fig. 6-26). Stimulation of the nerves would be provided by the region in which the displacement in the standing wave is near its maximum. While this theory continued to be of interest well into the twentieth century, there were many criticisms of it. The right answers had not yet been found.

A step in a new direction was provided by Alfred Mayer (1836–1897) [84]. By experimenting with the ticking of clocks, Mayer came to the conclusion that the sensation of one sound in the ear can be "obliterated" by the simultaneous action of a second, more intense, sound of lower frequency. At the same time, he also concluded that a sound, even when it is intense, cannot obliterate the sensation of another sound lower than it in pitch. This was the first study of the phenomenon now known as "masking," but it

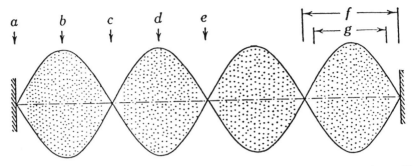

FIGURE 6-26. Standing waves on the basilar membrane. Ewald's pressure-pattern theory. The pattern shown here is that assumed by the basilar membrane when stimulated by a tone of 40⁻. Points *a, c, e* are nodes; *b, d* are loops. A loop covers half a wave length *f*, and is effective over only a part of its length *g*. (After Ewald, from Wever [83].)

would be nearly 50 years before further research was carried out in the field. It was an early step into the field of psychoacoustics.

There was considerable interest in determining the auditory threshold, i.e., the lowest intensity of sound that could be heard, and also the lowest frequency. Rayleigh and von Helmholtz concerned themselves with the lowest intensity, while Wilhelm Preyer conducted detailed experiments on the lowest audible frequency, which he maintained was below 15 Hz [85]. In 1890, C. Stumpf [86] set forth his theory of tonal fusion.

Preyer made an early study of the frequency discrimination of the average ear [87], a study that was followed up by those of Luft [88] and Meyer [89]. The basic conclusion of these works were that the "difference limen" (DL)—the relative frequency shift required to produce a just-noticeable-difference in the ear of the listener—was high at low frequencies, dropped to a plateau over much of the audio range, and rose again at high frequencies [90].

On the subject of deafness, some progress was made on the treatment of otosclerosis (see Chapter 1) by J. Kessel in 1878. He intervened surgically in attempts to mobilize the stapes (stirrup) which had become fixed due to the disease [91]. Somewhat later, C. Miot [92] performed over 200 similar operations.

6.11. Speech

One step forward in understanding the origins of speech was the invention of the laryngoscope by a famous singer and singing teacher, Manuel Garcia (1805–1906) [93] in 1855. This consisted of a small mirror inserted into the mouth, by which the vocal cords could be viewed in the glottis. He used this on singers and demonstrated that the tension in the cords varied and that the width of the slit between them changes as the frequency of the note or the rate of expulsion of the breath is changed. Stroboscopic methods were later used by Paul Musehold [94] to demonstrate that the cords executed simple harmonic motion even while the sounds emitted from the mouth were far more complex.

The study of speech in this period was scattered over many areas of research. As we have noted previously, von Helmholtz was interested in the synthesis of vowels by appropriate combinations of the harmonics of a fundamental tone (Chapter 3), while a set of his resonators was capable of separating out the various harmonics involved in all sounds, including human speech. A more direct analysis of speech was provided by the experiments of Ernest Merritt and E.L. Nichols [95], who employed the manometric flame technique developed by Koenig. Some of their results are shown in Fig. 6-27. One must confess that the reading of these signals must have been as complex an art as the reading of chicken entrails in the days of Greece and Rome. But they were the best that was available in 1900.

In Chapter 2, we mentioned the work of Robert Willis on speech in the early 1800s. From his researches on reed organ pipes, he concluded that the quality of a vowel depended on a fixed, characteristic pitch [96]. Charles Wheatstone gave support to this view, arguing that the larynx generated a complex signal that the various oral cavities acted upon to produce the final sound [97]. The next half-century saw a lively discussion of the nature of vowel sounds. Grassman [98] and others contributed to this fixed-pitch theory. Donders noted that the cavity of the mouth is tuned at different pitches for different vowels, and the theory received the support of von Helmholtz, who concluded that every vowel is characterized by a fixed region of resonances that are independent of the fundamental. An opposing theory—the relative-pitch theory—was developed, which maintained the view that the quality is determined by the set of overtones accompanying the fundamental, the ratio of overtone to fundamental remaining constant. The situation was summed up near the century's end by Rayleigh in the final paragraphs of his book. After noting that he thought the invention of the phonograph by Edison would have provided a decisive test of the fixed-pitch theory, Rayleigh commented on the work done by Ludimar Hermann [99] and others:

A general comparison of [Helmholtz's] results with those obtained by other methods has been given by Hermann, from which it will be seen that much remains to be done before the perplexities involving the subject can be removed. [100]

6.12. Music

Much of the research on music in the second half of the nineteenth century centered around the contributions of von Helmholtz and Rayleigh that were discussed in Chapters 3 and 4. But there were others working in the

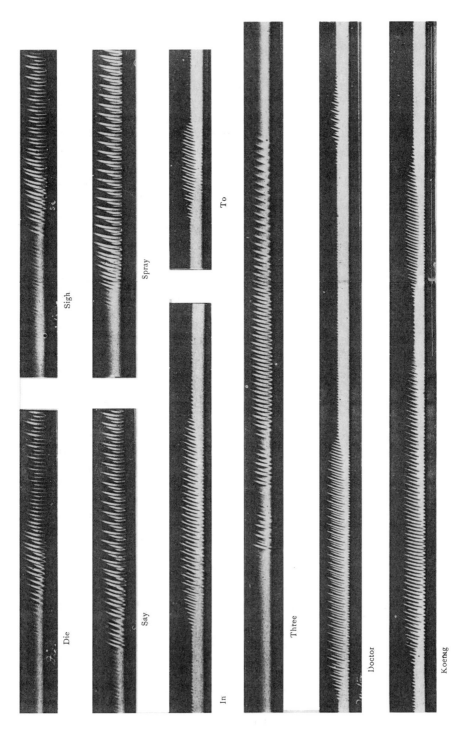

FIGURE 6-27. Vowel sounds by the flame technique. (From Merritt and Nichols [95].)

field, especially on the determination of the waveform of the vibrations of various musical instruments. In 1891, O. Krigar-Menzel and A. Raps used photography to update the work of von Helmholtz on violin strings, and published detailed studies of the behavior of such strings (Fig. 6-28) [101]. Later, they also studied the behavior of a plucked string [102], samples of which are shown in the book by Barton [103]. As a result of

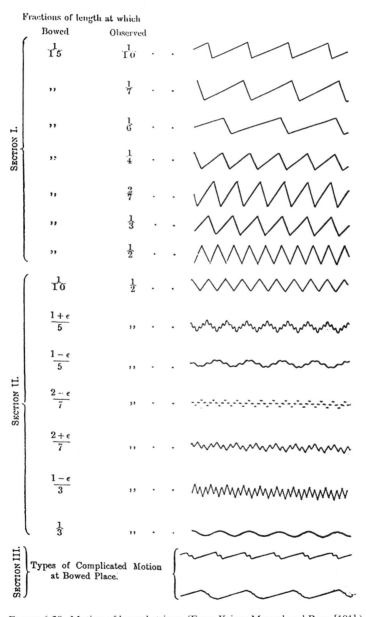

FIGURE 6-28. Motion of bowed strings. (From Krigar-Menzel and Raps [101].)

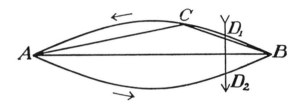

FIGURE 6-29. Shapes assumed by a bowed string. The arrow with a tail shows the position and direction of bowing. (From A.T. Jones [104].)

these researches, the behavior of the violin string was well established. As shown in Fig. 6-29 [104], as the string is bowed downward, the string takes on the shape of the two straight lines AC and CB, with the intersecting point C moving counterclockwise around the path indicated by the short arrows.

Progress was also made in this period on understanding the behavior of the hammer of the piano. Charles Kasson Wead (1848–1925), a professor at the University of Michigan, determined the length of time the hammer of the piano was in contact with the string [105]. He found that the hammer was in contact for only a small fraction of the period of the sound for lower notes, while the contact at higher tones (1082 Hz or c^5) could last an entire period.

Further piano research was done by Walter Kaufmann in Berlin. Kaufmann followed the method of Krigar-Menzel and Raps, and studied the vibration curves for the strings as a function of the various parameters, such as mass and hardness of the hammer, strength of the blow, and position at which the string is struck [106]. In so doing, he came into conflict with theories developed by von Helmholtz on the reason behind the choice of striking position of the hammer [107].

Other studies were carried out on the characteristics of organ pipes. The behavior of the air in the pipe was first well described by Hermann Smith in a series of short papers [108]. These were followed by works by F.W. Sonreck [109] and Heinrich Schneebeli [110], and finally by von Helmholtz himself [111]. To close this period, an experimental confirmation of the flow of the air was provided by Victor Hensen in 1900. The flow of the air, according to Hensen, is shown in Fig. 6-30 [112].

6.13. Architectural Acoustics

There was little progress in understanding architectural acoustics in this half-century. In 1853, the Boston physician J.B. Upham published two papers that summed up the state of knowledge in the field [113]. He recog-

FIGURE 6-30. Motion of the air mass in an organ pipe, shown in cross section. (From Hensen [112].)

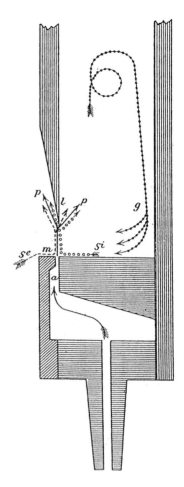

nized the significance of reverberation and reflection in room acoustics, and reported some bad rooms with roughly measured reverberation times of up to several seconds. (While the term "reverberation time" had not yet been defined, Upham spoke of the time during which sound continued after the original impulse had ceased.) He knew that a smooth and flat surface would give rise to longer reverberation times, but had little understanding of the effect of different materials in increasing or decreasing the size of that quantity. In fact, much of the two articles was a review of Herschel's encyclopedia article on sound from 1830 [114].

We have already mentioned the work of Henry on room acoustics in Chapter 3. The next great advance had to wait a half-century longer. We might remark that perhaps the biggest event in architectural acoustics in the second half of the nineteenth century was the birth of Wallace Clement Sabine in 1868!

6.14. Harbingers of the Future

There were great undercurrents in nineteenth-century physics that would lead to major advances in acoustics in the twentieth century. Some of these appeared at the time to have little to do with sound, but were later found to be enormously useful in expanding the field of acoustics.

Ultrasonics

Various experimenters in the early nineteenth century had found the upper frequency limit of human hearing to be in the region around 30 kHz. But instrumentation existed that could produce wave motions above these frequencies. Koenig continued to make smaller and smaller tuning forks, until he had some capable of vibrations as high as 90 kHz. His highest tone was f^9, which has a frequency of 87, 381 Hz [115]. These forks were of extremely small size (see Fig. 6-31), with the space between the prongs amounting to 1.0 mm and the length of the prongs to no more than 1.0 cm.

Koenig measured the presence of these oscillations by means of a Kundt tube. Since the wavelength is in air, the frequency of f^9 is 3.9 mm, the tubes

FIGURE 6-31. Koenig's ultrasonic signals of relatively low frequency [115].

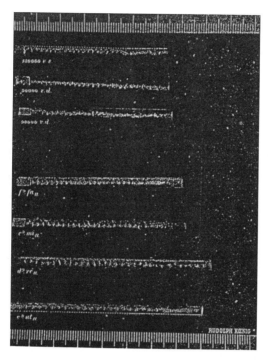

FIGURE 6-32. Koenig's ultrasonic signals of relatively high frequency [116].

were also very small. The clarity of the dust figures at lower frequencies is shown in Fig. 6-31, but the resolution at the highest frequency was quite poor (Fig. 6-32). Clearly, in both sound production and sound detection, Koenig had gone as far as was possible with the old techniques.

Another method for producing sound at frequencies above the audible range was developed by an English physician, Sir Francis Galton (1822–1911) [116]. Galton constructed a small whistle that was apparently capable of sounding frequencies as high as 84 kHz. He placed such a whistle in a cane and used it, in a very informal way, to test the high-frequency hearing of animals. As he put it,

I hold it (the whistle) as near as is safe to the ears of animals, and when they are quite accustomed to its presence and heedless of it, I make it sound; then if they prick their ears it shows that they hear the whistle. . . . Of all creatures I have found none superior to cats in the power of hearing shrill sounds; it is perfectly remarkable what a faculty they have in this way. Cats, of course, have to deal in the dark with mice, and to find them out by their squealing. [117]

Unfortunately, Galton gave no further details as to the upper audibility thresholds of the cats and other animals that he tested.

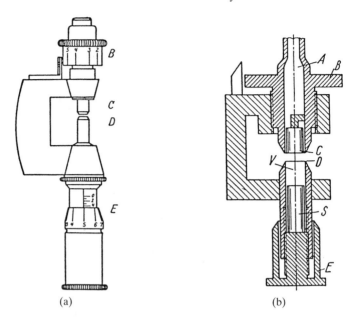

(a)   (b)

FIGURE 6-33. Edelmann's Galton whistle: (a) Galton whistle; and (b) section through Galton whistle. (From Bergmann [118].)

The Galton whistle was later improved by M.T. Edelmann in 1900 (Fig. 6-33) [118]. Air is blown through the tube A and emerges from a circular slit at C to fall on the knife edge D. The size of the chamber below the knife can be controlled by the drum E. The frequency of the sound emitted depends on the size of the chamber below the knife edge and the pressure at which the air enters the nozzle A. Frequencies as high as 100 kHz were obtained with this device.

Here again, however, the technique had been refined to the point of exhaustion. Something new was needed if the upper limits of sound vibrations were to be increased significantly. And that "something new" was to be found in magnetostriction and piezoelectricity.

Magnetostriction

The first of these two different discoveries that were ultimately to provide the means for raising this frequency limit was that of magnetostriction, i.e., the change in length of a ferromagnetic rod in the presence of a magnetic field. This effect was reported by the famous English brewer and amateur physicist, James P. Joule (1818–1889), whose name is more frequently connected with the measurement of the mechanical equivalent of heat. Joule first reported the change in length of ferromagnetic rods as early

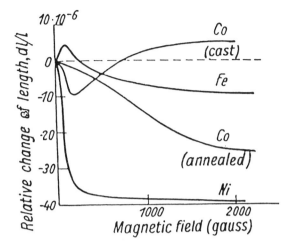

FIGURE 6-34. Change of length by magnetostriction. (From Bergmann [120].)

as 1842 [119]. A sample of the complexity of the behavior of magnetic materials in this effect is shown in Fig. 6-34 [120]. It can be seen that the rods can both increase and decrease in length. Joule also observed that the change in dimensions was independent of the direction of the magnetic field.

As has often been the case in science, a number of observers had preceded Joule with related results. In 1837, Dr. Charles Grafton Page (1812–1868), a physician of Salem, MA [121], observed something he called "galvanic music." He placed a coil of wire between or in front of the poles of a horseshoe magnet. When he interrupted the current, a ringing sound was heard [122]. As Hunt [123] later pointed out, it was not clear whether or not Page was observing a release of magnetostrictive stress or not. In the years following, a number of other observers reported similar phenomena. It is perhaps appropriate to quote Hunt's conclusion:

If it is just to concede to Joule the credit for "discovering" magnetostriction—and it probably is—then it can be said that this is one effect whose manifestations had been studied extensively for several years prior to its discovery. [124]

In 1884, Kirchhoff developed a theory to explain magnetostriction [125], and in 1898, two Japanese scientists, H. Nagaoka and K. Honda, reported experimental results in rough agreement with the theory [126].

Piezoelectricity

The second effect that was to bear much fruit in the development of ultrasonics was that of piezoelectricity (from the Greek *piezo*—pressure). Its discovery also has a lengthy history. In 1817, the French crystallographer René-Just Haüy (1742–1823) [127] observed that he could electrify a body "by pressing the body to be tested (any of a number of crystals, including Iceland spar) for a very short time between two fingers [128]." A few years later, Antonine César Becquerel (1788–1878) [129] confirmed these results, and pointed out a train of such observations dating back to Coulomb.

No immediate developments followed from these papers. In another area of crystal physics, it had been known for some time that opposite electric polarity appeared at the ends of certain crystals when heated. This phenomenon had been given the name of pyroelectricity (from the Greek word *pyro* for fire) by Sir David Brewster (1781–1868) in 1824, and on this basis, Lord Kelvin "postulated a state of permanent (electrical) polarization in every pyroelectric crystal [130]."

Begining with these ideas, two young French scientists, Jacques Curie (1855–1941) and his brother Pierre (1959–1906), reported a new discovery in 1880:

We have found a new mode of development of polar electricity in these same crystals (as had possessed pyroelectricity), which consists of submitting them to variations of pressure along their hemihedral axes.

The effects produced are entirely analogous to those caused by heat: during compression, the extremities of the axis of these crystals are charged with opposite electricity; once the crystal reaches a neutral state, if it is then subject to tension, the phenomenon is reproduced but with an inversion of sign; the extremity which is charged positively by compression becomes negative during tension, and conversely. [131]

Later that year, another Frenchman, Gabriel Lippmann (1845–1921) [132] carried out a thermodynamical study of the phenomenon and concluded that an opposite effect must also exist, namely, that the application of an electric field to these crystals should cause a change in their length. And, indeed, the Curie brothers, taking up this lead, quickly found this second effect [133].

The manner in which the electrical effects are produced is made clear in the schematic drawings of Fig. 6-35 [134]. In sharing electrons the silicon atoms of the quartz crystal take on a net positive charge and the oxygen a net negative one. Because of the crystalline structure of the quartz, the compression leads to an imbalance of atoms, as shown in the lower half of the figure, and to net positive and negative charges on the opposite faces. One can also see that the application of an electric field would have stretching or compressional effects, depending on the direction of the field.

The Curies found that a wide variety of crystalline substances could also exhibit this phenomenon, and the name "piezo-electricity" was soon proposed for the effect by W.G. Hankel [135].

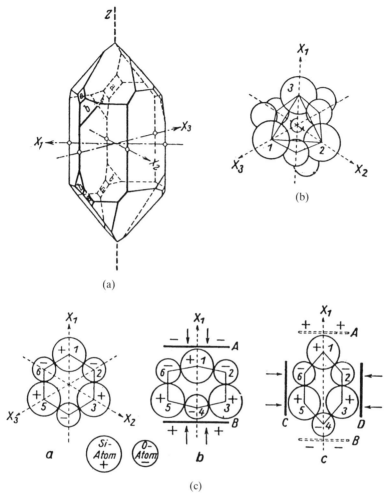

FIGURE 6-35. (a) Schematic of quartz crystal; (b) cell structure of quartz, viewed from above; and (c) on the theory of piezoelecticity. (From Bergmann [134].)

The mathematical formulation of the theory of the piezoelectric effect was developed first by Pierre Duhem (1861–1916) [136] and Friedrich Pockels (1865–1913) [137], and more completely by the eminent German crystallographer, Woldemar Voigt (1850–1919) [138]. We mentioned above the idea of Lord Kelvin on the existence of a permanent state of polarization in pyroelectric crystals. Other contributions were made by Wilhelm Conrad Röntgen (1845–1923), the discoverer of x-rays [139], by Charles Friedel (1832–1898) and Jacques Curie [140], and again, by Lord Kelvin [141]. Hunt notes that the work of Voigt was "not effectively challenged for more than half a century [142]."

Acousto-Optics

Another source of sound waves that had to wait even longer than magneto-striction and piezoelectricity before becoming a practical technique is the phenomenon known as acousto-optics, first observed by Alexander Graham Bell in 1880 [143]. Bell noted that, if an intermittent beam of light were incident on a wafer of a solid, a sound tone could be heard, the frequency of which was that of the intermittency. He later found that a wide variety of substances could be used for production of the sound. Subsequently, he placed his test material in a tube, and connected the mouth of the tube to his ear by means of a rubber tube. The apparatus is pictured in Fig. 6-36. Rays from the Sun were reflected onto a pair of perforated disks, one of which could be rotated, thus chopping the beam of light. The chopped beam was then focused by a parabolic reflector onto the test material, with the sound picked up by the earpiece. In a letter to his associate, Sumner Tainter, Bell wrote from Paris in 1880,

I have tried a large number of substances in this way (mostly, as described above) with great success, although it is extremely difficult to get a glimpse of the Sun here, and when it does shine the intensity of the light is not to be compared with that to be obtained in Washington. I got splendid effects from crystals of bichromate of potash, crystals of sulphate of copper, and from tobacco smoke. A whole cigar placed in the test-tube produced a very loud sound. [144]

Apparently, Bell, like Franklin and Edison, was prepared to try anything!

The phenomenon was studied by a number of Bell's contemporaries, including Tyndall [145], Röntgen [146], and W.H. Preece [147]. Sound could be produced from liquids and gases, as well as solids. There was some controversy at the time as to whether the effect was due to visible light or to radiant heat. The full solution of the problem, and its practical application had to wait nearly one hundred years and for the invention of the laser (see Chapter 10).

Density of States

Often, in this text, we have seen instances in which Rayleigh set forth an idea, but left it for others to follow up, for example, his analysis of sound absorption in gases. On other occasions, he picked up on the ideas of others, rephrased them, and extended them into new territory. One such case in point was his work on the theory of Chladni plates, wherein he developed the mathematical approximation that came to be called the Rayleigh–Ritz method in quantum mechanics [148], [149]. But there is also at least one case in which Rayleigh gave a derivation in his book, and then, long after, returned to that idea to apply it to a wholly different field. In the *The Theory of Sound* (Section 255), Rayleigh considered first the case of allowed vibrations in a closed tube of length l. In this case, the wave number

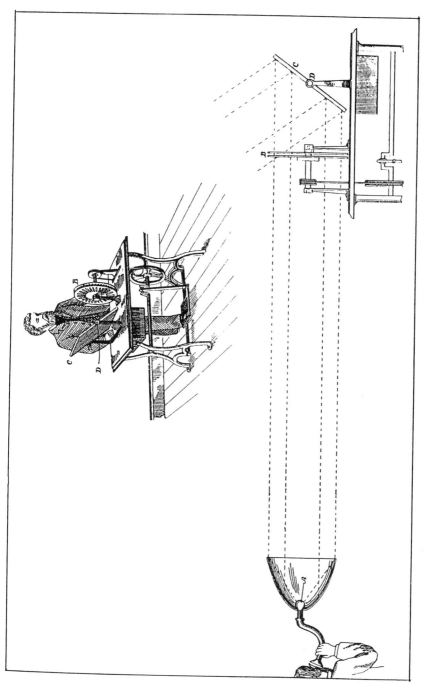

FIGURE 6-36. Bell's acousto-optic experiment [143].

k is equal to $(\pi/l)p$, where p is an integer. In Section 267 of his book, he extended this analysis to the three-dimensional case, where, if the box were a cube, the wave number k would be given by

$$k^2 = \left(\frac{\pi}{l}\right)^2 (p^2 + q^2 + r^2),\qquad\qquad(6.6)$$

where p, q, r are integers. Rayleigh let this sit on the shelf for nearly a quarter of a century. Then, in 1900, he returned to it [150], making use of it in calculating classically the radiation from a black body. He wrote

If we regard p, q, r as the coordinates of points forming a cubic array, k is the distance [151] of any point from the origin. Accordingly, the number of points for which k lies between dk and $k + dk$, proportional to the volume of the corresponding spherical shell may be represented by $k^2\,dk$, and this expresses the distribution of energy according to the Boltzmann–Maxwell law, so far as regards the wavelength or frequency. [150]

This, of course, was the development of Rayleigh's portion of the Rayleigh–Jeans law of black body radiation. It also marked the use of a reciprocal, or phase space. His calculation was a calculation of the density of states, which became, together with the reciprocal lattice, a standard of solid state physics of the next century.

Response to Stimuli

In the early days of experimental psychology, two German scientists made enduring contributions to psychology that would prove to be of great significance in acoustics. Ernst Heinrich Weber (1795–1878), professor first of anatomy and then of physiology at the University of Leipzig, published a book on *Touch and Sensibility* (in German, *Der Tastsinn und das Gemeingefühl*) in 1846, clearly an ancestor of our modern acoustic field of "bioresponse to vibration [152], [153]." In it he enunciated the principle that, if R is the magnitude of a stimulus, then the ratio of the increase in stimulus ΔR necessary for a just noticeable difference (written *jnd* in much of the literature) in the sensation (S) is given by

$$\frac{\Delta R}{R} = \text{constant.}\qquad\qquad(6.7)$$

In 1860, Gustav Theodor Fechner (1801–1887), published his book *Elements of Psychophysics* [154]. Fechner spent his long career as a professor at Leipzig, being variously a physiologist, a physicist, an experimental psychologist and a philosopher. The breadth of his interests rivaled that of von Helmholtz. In his book, Fechner stated a law that now bears his name:

$$S = k \log R.\qquad\qquad(6.8)$$

Fechner later referred to Eq. (6.8) as Weber's law, but it is more appropriate to call Eq. (6.7) by that name, and apply the name Fechner to Eq. (6.8). Of Fechner, Boring later wrote

The great thing that he accomplished was a new kind of measurement. The critics may debate the question as to what it was that he measured; the fact stands that he conceived, developed and established new methods of measurement, and that, whatever interpretation may later be made of their products, these methods are essentially the first methods of mental measurement and thus the beginning of quantitative experimental psychology. [155]

In 1888, Max Carl Wien (1866–1938) [156] used a telephone receiver located at the mouth of a von Helmholtz resonator, with a rubber tube connecting the resonator to the observer's ears, and made measurements of the Fechner constant for different musical tones. He found that the "constant" k was in fact not constant, but varied from 0.22 at 220 Hz down to 0.13 an octave higher (440 Hz). He also noted that k increased at very low and very high sound intensities.

At almost the same time, A. Deenik made similar measurements of the relative increase in stimulus necessary for a just-noticeable difference in sensation, using tuning forks and organ pipes. As did Wien, Deenik found that the value of k decreased somewhat asymptotically with increase in frequency [157].

From the viewpoint of acoustics we might observe that Fechner made out the connection between biological response to stimulation to be a logarithmic one. When the time came for developing a scale for sound pressure levels, it was therefore appropriate to take Fechner's law as a model and introduce the decibel (see Chapter 7).

Overview

The development of new apparatus and new directions promised a bright future for acoustics in the twentieth century. The different advances described in Section 6.14 were especially inviting. Eventually, all of them would come to fruition, but most of them took much longer than the next quarter century to bring that about.

Notes and References

[1] Regnault, *C.R. Acad. Sci. Paris*, **66**, 209 (1868). See also the discussion in E.G. Richardson, *Sound*, 3rd ed., Edward Arnold, London, UK, 1940, pp. 6–7. The sewers of Paris were made famous by Hugo's *Les Miserables*, but the water pipes of the city belong to acoustics—from Biot to Regnault.

[2] *Collected Works of P.N. Lebedev* (in Russian). Acad. Sci. USSR Press, Moscow, 1963, p. 51.

[3] A. Kundt, *Ann. Physik* **127**, 497–523 (1866). Kundt first worked at Zürich and Würzburg, then was Professor at Strasbourg, where he had the famous Russian physicist Petr Lebedev as a student, and finally at Berlin, where he succeeded von Helmholtz. The irritating sounds from *Kundt's tube* have provided an enduring memory for generations of elementary physics students.

[4] H. von Helmholtz, *Wiss. Abhandl.* **1**, 338 (1882); Rayleigh, *The Theory of Sound*, 2 vols. 2nd ed. Reprinted by Dover, New York, NY, 1945, vol. 2, pp. 318–319. See also G.W. Stewart and R.B. Lindsay, *Acoustics*, Van Nostrand, New York, NY, 1930, pp. 68–69.

[5] G. Kirchhoff, *Gesammelte Abhanbdlungen* (*Collected Works*), Barth, Leipzig, 1882, Frontispiece.

[6] G. Kirchhoff, *Ann. Physik* **134**, 177–193 (1868).

[7] W. Duff, *Phys. Rev.* **6**, 129–139 (1898).

[8] Rayleigh, *Phil. Mag.* **97**, 308–314 (1899).

[9] Ibid., p. 314.

[10] F.E. Melde, *Ann. Physik* **109**, 193–215 (1860); **111**, 513–537 (1860); (new series) **21**, 452–470 (1884); **24**, 497–522 (1885). Figure 6-5 is taken from E.G. Richardson, *Sound*, Edward Arnold, London, UK, p. 118. See also the discussion in A.T. Jones, *Sound*, Van Nostrand, New York, NY, 1937, pp. 204–208.

[11] Sir John Herschel, *Phil. Mag.* **3**, 401–412 (1835).

[12] G.H. Quincke, *Ann. Physik* **128**, 177–192 (1866). Figure 6-6 is taken from E.G. Richardson, *Sound*, p. 58. The Herschel–Quincke systems are described in A.T. Jones, *Sound*, pp. 94–95, 202–203.

[13] In the 1920s. George W. Stewart found that the sound also vanishes when the sum of the lengths of the two parallel tubes amounted to an integral number of wavelengths, while at the same time, the difference in their lengths was not also an integral number of the wavelengths. See G.W. Stewart, *Phys. Rev.* **31**, 696–698 (1928).

[14] G.W. Stewart and R.B. Lindsay, *Acoustics*, Chapter 5.

[15] *Benchmark/8*, p. 54. The book provides a useful study of the development of this field through the past one hundred years.

[16] Rayleigh, *The Theory of Sound*, vol. 1, p. 251.

[17] Rayleigh, *The Theory of Sound*, vol. 1, pp. 395–432. See also Rayleigh, *Proc. London Math. Soc.* **13**, 4–16 (1881).

[18] J. Pochhammer, *J. Reine. Angew. Math.* (*Crelle's*) *J.* **81**, 324–336 (1875).

[19] Rayleigh, *Proc. London. Math. Soc.* **20**, 225–234 (1889). Lamb himself contributed additional comments at this time: *Proc. London Math. Soc.* **21**, 70–85 (1889).

[20] A.E.H. Love, *Phil. Trans. Roy. Soc.* (*London*) **179A**, 491–546 (1888).

[21] A. Clebsch, *Crelle's J. Reine Angew. Math.* **61**, 195–243 (1863). For a detailed historical review of this subject, see Y.-H. Pao and C.-C. Mow, *Diffraction of Elastic Waves and Dynamic Stress Concentrations*, Crane–Russak, New York, NY, 1973, Chapter 1. Clebsch was even more famous for his book, *Theorie der Elasticität fester Körper*, Leipzig, 1862. A French translation by Barré de Saint Venant and Alfred Flamant appeared in 1885.

[22] Rayleigh, *The Theory of Sound*, vol. 2, pp. 244–234.

[23] P.L. Rijke, *Ann. Physik*, **107**, 339–343 (1859); *Phil Mag.* **17**, 419–422 (1859).

[24] Rayleigh, *Phil. Mag.* **7**, 155 (1879). See also Rayleigh, *The Theory of Sound*, vol. 2, pp. 232–233.

[25] E.G. Richardson, *Sound*, Edward Arnold, London, UK, 1945, p. 202.

[26] *Encyclopaedia Britannica*. William Benton, Chicago, IL, 1958, Vol. 20, p. 289.

[27] Cited in Benjamin F. Howell Jr., *An Introduction to Seisomological Research*, Cambridge University Press, Cambridge, UK, 1990, pp. 57–58.

[28] F. Cecchi, *Nuovi Lincei Atti* **29**, 421–428 (1876). For this and other historical facts on seismology, the author is indebted to the text by Thorne Lay and Terry C. Wallace, *Modern Global Seismology*, Academic Press, San Diego, CA, 1995, especially Chapter 1, as well as to the book by Howell mentioned in Ref. [24].

[29] T. Lay and T.C. Wallace, *Modern Global Seismology*, p. 174.

[30] Alfred M. Mayer, *Am. J. Sci. Arts*, **8**, 21–42 (1874).

[31] D.C. Miller, *An Anecdotal History of the Science of Sound*, Macmillan, New York, NY, 1935, opposite p. 84.

[32] D.C. Miller, *An Anecdotal History*, pp. 85–92. See also, R. Koenig, *Quelques Expériences d'Acoustique*, A, Lahurel, Paris, 1882, p. 174.

[33] Ibid., pp. 91–92.

[34] E.L. Scott, *Cosmos*, **14**, 314 (1859); *C. R. Acad. Sci. Paris* **53**, 108–111 (1861).

[35] D.C. Miller, *An Anecdotal History*, p. 88.

[36] M. Josephson, *Edison*, McGraw-Hill, New York, NY, 1959, p. 159.

[37] E.W. Blake, Jr., *Am. J. Sci. Arts*, Ser. 3, **16**, 54–59 (1878). Blake was given the title of Hazard Professor of Physics, a chair that was financed by a well-to-do Rhode Island family (which included Oliver Hazard Perry as a member). Later (acoustical) holders of the chair included R. Bruce Lindsay in 1934, A.O. Williams, Jr. in 1970, and this author in 1984. The current holder is C.E. Elbaum.

[38] *Ostwalds Klassiker der Exakten Wissenschaften*. W. Engelmann, Leipzig, 1906, No. 157, Frontispiece.

[39] A. Toepler, *Ann. Physik*, **127**, 556–580 (1866); **128**, 108–125, 126–129 (1866).

[40] A. Toepler, *Ann. Physik*, 5th ser., **131**, 33–55, 180–215 (1867). Toepler gave his reason for using the name in the first paper, and continued a further defense of it in the second. Apparently, he was not very happy about the choice of the name, but it not only survived in German, but has also entered the English language.

[41] R.W. Wood, *Phil. Mag.* 5th ser., **48**, 218–227 (1899).

[42] Cited in R.W.B. Stephens and A.E. Bate, *Acoustics and Vibrational Physics*, Edward Arnold, London, 1966, UK, p. 486.

[43] F.A. Journée, *C.R. Acad. Sci. Paris* **106**, 244–246 (1888).

[44] E. Mach and P. Salcher, *Sitzber. Akad. Wiss. Wien* **95**, part 2, 764–780 (1887).

[45] E. Mach and L. Mach, *Sitzber. Akad. Wiss. Wien* **98**, part 2a, 1310–1326 (1889).

[46] S. Earnshaw, *Phil. Trans. Roy. Soc. (London)* **150**, 133–148 (1858).

[47] The derivation in more modern terms may be found in R.T. Beyer, *Nonlinear Acoustics*, Naval Sea Systems Command, 1974, p. 103. Reprinted by the Acoustical Society of America, Woodbury, NY, 1997.

[48] And when the problem was solved, it was reported in an Italian journal [E. Fubini-Ghiron, *Alta Freq.* **4**, 532–581 (1935)], where it did lie unread by most of the acoustical world.

[49] F.W. Bessel, Collected Work of *Abhandlungen von F.W. Bessel*, W. Engelmann, Leipzig, 1875. Band I, pp. 17–21, 84–109.

[50] The history of the rediscovery of the Bessel connection is given by D.T. Blackstock, *J. Acoust. Soc. Am.* **34**, 9–30 (1962).

[51] G.F.B. Riemann, *Gött. Abhand.*, 1860; reprinted in *Collected Works*, Dover, New York, NY, 1953, p. 156. An English translation of part of the article appears in *Benchmark/19*, pp. 42–60.

[52] R.T. Beyer, *Nonlinear Acoustics*, p. 107.

[53] W. Preyer, *Akustische Untersuchunge (Acoustical Investigations)*, and *Ueber die Grenzen der Tonwahrnehmung (The Limits of the Perception of Tone)*, in *Physiologische Abhandlungen (Physiological Essays)* (W. Preyer, ed.), Jena, 1876.

[54] R.M. Bosanquet, *Proc. Phys. Soc. (London)* **4**, 221–256 (1881).

[55] R. Koenig, *Ann. Physik* **12**, 335–349, 350–353, **14**, 369–393 (1881).

[56] A.J. Ellis, in Appendix XX, part L, to his English translation of H. von Helmholtz, *On Sensations of Tone*, 1885. Reprinted by Dover, New York, NY, 1954.

[57] A.W. Rücker and E. Edser, *Phil. Mag.* Ser. 5, **39**, 342–357 (1895).

[58] Ibid., pp. 343–344.

[59] E.H. Barton, *A Textbook on Sound*, Macmillan, London, UK, 1908. Reprinted, 1932.

[60] A.T. Jones, *Sound*, Van Nostrand, New York, NY, 1937.

[61] E.G. Richardson, *Sound*, 3rd ed., Edward Arnold, London, UK, 1949.

[62] A. Corti, *Wiss. Zool. Z.* **3**, 109–165 (1851).

[63] An excellent discussion of the work on the cochlea in this period is given in *Benchmark/15*, pp. 98–101.

[64] A. Kölliker, in *Festschrift für F. Tiedemann*, Würzburg, 1854.

[65] Ernest Reissner, *De Auris Internae Formatione*, Dorpat, Estonia (Livonia), 1851.

[66] Adam Politzer, *Wandtafeln zur Anatomie des Gehörorganes*, 1873.

[67] G. von Békésy, *Experiments in Hearing*, McGraw-Hill, New York, NY, 1960. Reprinted by the Acoustical Society of America, Woodbury, NY, 1989, pp. 12, 13.

[68] H. von Helmholtz, *On Sensations of Tone*, p. 134.

[69] E.G. Wever and M. Lawrence, *Physiological Acoustics*, Princeton University Press, Princeton, NJ, 1954, p. 418.

[70] H. von Helmholtz, loc. cit.

[71] A detailed consideration of Helmholtz's work, as well as work in testing their theories is given in Ernest Glen Wever and Merle Lawrence, *Physiological Acoustics*, pp. 90–114.

[72] A. Lucae, *Arch. Ohrenheilk.* **1**, 303–317 (1864).

[73] E. Mach, *Sitz. Ber. Akad. Wiss. Wien, Math.-Phys. Cl.* part 2, **48**, 283–300 (1863); part 2, **54**, 11 (1864); part 2, **57**, 11 (1864).

[74] Friedrich Petzold, *Z. Ohrenheilk.* **48**, 107–171 (1904).

[75] A discussion of this problem is given in E.G. Wever and M. Lawrence, *Physiological Acoustics*, p. 224.

[76] E.F. Weber, *Amtl. Ber. Versamml. Deutsch. Naturf. Aerzte*, 1841, pp. 83–84.

[77] An excellent critique of early cochlear theories is given by G. von Békésy in his *Experiments in Hearing*, pp. 404–405.

[78] W. Rutherford, *J. Anat. Physiol.* **21**, 166–168 (1886).

[79] E. Hurst, *Trans. Liverpool Biol. Soc.* **9**, 321–353 (1895). A detailed comparison of the various traveling wave theories (up to 1951) is given by Wever and Lawrence, *Physiological Acoustics*, Chapter 13.

[80] E.G. Wever and M. Lawrence, *Physiological Acoustics*, p. 249.

[81] Pierre Bonnier, *Bull. Sci. France Belgique* **25**, 367 (1893).

[82] E. ter Kuile, *Pflüg. Arch. Ges. Physiol.* **79**, 484–519 (1900).

[83] J.R. Ewald, *Wien. Klin. Wochenschr.* **11**, 721 (1898); *Pflüg. Arch. Ges. Physiol.* **76**, 147–188 (1899).

[84] A.M. Mayer, *Phil. Mag.* **2**, 500–507 (1876).

[85] W. Preyer, *Akustische Untersuchungen*, Jena, 1879. A detailed description of Preyer's work is given in H. von Helmholtz, *On Sensations of Tone*, p. 176, including a repeat experiment by A.J. Ellis on Preyer's apparatus.

[86] C. Stumpf, *Tonpsychologie*, vol. 2, S. Hirzel, Leipzig, 1895, p. 1525; *Z. Phys.* **38**, 745–758 (1926).

[87] W. Preyer, *Über die Grenzen der Tonwahrnehmung*, H. Dufft, Jena, 1876, pp. 24–38.

[88] E. Luft, *Phil. Stud.* No. 4, 1888, p. 511.

[89] M. Meyer, *Z. Psychol. Physiol. Sinn.* **16**, 352–372 (1898).

[90] An extensive history of these pitch studies may be found in T.F. Vance, *Psychol. Mongr.* **16**, 115–149 (1914).

[91] J. Kessel, *Arch. Ohrenheilk.* **13**, 69 (1878).

[92] C. Miot, *Rev. Laryngol.* **10**, 49 (1890).

[93] M. Garcia, *Phil. Mag.* **10**, 218–226 (1855). Among Garcia's students was the great Swedish soprano, Jenny Lind.

[94] P. Musehold, *Arch. Laryngol.* **7** (1898).

[95] E.L. Nichols and E. Merritt, *Phys. Rev.* **2**, 93–100 (1894).

[96] R. Willis, *Trans. Cambridge Philos. Soc.* **3**, 233–234 (1830).

[97] C. Wheatstone, *London and Westminster Rev.* Oct. 1837. The work of both authors is discussed in H. von Helmholtz, *On Sensations of Tone*, pp. 103–104, 117–118.

[98] A brief discussion of the works of these various authors is given by D.C. Miller, *The Science of Musical Sounds*, pp. 215–217.

[99] L. Hermann, *Pflüg. Arch. Ges. Physiol.* **47**, 351 (1890).

[100] Rayleigh, *The Theory of Sound*, vol. 2, pp. 478. As we say in the twentieth century, in concluding our research reports to contractors, "further research is necessary."

[101] O. Krigar-Menzel And A. Raps, *Sitz. Berlin Akad. Wiss.* **44**, 613–625 (1891).

[102] O. Krigar-Menzel and A. Raps, *Ann. Physik* **50**, 444–455 (1893).

[103] E.H. Barton, *A Textbook on Sound*, p. 432.

[104] A.T. Jones, *Sound*, p. 204.

[105] C.K. Wead, *Am. J. Sci.* **32**, 366–368 (1886).

[106] Walter Kaufmann, *Ann Physik* **54**, 675–712 (1895). A good discussion of Kaufmann's work is given by A.T. Jones, *Sound*, pp. 302–304.

[107] A discussion of this point is given in A.T. Jones, *Sound*, pp. 302–303.

[108] Hermann Smith, *Nature* **8**, 25 (1873); **10**, 161–163, 481–482 (1974). The writing style of Hermann Smith is particularly droll. A brief sample follows:

This same crisp little bit of paper will reveal to your eyes the treasured secret of the organ-pipe, tell you how its wealth of varied tone is wrought, show you its fine arc of flexure ... and how its free curves are moulded to your will; listen and you shall hear the domestic wrangle of the reed and pipe; (*Nature*, p. 163).

[109] F.W. Sonreck, *Ann. Physik* **158**, 129–147 (1876); **159**, 666–667 (1876).

[110] Heinrich Schneebeli, *Ann. Physik* **153**, 301–305 (1874).

[111] H. von Helmholtz, Ref. [1, pp. 92, 390, 394].

[112] Victor Hensen, *Ann. Physik* **2**, 719–741 (1900).

[113] J.B. Upham, *Am. J. Sci.* **65**, 215–226 (1853); **66**, 21–33 (853).

[114] J. Herschel, *Sound*, Encylopaedia Metropolitana, London, UK, 1830.

[115] R. Koenig, *Ann Physik* **69**, 626–660, 721–738 (1899). Koenig was using the von Helmholtz designation of the scales. According to the modern American standard, this would be called f^{12}. See the discussion of various scale notations in J. Backus, *The Acoustical Foundations of Music*, W.W. Norton, New York, NY, 1969, pp. 134–135.

[116] Francis Galton, *Inquiries into Human Faculty and its Development*, 1st ed., 1883; 2nd ed., 1907. Reprinted in *Everyman's Library* (Ernest Rhys, ed.), E.P. Dutton, New York, NY, 1911.

[117] Ibid., p. 27.

[118] M.T. Edelmann, *Ann Physik* **1**, 469–482 (1900). See also L. Bergmann, *Ultrasonics* (English translation), G. Bell and Sons, London, UK, 1938, pp. 1–2.

[119] J.P. Joule, *Phil. Mag.* ser. 3, **30**, 76–87, 225–241 (1847). Joule observed in this paper that he first reported the measurements in a "Conversatione" in Manchester, England, in (Sturgeon's) *Annals of Electricity, Magnetism and Chemistry* **8**, 219–224 (1842).

[120] L. Bergmann, *Ultrasonics* (English translation), G. Bell, London, UK, 1938, p. 6.

[121] C.G. Page, *Am. J. Sci.* **32**, 396–397 (1837); **33**, 118–120 (1837).

[122] C.G. Page, *Am. J. Sci.* **32**, 196 (1837).

[123] F.V. Hunt, *Electroacoustics*, McGraw-Hill, New York, NY, 1954. Reprinted by the Acoustical Society of America, Woodbury, NY, 1982, pp. 18–20. Hunt devotes some attention to the various other "discoverers" of magnetostriction.

[124] F.V. Hunt, *Electroacoustics*, p. 20.

[125] G. Kirchhoff, *Ann. Physik* **24**, 52–74 (1884).

[126] H. Nagaoka and K. Honda, *Phil. Mag.* Ser. 5, **46**, 261–290 (1898).

[127] René-Just Haüy, *Mém. Muséum Histoire Naturelle (Paris)* **3**, 223–228 (1817). Reprinted in *Ann. Chim. Phys.*, Ser. 2, **5**, 95–101 (1817). Accounts of these developments are given in Walter Cady, *Piezoelectricity*, 2 vols, McGraw-Hill, New York, NY, 1946. Reprinted by Dover, New York, NY, 1964, vol. 1, p. 1, and in F.V. Hunt, *Electroacoustics*, McGraw-Hill, New York, NY, 1954. Reprinted by the Acoustical Society of America, Woodbury, NY, 1982.

[128] R.-J. Haüy, *Ann. Chim Phys.*, Ser. 2, **5**, 96 (1817).

[129] A.C. Becquerel, *Bull. Sciences, Paris*, Ser. 3, **7**, 149–155 (1820); *Ann. Chim. Phys.*, Ser. 2, **22**, 5–34 (1823).

[130] Sir William Thomson (Lord Kelvin), *Phil. Mag.*, Ser. 5, **5**, 4–27 (1878).

[131] J. Curie and P. Curie, *C.R. Acad. Sci. Paris*, **91**, 294–295 (1881). An earlier report appeared in 1880 in *Bulletin de la Societé Minéralogique*.

[132] G. Lippmann, *C.R. Acad Sci. Paris* **92**, 1049–1051, 1149–1152 (1881); *Ann. Chim. Phys.* Ser. 5, **24**, 145–178 (1881).

[133] J. Curie and P. Curie, *C.R. Acad. Sci. Paris* **93**, 1137–1140 (1881).

[134] The figures are taken from S. Bergmann, *Der Ultraschall*, 3rd ed., VDI Verlag, Berlin, 1942, pp. 20–21. Bergmann's book, in its various editions, provides a wealth of information on the early development of ultrasonics.

[135] W.G. Hankel, *Abh. Sächs.* **12**, 457–548 (1881); **12**, 549–596 (1882).

[136] P. Duhem, *Ann. Sci. École Norm. Sup.* **2**, 405–424 (1885); **3**, 263–302 (1886); **9**, 167–176 (1892); *Compt. Rend.* **112**, 657–658 (1981).

[137] F. Pockels, *Neues Jahrbuch für Mineralogie Geologie und Palaeontologie* **7**, 201–231 (1890).

[138] W. Voigt, *Abh. Gesell. Wiss. Gött. (Math Classe*, Part 2) **36**, 1–99 (1890). *Lehrbuch der Krystallophysik*, B.G. Teubner, Leipzig, 1910, 1928.

[139] W.K. Röntgen, *Ann. Physik* **18**, 213–228 (1883); **19**, 513–518 (1883); **39**, 16–24 (1890).

[140] C. Friedel and J. Curie, *Compt. Rend.* **96**, 1262–1269, 1389–1395 (1883).

[141] Kelvin, *Phil Mag.* **36**, 331–342, 342–343, 384, 453–459 (1893).

[142] F.V. Hunt, *Electroacoustics*, p. 44.

[143] A.G. Bell, *Am. J. Sci*, **20**, 305–314 (1880). S. Tainter and A.G. Bell, *Am. J. Sci.*, **20**, 314–324 (1880). A.G. Bell, *Upon the Production of Sound by Radiant Energy*, Gibson, Washington, DC, 1881.

[144] Quoted by A.G. Bell, *Phil. Mag.* **11**, 510–528 (1881).

[145] John Tyndall, *Proc. Roy Soc. (London)* **31**, 307–317 (1881).

[146] W.C. Röntgen, *Phil. Mag.* **11**, 308–311 (1881).

[147] W.H. Preece, *Proc. Roy. Soc. (London)* **31**, 506–520 (1881).

[148] Rayleigh, *Proc. London Math. Soc.* **4**, 357–368 (1873); *The Theory of Sound*, p. 109ff.

[149] W. Ritz, *J. Reine Angew. Math.* **135**, 1–61 (1909).

[150] Rayleigh, *Phil. Mag.* **49**, 539–540 (1900).

[151] In his article, unlike the case treated in his book, Rayleigh has nondimensionalized his k, dividing it by π/c.

[152] E.H. Weber (1795–1878), *Die Lehre vom Tastsinn und Gemeingefühl* F. Viewig, Braunschweig, 1851. English translation *The Sense of Touch*, London, 1978.

[153] For a general discussion of the work of Weber and Fechner, see E.G. Boring, *A History of Experimental Psychology*, 2nd ed. Appleton-Century-Crofts, New York, NY, 1950, Chapter 10 and 14. A leisurely and detailed discussion of stimuli, sensations and the Weber–Fechner law is contained in Wilhelm Wundt, *Lectures on Human and Animal Psychology*. A translation of the second German edition, Macmillan, New York, NY, 1894, Lectures 1–5.

[154] G.T. Fechner, *Elemente der Psychophysik*, 1860.

[155] E.F. Boring, *A History of Experimental Psychology*, p. 293.

[156] M. Wien, *Ann. Physik* **36**, 834–857 (1888).

[157] A. Deenik, *Verslag Akad. Wet. Natuurk.* **14**, 396–400 (1905).

7
The Twentieth Century:
The First Quarter

> ... a rather stagnant state of acoustics during the first two decades of the twentieth century ... academically, acoustics became, by and large, an uninteresting subject
>
> R. Bruce Lindsay [1]

The gloomy comments on this period in the passage quoted above have been repeated by a number of authors [2]. The early part of the twentieth century was certainly one of amazing growth in most of physics, so that progress in acoustics did suffer by comparison. Of course, von Helmholtz was dead, and Lord Rayleigh was spending more of his time on non-acoustical research, but there was still great work being done, especially in architectural acoustics, electroacoustics, and ultrasonics. It was, however, for acoustics, a time when the field as a whole had to wait while the supporting disciplines of electricity and electrical devices moved forward, so that virtually all of acoustics could be redone with new, more powerful, and more accurate equipment.

In his book *Electroacoustics*, F.V. Hunt [3] has written a masterful essay covering the historical aspects of that field. We shall give a much briefer coverage, setting the scene for the applications of electroacoustics to the rest of the field, and leaving the details to the book by Hunt.

In the first part of the century, the various subbranches of acoustics (and the workers in them) began to sort themselves out. The days in which a single individual might write papers on speech, hearing, music, electroacoustics, and acoustical propagation, generally, were drawing to a close. By 1925, the major subdisciplines were pretty well established. But everyone was interested in the possibilities of application of the new tools from electrical science.

7.1. Electroacoustics

By 1900, the use of household electricity was becoming widespread, and so was the telephone. The early forms of radio were already on the scene. Electrical instruments were coming into the acoustical laboratory, ready to improve the taking and the interpretation of measurements.

Six important electrical instruments were to speed the early acousticians on their way: the microphone, the loudspeaker, the amplifier, the vacuum tube, the oscillator, and the oscilloscope. We have already reported on the invention of the carbon microphone by Edison and others in the 1870s in the development of the telephone (Chapter 5). Similarly, the first loudspeakers were the earpieces on these same instruments [4]. In the years following the Bell invention, numerous improvements were made on microphones and loudspeakers, but the others on our list of instruments were largely products of the new century.

The Microphone

The first microphones were the loosely packed carbon microphones of Edison and Berliner (Chapter 5). Various improvements in this device were recorded over the next 50 years [5]. Another type of microphone was the hot-wire microphone invented in 1880 by Sir William Henry Preece (1834–1913) [6] a device that was successively improved by Forbes (1849–1936) [7], Karl Ferdinand Braun (1850–1918) [8], and Tucker (b. 1877) [9]. In the Forbes instrument, a hot wire was placed in a resonator. The wire was periodically cooled by the "local wind" which was dependent on the particle velocity, Tucker refined this instrument (1921) by mounting the wire in the neck of a von Helmholtz resonator. Braun's instrument was later applied by Harold Arnold (1883–1933) and Irving Crandall (1890–1917) [10] for the absolute calibration of microphones and their instrument was given the name of thermophone (Fig. 7-1).

The greatest step forward, however, was made by Edward Wente (1889–1972) who invented the condenser microphone in 1917 (Fig. 7-2) [11]. Since this microphone had a resonant frequency in the high audible range, its sensitivity was not uniform over the audio frequency range, as is seen from Fig. 7-3 [12], but it was a beginning for what was to be a great instrument.

Another device of this period was the phonometer, invented by Arthur Gordon Webster (1863–1923) [13] (Fig. 7-4). Designed by him to measure the absolute sound intensity, it was one of the last of the old, non-electroacoustic instruments [14]. Still another was the tonoscope, the invention of Carl E. Seashore [15] (Fig. 7-5), which used perforations in a large rotating drum, through which one could view a manometric flame. It was a spectrum analyzer, but one whose time came and went rapidly.

The Loudspeaker

In the book on *Technical Aspects of Sound*, in 1953, F. Spandöck observed

At the end of every electroacoustic line there is a loudspeaker or telephone as acoustic source, whose function is to transform the electrical into sound energy . . . . This is in fact the weakest element in the system and so determines the overall

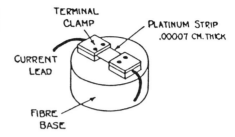

FIGURE 7-1. The thermophone. Sectional drawing of transmitter. (From Arnold and Crandall [10].)

equality of production. The problem of getting electroacoustic radiation free from distortion is in fact a rather difficult one. [16]

The story of the early development of the telephone, recounted in Chapter 5, indicates that this distortion was a problem from the very beginning. A great improvement on the original Bell and Edison devices was that of the moving-coil electroacoustic transducer, first patented in America in 1877 by Charles Cuttriss (1849–1905) and Jerome Redding (1841–1939) in 1877 [17], and in Germany and Great Britain by Ernst Werner Siemens [18] (1816–1892) in the same year. The Siemens patents became the dominant ones. In both of them, a permanent magnet produced a radial magnetic field and a nonmagnetic parchment diaphragm was used as the sound radiator. Hunt notes that Siemens

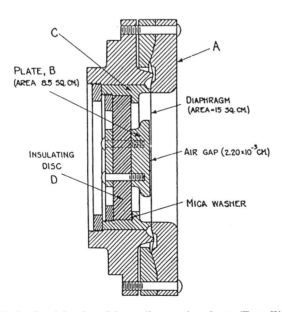

FIGURE 7-2. Sectional drawing of the condenser microphone. (From Wente [11].)

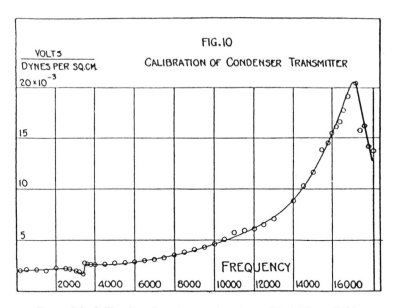

FIGURE 7-3. Calibration of condenser microphone. (From Wente [12].)

went further and specified that the diaphragm was to have the general form of the frustum of cone, but that it should have an exponentially flaring profile of "trumpet form"—a shape sometimes characterized as "morning glory" form. [19]

Siemens' drawing for the British patent is reproduced in Fig. 7-6. The moving-coil type of loudspeaker soon became standard, as did the shape of the Siemens loudspeaker.

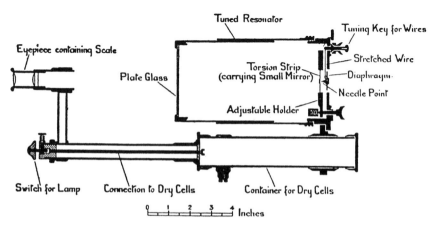

FIGURE 7-4. Webster's portable selective phonometer [13].

FIGURE 7-5. Early model of tonoscope. (From Seashore [15].)

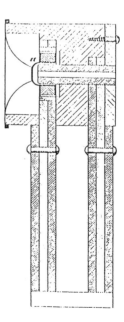

FIGURE 7-6. Siemen's loudspeaker [18].

Amplifiers, Vacuum Tubes, and Vacuum Tube Amplifiers

Perhaps the greatest problem of both the telegraph and the telephone was the weakness of the signal. To convey it over long distances, some sort of amplification was always needed. For the telegraph, this was generally solved by the use of a "repeater," namely, a duplicate telegraph circuit that used the incoming signal to trigger the duplicate and send forward a signal of renewed strength.

The electron tube amplifier has a lengthy history. The fact that a heated metallic body produced some electric charge was known from the observations of Frederick Guthrie in 1873 [20], and of Julius Elster (1854–1920) and Hans Geitel (1855–1923) in 1882 [21]. Somewhat later (1883), Edison observed the emission of negative charge from a heated wire, collecting a current at a second wire in a vacuum container that was used in the development of the incandescent light [22]. Edison made no further use of his discovery, which was in fact a primitive two-electrode vacuum tube (diode). The effect was later given the name of the Edison effect or thermionic emission.

The development of radio by Guglielmo Marconi (1874–1937) brought about the need for a device that could "rectify" the high-frequency electrical currents involved in the radio circuits so as to allow low-frequency alterations in these currents, produced by sparks, voice, or music, to drive a loudspeaker. In 1897, J.J. Thomson (1856–1940) identified the charge carriers in the Edison effect as electrons, and in 1904, John Ambrose Fleming (1849–1945) made use of the discovery of Edison to construct a diode that could perform this rectification (Fig. 7-7) [23]. Two years later, Lee De Forest (1873–1961) added a grid of wires between the filament and the receiving plate. By setting the electric potential level of the grid below that of the heated filament, he was able to control the current to the plate

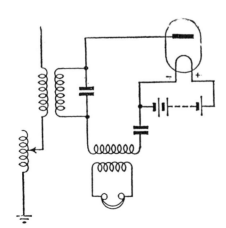

FIGURE 7-7. Fleming's diode, 1904. Early radio receiver. (From Dunsheath [23].)

FIGURE 7-8. De Forest's triode, 1906. (From Dunsheath [23].)

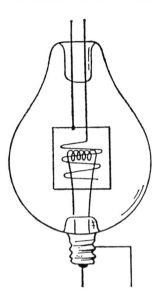

(Fig. 7-8). Thus the triode was born [24]. Here also, a long and bitter patent dispute took place that was not ended until 1953.

The triode could be used as an amplifier (Fig. 7-9), since a small voltage change at the grid could result in a large voltage change at the plate. In 1915, Edward Howard Armstrong [25] (1890–1954) combined the operations of amplification and rectification by operating a triode with the electrical potential of its grid at such a level that current from the negative swings of the voltage applied to the grid could be cut off, thus producing one-way currents (Fig. 7-10).

The Oscillator

The work of Heinrich Rudolf Hertz (1857–1894) in his discovery of radio waves involved a high-frequency oscillator, in which first a Leyden jar and

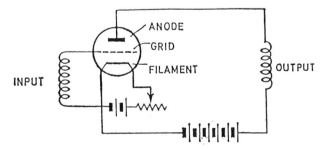

FIGURE 7-9. The triode amplifier. Simple amplifying circuit. (From Dunsheath [23].)

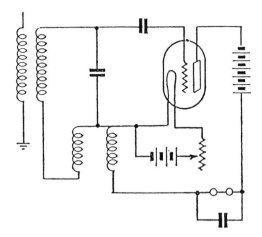

FIGURE 7-10. Armstrong's circuit for simultaneous rectification and amplification, 1915. (From Dunsheath [23].)

then an induction coil was used [26], but there was no self-sustained electrical oscillator at audible frequencies. This situation changed in 1913, with the invention of the feedback circuit by Charles Samuel Franklin (b. 1879) [27]. In this circuit, electrical energy was "fed" from the plate circuit (Fig. 7-11) back to the grid with such a phase relation that it amplified the signal at the grid, which in turn amplified the signal at the plate. When sufficient energy was so transferred, the circuit set into oscillations of it own, with the frequency dictated by the values of the inductance and capacitance in the grid circuit [28].

The Oscilloscope

In 1897, Ferdinand Braun constructed the first electronic oscillograph [29], in which a beam of electrons was controlled by electrical and magnetic fields

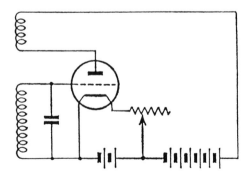

FIGURE 7-11. Franklin's feedback, used to generate continuous oscillation, 1913. (From Dunsheath [23].)

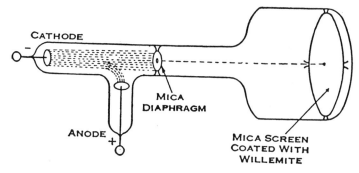

FIGURE 7-12. The Braun oscillograph. (From *Encyclopaedia Britannica* [29].)

applied along its passage. The beam was incident on a fluorescent screen at the end of the tube, so that variations in the applied voltage could be measured by motions of the electronic spot (Fig. 7-12). The improvements made by 1911, for the basic form of the instrument, now known as the oscilloscope, are shown in Fig. 7-13 [30].

Acoustic Impedance

While not an instrument, the concept of acoustic impedance and its application has played a very important role in electroacoustics, and some discussion of its origin is appropriate. In the 1890s Oliver Heaviside [31] (1850–1925) introduced the concept of electrical impedance and the use of complex exponentials in treating it. In 1914, Arthur Gordon Webster [32],

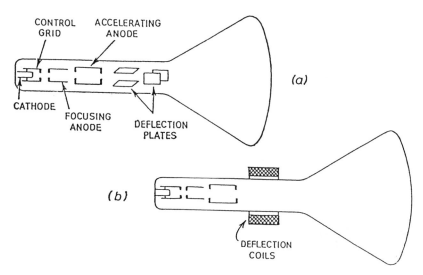

FIGURE 7-13. Cathode ray oscillograph, c. 1911. (From Dunsheath [23].)

presented a paper at a meeting of the American Physical Society, and, making use of the ideas of Heaviside, presented the idea of an acoustical impedance. Arthur Edward Kennelly (1861–1935) had been working with the electrical impedance of telephones [33], and made use of a motional impedance [34].

This growth of electronics and electroacoustics revolutionized the study of acoustics in the life sciences. It supported the growth of architectural acoustics, and led the way to the widespread study of the acoustics of listening spaces, the study of the composition of sounds of both speech and music, as well as the development of microphones and loudspeakers, of physiological and psychological acoustics, and physicists and engineers were mixed up in all of them. It is not surprising that the Acoustical Society of America was created just after the end of this period. There must have been an increasing sense of the unity of acoustics among the researchers from the many separate disciplines. We shall witness this interplay of the different disciplines again and again during the course of this chapter.

7.2. Architectural Acoustics

The most momentous lines in architectural acoustics are those beginning in a paper by Wallace Clement Sabine (Fig. 7-14) [35] in 1990:

The following investigation was not undertaken at first by choice but devolved on the writer in 1895 through instructions from the Corporation of Harvard University to propose changes for remedying the acoustical difficulties in the lecture-room of the Fogg Art Museum, a building that had just been completed. [36]

FIGURE 7-14. Wallace Clement Sabine. (From D.C. Miller [35].)

That sentence marked the beginning of the distinguished career of Professor Sabine and of the field of architectural acoustics itself. As we have mentioned before, there was little organized scientific knowledge of the subject before Sabine. The basic difficulty of the overlapping of signals due to multiple reflections had been noted by Mathews (Chapter 2), the duration of sound surviving, after it had originally been uttered, has been noted by Upham (Chapter 6), but there was little effective advice as to how to design a room, or to improve its characteristics after it had been built.

Sabine began his study in systematic fashion. He noted three basic problems: loudness, i.e., getting enough sound energy to the ear of the listener; distortion of complex sounds by interference and resonance; and confusion—reverberation, echo, and extraneous sounds. Sabine's first problem was to determine what he called the reverberation time—the time required for the level of a sound to become inaudible after its source has been shut off. His apparatus is shown in Fig. 7-15. An organ pipe of known frequency was driven by air from a reservoir, and a chronograph marked the passage of time. The observer could control the air flow and the chronograph simultaneously, and thus record the appropriate time interval.

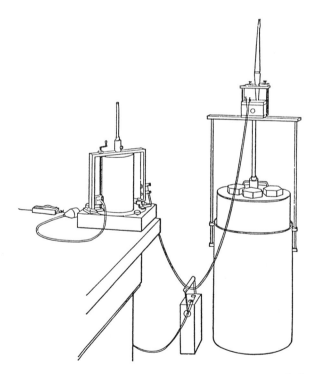

FIGURE 7-15. Sabine's apparatus for measuring the reverberation time [36]: Chronograph, battery, and air reservoir, the latter surmounted by the electro-pneumatic valve and organ pipe.

Reverberation times of five or six seconds were obtained for the lecture hall, indicating the almost hopeless confusion that must result for any spoken words. Sabine made these measurements in an empty room. Reasoning that people, or cloth material, would absorb some of the sound energy, and therefore cut down on the reverberation time, Sabine began moving in seat cushions from nearby Sanders Theater auditorium, and plotted the graph shown in Fig. 7-16 for the reverberation time as a function of the number of meters of cushions put in place [37].

Sabine clearly recognized two new directions for his research. What sort of theory might be developed to predict the relation between reverberation time and the dimensions of the room, and what were the relative effects of various absorbing materials.

In response to the first problem, Sabine developed an empirical formula for the reverberation time, $T = 0.164 V/a$, where V is the volume of the room (in metric units) and a is the average value of the absorption coefficient, i.e., the average aborbing power per unit area of exposed surface times the total surface area [38]. In 1903, W.S. Franklin (1863–1930) derived an expression for the decay of sound energy and obtained the expression $E = E_0 \exp[-(c\alpha S/4V)t]$ [39]. Sabine had recognized that the sound energy density at the threshold of audibility would be lower than the initial energy density by a factor of 10^6. Franklin's formula would then yield a reverberation time $T = 0.161 V/a$, i.e., an expression almost identical with Sabine's result.

In his first work, Sabine used the seat cushions of Sanders theater as the unit of absorption, but recognized that a more universal unit had to be developed. Sabine describes both the problem and his solution:

For the purposes of the present investigation, it is wholly unnecessary to distinguish between the transformation of the energy of the sound into heat and its transmission into outside space. Both shall be called absorption. The former is the special accomplishment of the cushions, the latter of open windows. It is obvious, however, that if both cushions and windows are to be classed as absorbents, the open window, because the more universally accessible and the more permanent, is the better unit .... It is necessary [for reasons of convenience] to work with cushions, but to express the results in open-window units. [40]

Wallace Sabine continued to dominate the field of architectural acoustics throughout his career. In 1908, recognizing that the ultimate merit of a room for music is the approval of the listeners, he carried out a study of five rooms, varying in size from 2500 to 7500 cubic feet, changing the amount of absorption material in each of them, and found the optimal reverberation times, based on the preferences of a panel of musical authorities. The average results for the rooms clusters around 1.1 s [41].

In a later paper (1913) [42], he made use of the Toepler method of visualizing sound waves to follow the wavefronts of a sound emitted from the stage of a model of a theater (Fig. 7-17(a)). This modeling was to become a powerful tool in architectural design. A similar technique, using water waves was developed about the same time [43].

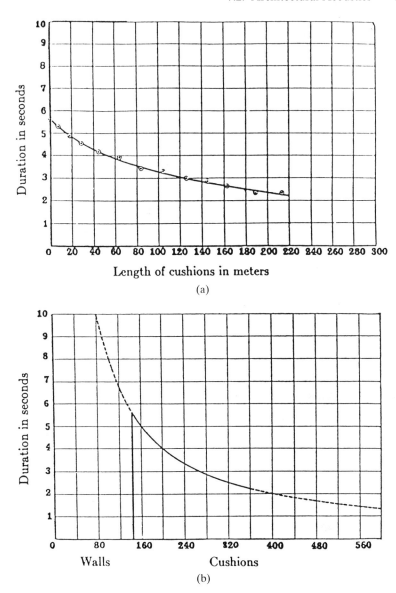

FIGURE 7-16. Sabine's results for the Fogg Art Museum auditorium [36]. (a) Curve showing the relation of the duration of the residual sound to the added absorbing material. (b) Curve 5 plotted as part of its corresponding rectangular hyperbola. The solid part was determined experimentally; the displacement of this to the right measures the absorbing power of the walls of the room.

Sabine had indeed related the reverberation time to the dimensions of the room, but what would be the best value for this reverberation time? Paul Sabine (1879–1958), a relative of W.C. Sabine and a distinguished architectural acoustician in his own right, attempted to answer that in a

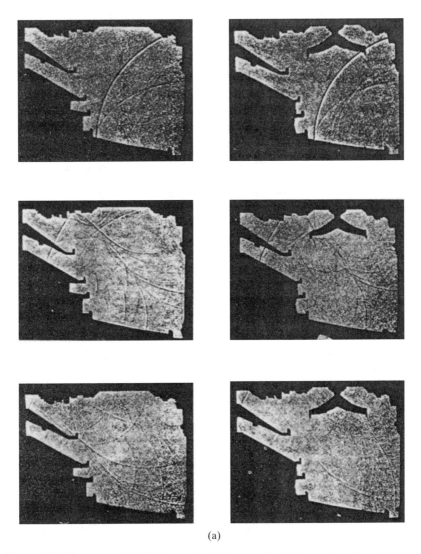

(a)

FIGURE 7-17. Theater models. (a) Two series of photographs of the sound and its reflection in the New Theater, Sabine [42]. (b) Geometrical, sound-pulse, and ripple-tank studies of the reflection of sound in a section of an auditorium. (From Davis [43].)

rough way in 1924, observing that "the time of reverberation for an auditorium with its maximum audience . . . should lie between 1 and 2 seconds. For speech and light music it should fall in the lower half of this range, while for music of the larger sort it may lie nearer the upper limit [44]."

A little later, Samuel Lifshitz, in Moscow, attempted to answer this question in more detail. Using a board of "highly qualified musicians," he concluded that, for small rooms, the ideal reverberation time was 1.06 s, but

FIGURE 7-17. *Continued*

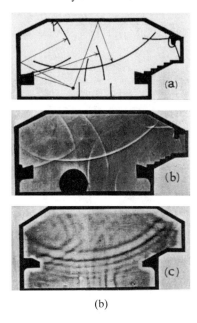

(b)

larger rooms required a more complicated relation connecting the time and the volume. His optimal curve is shown in Fig. 7-18 [45].

7.3. Physical Acoustics

Ultrasonic Propagation Measurements

The ingenious experiment of Wilmer Duff at the end of the nineteenth century (Chapter 6) indicated that the absorption of sound in air was considerably greater than that predicted by the theories of Stokes and Kirchhoff, and we have made reference to Rayleigh's remark that one should look for the explanation of this extra absorption in terms of the slowness of exchange of the energy between the sound wave and the internal degrees of freedom of the molecules. This idea was picked up by Sir James Jeans (1977–1946) in his *Dynamical Theory of Gases* in 1904 [46]. Jeans treated molecules as loaded spheres and considered the delay in getting the energy of the sound wave into the rotational energy of the molecules. He found (correctly) that the velocity of sound would vary from the value for a diatomic gas to that of a monatomic gas, and also found (incorrectly) that the resultant absorption would be smaller than that due to viscosity as the frequency is increased or the gas pressure decreased. He was on the verge of describing molecular relaxation, but didn't quite make it.

According to the classical theories of Stokes and Kirchhoff, the sound absorption coefficient increases with the square of the sound frequency.

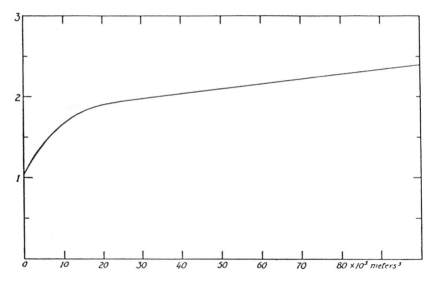

FIGURE 7-18. Optimum reverberation time in seconds as a function of volume of auditorium. (From S. Lifshitz [45].)

Since the absorption of audible sound was extremely small, it appeared that the first direct measurement of the sound absorption coefficient would have to be made at higher (ultrasonic) frequencies, so a number of investigators turned their attention to the production of these higher frequencies.

One of the first to be successful in this area was Woldemar Yakovlevich Altberg (1877–1942), working in the laboratory of Petr Lebedev (1866–1912) in Moscow in 1907 [47]. An electric spark was generated in his apparatus, producing bursts of sound energy, which contained many frequencies. By using a Fraunhofer diffraction grating (for sound), he was able to isolate a single sound frequency. He obtained frequencies in air up to 300 kHz.

In 1911, Nikolai Pavlevich Neklepaev (1886–1927), working in the same laboratory as Altberg, and using some of the same apparatus (Fig. 7-19), measured the absorption coefficient for air at room temperature in the frequency range of hundreds of kilohertz, and found that the size of the coefficient was more than twice the value calculated from the theories of Stokes and Kirchhoff [48]. Thus the earlier results of Duff were confirmed. In a note published in the same journal issue as Neklepaev's article, Lebedev reviewed the various possible sources of sound absorption in a gas, and observed:

it remains uncertain in what respect it is justified to operate according to Stokes and Kirchhoff with constants that are valid for static measurements, and transfer them to processes that are propagated in space with the velocity of sound. [49]

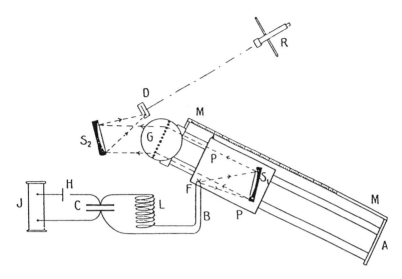

FIGURE 7-19. Neklepajew's apparatus for measuring the absorption coefficient in air [48].

The next advance came from a most unexpected source. In 1920, Albert Einstein (1879–1955) published a paper in which he considered the problem of the passage of a sound wave through an interracting gas mixture, such as the reaction $2NO_2 \rightarrow N_2O_4$. He was at the time interested in determining the rate constants for such chemical reactions, and reasoned that they could be found by measuring the dispersion that would occur in the sound velocity. He therefore derived an expression for the velocity of sound in such mixtures, and, in so doing, came up with what would later be called the molecular relaxation relation [50].

Work on the improvement of the production of sound at ultrasonic frequencies continued [51]. Walter Guyton Cady (1874–1974), for many years a professor of physics at Wesleyan University, was primarily interested in the piezoelectric properties of crystals, and of their use as frequency stabilizers in radio circuits and as frequency standards [52]. As so often was the case in electroacoustics, there was a long patent fight between Cady, who had assigned his rights to the crystal-controlled oscillator to the Radio Corporation of America, and the Western Electric Co., which held the patent of Alexander McLean Nicolson (1880–1950). The situation was further complicated by the association of Cady with George Washington Pierce (1872–1956), a Harvard Professor, who made many advancements in the frequency-stabilized oscillator [53].

Pierce also turned his attention to applications in acoustics. He cut his quartz crystal (Fig. 7-20), so that an electric field applied in the E direction would result in mechanical vibrations in the B direction. The resultant sound waves propagated through the air and were reflected back onto the transducer (Fig. 7-21). The presence of a standing wave could be detected

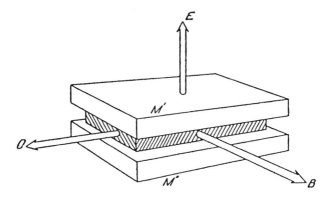

FIGURE 7-20. The piezoelectric "sandwich." Electrodes M' and M″ with plate of crystal between. (From G.W. Pierce [54].)

by measuring the current in the plate circuit of the oscillator as a function of the distance been the crystal and the reflector (Fig. 7-22). The distance between such peaks therefore gave an accurate measurement of the wavelength of the sound in air, and hence of the sound velocity [54].

At about the same time as Pierce, D.L. Rich and W.H. Pielemeier [55], working at the then Pennsylvania State College, now Pennsylvania State University, began a long and illustrious association of that institution with the field of acoustics [56]. In a paper given at a meeting of the American Physical Society in 1924, they reported that the absorption in air was larger than the theoretical value given by Lebedev, but less than that reported experimentally by Neklapaev. These observers further

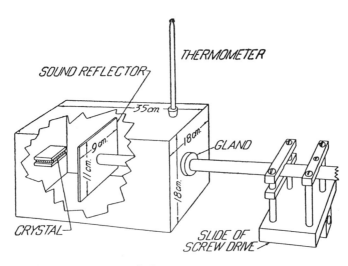

FIGURE 7-21. Pierce's interferometer [54]. Crystal and mirror in brass box for containing gas.

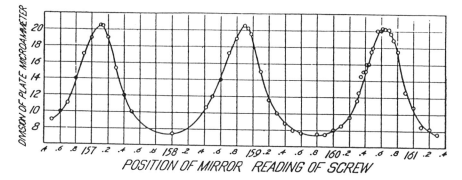

FIGURE 7-22. Standing waves. Plot of readings of microammeter in plate circuit of piezoelectric oscillator against readings of position of reflecting mirror. (From Pierce [54].)

noted the great sensitivity of the results to the amount of carbon dioxide in the air, thus opening up the large field of ultrasonic absorption in gas mixtures.

Atmospheric Acoustics

The studies of Tyndall (Chapter 3) and Henry (Chapter 5) on propagation of sound in the atmosphere, especially in the presence of fogs, marked the beginning of atmospheric acoustics [57]. Our knowledge of the atmosphere was given a strong boost by the use of balloon measurements. In 1902, M.L. Teisserenc de Bort (1858–1913) discovered the "stratosphere," a layer above the turbulence-filled air of the first 20 km from the Earth, which layer we call the troposphere [58]. In 1910, van den Borne reported the existence of a "dead zone" or "zone of silence" around an explosion, at distances of 100 km or so from the source of the explosion, in which little sound is heard, followed by a second, outer zone where the explosion could also be heard [59].

The behavior of the sound from explosions was given great study during World War I. In 1916, Emde [60] noted that the zones of silence could be understood if there was a region of altitudes in which the temperature increased with height. Such an inversion does take place, with a temperature maximum at an altitude of about 50 km, but this fact was not known until Whipple deduced the temperature dependence on altitude from a study of meteor trails in 1923 [61]. Lindemann and Dobson came to a similar conclusion in the same year [62].

In 1917, another great figure of modern physics, Erwin Schrödinger (1887–1961), contributed a paper on the acoustics of the atmosphere, studying the effect that absorption differences played in the strength of the signals refracted back to the Earth [63]. He concluded that the low-frequency signals, with their lesser absorption, could be refracted back to the Earth from higher altitudes than the high-frequency signals and their roles in the refracted sound would be accentuated.

The problem of the variability of sound signals from fog horns and sirens continued to be of great interest. In attempts to settle the difference of opinion between Henry and Tyndall, Louis Vessot King (1886–1956), a Canadian scientist, carried out extensive measurements of the sound detected from a fog horn under varying conditions [64]. Some of his results, taken from experiments performed just after World War II, are shown in Fig. 7-23. The results were highly variable, and depended in a complicated way on the air temperature, wind speed, and state of the atmosphere.

A partial answer to the problem was advanced by Sir Geoffrey Ingram ("G.I.") Taylor (1886–1975) [65]. Taylor considered vortices in a two-dimensional plane and concluded that such vortices in air may have diameters as large as 40 m. King employed Taylor's ideas and hypothesized even

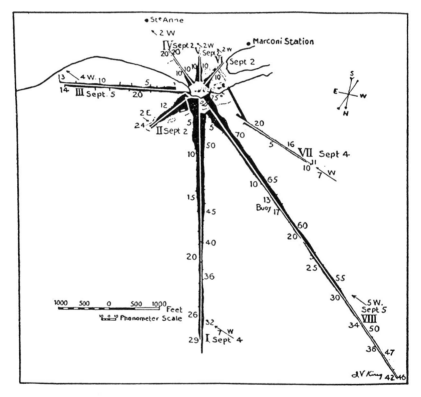

FIGURE 7-23. Sound reception in the atmosphere. Intensity of sound from fog signal at Father Point [King]. The phonometer readings are indicated by the thickness of the black trace. Observations plotted on one side of the line were taken on the journey away from Father Point, and those plotted on the other side were taken on the return trip. The numbers refer to observations of the position of the boat. (From L. King [64].)

large vortices. Such vortices could account for much of the variablity in air-borne sound signals [66].

Aeroacoustics

Although the name aeroacoustics is almost the same as atmospheric acoustics, it is a quite different subject. While atmospheric acoustics limits itself to the propagation of sound in the atmosphere and to what we might learn about the atmosphere and sound through their interplay, aeroacoustics is concerned with the interaction of gaseous flows with solid bodies in the production of sound, and with sound waves themselves. Thus the Aeolian tones (Chapter 4), sensitive flames (Chapter 4), and edge tones (Chapter 4) come within its scope. The famous nondimensional numbers—Stokes, Strouhal, Mach, and Reynolds also are part of its history [67]. During the last 20 years of Lord Rayleigh's life, he devoted much time to this subject, writing at least eight papers. In 1916, he set forth his principle of similitude [68], which today we would call dimensional analysis, demonstrating, for example, how the dependence of the frequency of Aeolian tones on flow velocity and string diameter can be found through its use. We shall encounter much more of aeroacoustics in the next two chapters.

7.4. Underwater Sound

The first interest in underwater sound seems to have come from efforts to improve marine signaling systems, designed to warn navigators of icebergs and ship channels. In 1889, the American Lighthouse Board mentioned the possibility of use of an underwater bell and microphone, devised by Lucien Blake [69]. However, the actual implementation of these ideas had to await waterproofing of telephone transmitters (work done by Elisha Gray), and in 1901, the underwater signaling system shown in Fig. 7-24 was developed [70]. It was widely used in the next decade, but apparently was being replaced (by 1912) by radio directional finders.

The study of underwater sound was increased considerably as a result of two modern catastrophes—the sinking of the *Titanic* in 1912 and the onset of World War I in 1914. Five days after the sinking of the *Titanic* by an iceberg, Lewis Fry Richardson (1881–1953) applied for two British patents, one for an acoustic airborne echo ranger and the other for a similar device for use under water, both designed to detect the presence of nearby icebergs [71]. Shortly thereafter, Reginald Aubrey Fessenden (1866–1932) devised an oscillator with a reciprocating induction motor that operated at 540 cycles per second and drove a large plate at its resonance frequency again, 540 Hz) (Fig. 7-25) [72], [73]. The detection was provided by shipboard listening. According to Hunt, Fessenden demonstrated this underwater detection in April, 1914 [74].

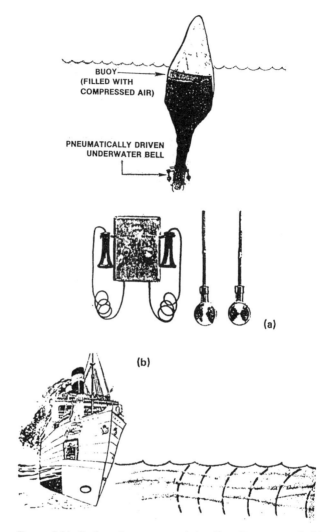

FIGURE 7-24. Early underwater sound signaling. (From Lasky [70].)

The next step forward came from abroad. Constantine Chilowsky (1880–1958), a Russian engineer living in Switzerland, developed ideas for a high-frequency echo-ranging device, involving a moving-armature magnetic transducer, and communicated this to the French physicist Paul Langevin (1872–1946) sometime in 1915 [75], and the two began a collaboration. Various means of sound production and reception were attempted, including the use of a carbon microphone, with only modest success until Langevin in 1917 obtained a large crystal of natural quartz, from which 10 × 10 × 1.6 cm slices could be cut. His resultant transmitter sent out a sufficiently powerful beam as to kill fish in its near field [76]. Later,

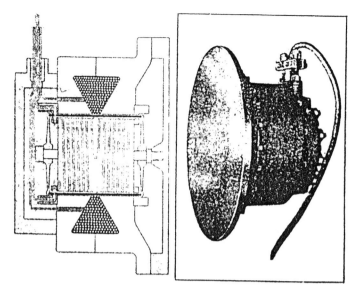

FIGURE 7-25. The Fessenden oscillator [72]. In 1913 this oscillator transmitted Morse Code signals underwater to ranges of more than 30 miles.

Langevin found that he could employ a mosaic of smaller pieces of quartz as his transmitter and by early 1918, he was able to send signals and detect echoes from a submarine at distances up to 1500 m.

Langevin gave some of his natural quartz to Robert W. Boyle (b. 1883) and British work on underwater sound began. Like Langevin, Boyle was on the hunt for natural quartz and soon discovered a warehouse-full in Bordeaux, from which he obtained enough quartz to equip ten naval vessels with the apparatus which, by now, the British were calling ASDIC (AntiSubmarine Detection Investigation Committee) [77].

Early American (and British) work on submarine detection concentrated strongly on underwater listening. A wide variety of microphones were developed for this purpose (Figs. 7-26 and 7-27) [76], [77], and sophisticated systems of binaural listening were constructed (Fig. 7-28). While these were primitive by later standards, a succession of improvements paved the way for later use. In addition, arrays of quartz and, later, Rochelle salt transducers were developed, housed in protective shells (Fig. 7-29) [77]. American work on echo-ranging did not develop soon enough to be used in World War I, but, as Professor Hunt points out, "There was to be, alas, ample opportunity for its military value to be demonstrated before another quarter century had passed [78]."

This collaboration with the British waned after the end of the war, but not before the British had given American scientists extensive demonstrations of their ASDIC systems. Research in underwater sound suffered a

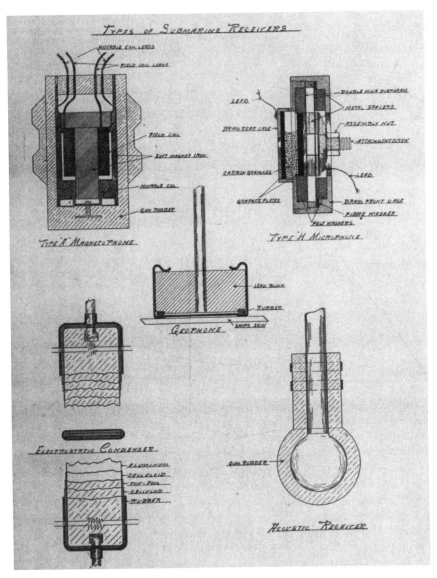

FIGURE 7-26. Early underwater microphones. (From Hayes [76].)

great loss of interest and funding in the years following that war [79]. The major gains in the 1920s were developments in fathometry, or what was called "pinging off the bottom." This work was carried on by Harvey Hayes (1878–1968), Fessenden, and Herbert Grove Dorsey (1876–1967) in the United States. A significant development was the founding of the US Naval Research Laboratory in Washington, DC, in 1923.

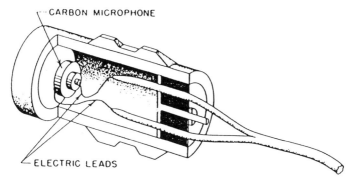

FIGURE 7-27. The RAT, a British listening device. (From Klein [77].)

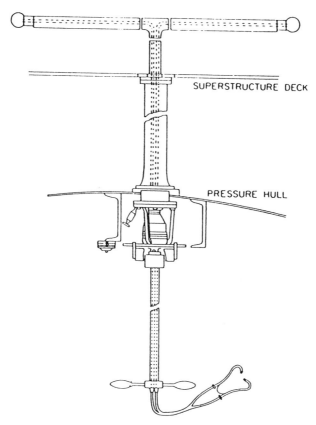

FIGURE 7-28. Binaural submarine listening apparztus installed on a US submarine. Type SE-4214 (SC) sound receiver as installed on a US submarine. (From Klein [77].)

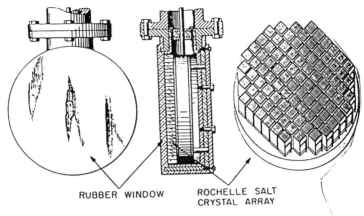

FIGURE 7-29. Use of a Rochelle-salt crystalline array in underwater listening. (end and side views, diameter—15 in.). (From Klein [77].)

One of the great advances of the immediate postwar period was the development of magnetostrictive oscillators by G.W. Pierce [80]. These operated at 25 Hz and were free of the element of fracture inherent in the quartz crystal systems.

Cavitation

The term cavitation was introduced in the literature in describing the formation of holes or bubbles in a liquid behind a screw-propeller, and the phenomenon of the great pressures that develop when such bubbles collapse was studied as early as 1859 [81]. More modern developments began with the paper by Osborne Reynolds on the origin of steam noise from a kettle, which he attributed to the complete collapse of the bubbles of steam as they passed through cooler water [82]. Then, in 1917, Rayleigh published his analysis of the pressures produced upon the collapse of a spherical bubble, deriving a relation between the pressures produced and the original size of the bubble, as well as calculating the time for the bubble collapse to occur [83]. As in many other theories of Rayleigh, this had to wait for experimental confirmation at a much later time.

7.5. Bioacoustics

In the first decade of the twentieth century, there was a stirring of research on the ability of bats to navigate in the dark that repeated work done one hundred and more years earlier. This long interval of dormancy was largely due to the unwillingness of the scientists involved to accept experimental results that seemed to them to be unreasonable and counter to common sense. Let us review these events [84].

Toward the end of his distinguished career as a physiologist, the Abbe Lazaro Spallanzani (1729–1799) became interested in the ability (or inability) of flying animals to move about in the dark. Spallanzani served as a professor at the Italian universities of Reggio, Modena, and finally, Pavia. He became famous for his work on respiration and digestion, and his refutation of the theory of spontaneous generation of minute animal life. Near the end of his career, in 1793, he found that owls were helpless in flying in the dark and would collided with walls, floors, or other objects, but that bats had no such problem. Spallanzani performed many experiments with the bats, including covering their eyes and even blinding them. The bats would still navigate with ease [85].

The idea occurred to Spallanzani that the cause of this ability might be a heightened tactile sense or possibly, a sixth sense. In 1794, however, a French surgeon, Charles Jurine, learning of Spallanzani's work, performed an experiment in which the bat's *ears* were plugged. The result was that the bats became as helpless in the dark as owls [86].

Jurine's work was confirmed by further researches by Spallanzani, but no effective explanation was found for what became known as Spallanzani's bat problem. The seal of disapproval on the work of Spallanzani and Jurine was placed by the eminent French zoologist, Georges Cuvier (1769–1832), who wrote

Spallanzani has concluded that bats possess a sixth sense of which we have no idea. Jurine has performed still other experiments which appear to prove that it is by means of the ear that bats direct themselves; but it seems to us that the operations performed on those individuals which were deprived of their ability to direct themselves were extremely cruel and have done much more than simply lessen their powers of hearing. To us the organs of touch seem sufficient to explain all the phenomena [avoidance of obstacles] which bats exhibit. [87]

And that put the quietus on any further developments in bat research for the next century.

In 1900, a French team of zoologists, R. Rolinat and E. Trouessart, repeated some of Spallanzani's experiments, and an American, W.L. Hahn, of Indiana University, repeated the work of both Spallanzani and Jurine in 1908, although, apparently, these researchers were unaware of the work of Jurine [88]. A Dutch scholar, S. Dijkgraaf, in working through the unpublished papers of Spallanzani, found that the Italian had suggested that the sounds of the beating wings of the bat may might be picked up by its ears and they were reflected from objects [89]. This same idea was picked up by the famous American-born inventor, Sir Hiram Maxim (1840–1916), inventor of the Maxim machine gun, and the Maxim silencer) [90], who, after the sinking of the *Titanic* in 1912, suggested the use of low-frequency sound (inaudible to human ears, in anticipation of sonar), to protect ships against icebergs. It is interesting that attention was paid to the possible use of low-frequency waves, whose enormous wavelengths would rule out significant scattering from anything but the largest of obstacles, while no one suggested the possibility that bats might generate sounds above the audible range (for humans), even though Koenig and others had been aware of the existence of ultrasound since the mid-nineteenth century, and Galton had demonstrated that some animals could hear sounds in this high-frequency range (see Chapter 6, including the verse at its beginning).

In 1920, the suggestion was finally made, by G. Hartridge, that bats might produce and hear high-frequency sounds [91]. Griffin notes [92] that Hartridge performed no experiments, and might even have thought only in terms of high, humanly audible, sounds of frequencies such as 25 kHz. Nevertheless, the right connection had finally been made. The experimental verification had to wait another generation (see Chapter 8).

7.6. Structural Vibrations

In 1921, Timoshenko (1878–1972) wrote a short paper that became a classic. In it he incorporated "the effects of transverse shear deformation into a model of a bent beam that had already included the effects of rotatory inertia [93]." Timoshenko found that the effects of rotatory inertia and transverse shear deformation were of the same magnitude [94]. In this paper, Timoshenko also introduced the shape factor, which enabled him to assume a shear stress that was uniform across the thickness of the beam [95].

We met up with the vibrations of plates in both Chapters 1 and 2. The subject continued to have its investigators. In 1911, John R. Airey (1868–1937) worked out the mathematical solution of the eigenvalue problems of free vibrations of plates, while, a few years later, Horace Lamb [96] carried forward the work on elastic waves in plates, a work that had been started in the latter part of the nineteenth century by Pochhammer [97] and Rayleigh [98]. Lamb discussed, for the first time, many of the features of such wave propagation, and included a treatment of Rayleigh surface waves.

Surface Waves in Solids

In 1911, Augustus Edward Hough Love (1863–1940) described a second type of surface wave, differing from the Rayleigh surface waves in that the displacements lie in the plane of the surface [99]. The displacements of the set of waves that can exist in solids are sketched in Fig. 7-30 [100]. The presence of the whole set of these waves in seismic signals is shown in Fig. 7-31 [101].

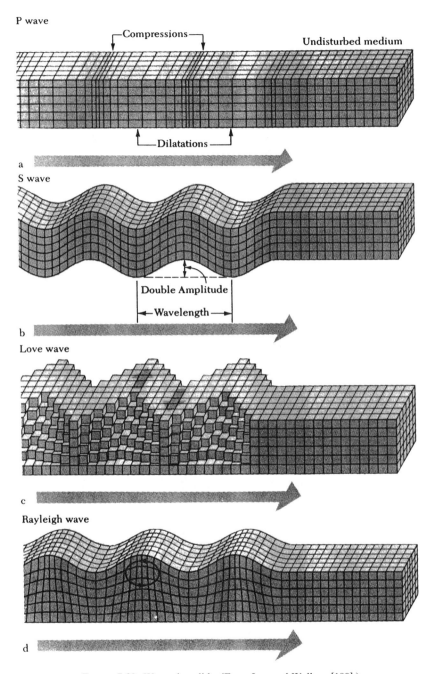

FIGURE 7-30. Waves in solids. (From Lay and Wallace [100].)

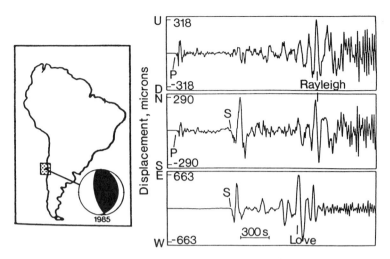

FIGURE 7-31. P and S waves. Recordings of the ground displacement history at station HRV (Harvard, Massachusetts) produced by seismic waves from the March 3, 1985, Chilean earthquake, which had the location shown in the inset. The three seismic traces correspond to vertical (U–D), North–South (N–S), and East–West (E–W) displacements. The direction to the source is almost due South, so all horizontal displacements transverse to the raypath appear on the east–west component. The first arrival is a *P* wave that produces ground motion along the direction of wave propagation. The *S* motion is large on the horizontal components. The Love wave occurs only on the transverse motions of the E–W component, and the Rayleigh wave occurs only on the vertical and North–South components. (From Lay and Wallace [101].)

7.7. Noise

Samuel Johnson (1709–1784) defined noise in his dictionary as "Any kind of sound [102]," which is hardly an effective definition. Even as late as Harvey Fletcher's book on *Speech and Hearing* (1928), the subject noise merited only a short chapter in a section labeled "music and noise [103]." (Fletcher (1884–1980) was the first President of the Acoustical Society of America, and also served as president of the American Physical Society.) von Helmholtz did mention the sounds of scraping and breathing in connection with the playing of musical instruments as noise [104], and that perhaps accounts for the classification due to Fletcher. However, other noises were manifesting themselves. Fletcher first defined noise as "sound with no definite pitch," and included hand-clapping, typewriter hammering, and street noise, showing an example of the latter in Fig. 7-32 [105]. In dealing with telephone reception, the engineers of the day found that the presence of other sounds interfered with such reception and began calling these extraneous sounds noise also, so that the modern definition of noise, simply as unwanted sound, was on its way [106].

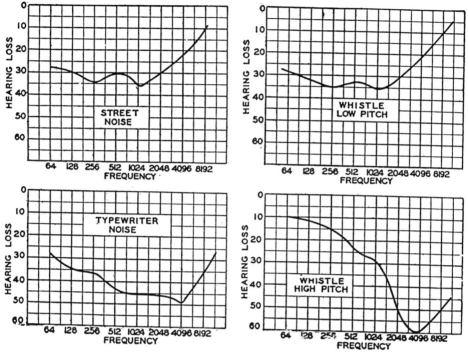

FIGURE 7-32. Noise measurements. (From Fletcher [105].)

Next to the existence of noise came the desire for its control and abatement. Complaints about noise extend far back in history. In an amusing tabulation of ordinances against noise passed by the city of Bern, Switzerland, R. Murray Schafer [107] lists one in 1628 against singing and shouting in the streets, one against noisy conduct at night in 1763, and many others, including noise of the woodworking industry (1886) and motor vehicles (1913) [108].

7.8. Hearing

The first quarter of the twentieth century saw little development in our knowledge of the cochlea. Henry J. Watt [109] proposed a modification of ter Kuile's [110] traveling wave theory with slightly different assumptions, while H. Oort studied the crossed and uncrossed portions of the nerve fibers in connection with the vestibular and cochlea nerves. Oort's discoveries were pioneering, but they did not bear fruit until much later (see Chapter 9).

This period did see substantial growth of research in what later would be called psychological acoustics. It was then, as it is today, a meeting ground of physicians, psychologists, electrical engineers, and physicists. In work done on the threshold of hearing or audibility, Harvey Fletcher pioneered in using electronic equipment in this research, having developed an early form of the audiometer [111]; he had also written a number of papers on hearing loss [112]. He was joined in this work by the physician Edmund P. Fowler, Jr., who plotted Fletcher's hearing loss diagrams (Fig. 7-33) into the "audiogram" that we know today [113], [114] (Fig. 7-34). Fowler is also noted for discovering and naming the phenomenon of recruitment of loudness, in which "the sensation of loudness grows more rapidly than normal as a function of intensity [115]."

Thomas F. Vance continued the various studies of the difference limen for frequency. A graph of his results, along with those of the nineteenth-century researcher Luft [116] and later work of Shower and Biddulph [117] is shown in Fig. 7-35 [118].

Two other aspects of tones whose study dates from early in the century are volume and localization. The concept of volume—one of the sensation of "largeness" or "smallness" of a tone, was first exploited by G.L. Rich [119], in measuring the just-noticeable-difference (in the sense of volume) as a function of frequency, while Halverson [120] measured it as a function of intensity. It was their intent to establish volume as an entity, separate

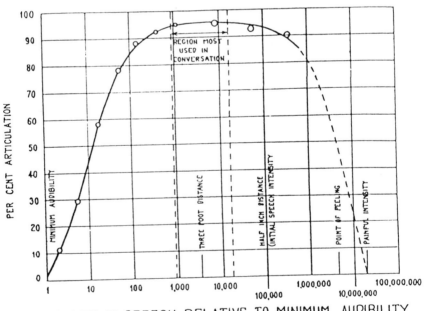

FIGURE 7-33. Fletcher's hearing loss diagram. (From Wegel and Lane [114].)

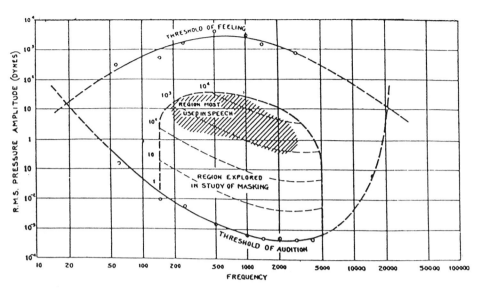

FIGURE 7-34. The audiogram. (From Wegel and Lane [114], [115].)

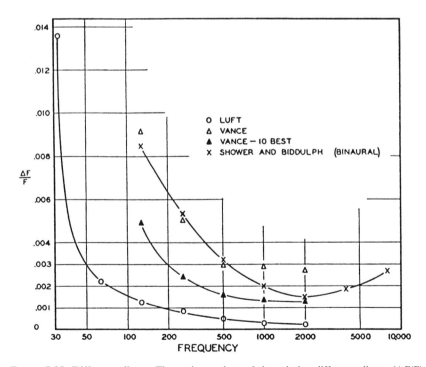

FIGURE 7-35. Difference limen. The various values of the relative difference-limen ($\Delta F/F$) obtained by various experimenters under different experimental conditions. (From Stevens and Davis [116].)

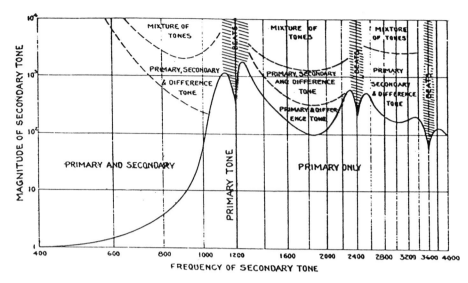

FIGURE 7-36. Masking. Sensation caused by two peere tones. (From Wegel and Lane [123].)

from pitch and loudness, that is characteristic of tones, but they did not fully succeed.

George W. Stewart followed up on Rayleigh's work on binaural hearing (Chapter 4), studying the phase relations in the acoustic shadow of a rigid sphere (which served as a model for the human head) [121]. Rayleigh had first thought that the directionality of binaural hearing was due to intensity differences, but concluded that the effect was actually due to phase differences. Stewart's work confirmed that idea.

The phenomenon of masking was also given wide study at this time. As noted in Chapter 6, Mayer had noted that low-frequency sounds could completely "obliterate" higher-frequency sounds, but not vice versa [122]. The next significant work was that done by Raymond Lester Wegel (b. 1888) and Clarence Edward Lane (1892–1951) [123], who studied the masking of tones of frequency 250–3000 Hz by tones both below and above that range, and who introduced the masking audiogram (Fig. 7-36). Of their paper, Earl Schubert later wrote: "Scarcely any serious student of audition will be unaware of the content of this paper, but it is well worth reading for didactic purposes in addition to its historical interest [124]."

7.9. Speech

Research in the nineteenth century and given the world an accurate knowledge of the structure and behavior of the larynx, and a basic understanding of the air flow from the lungs through the larynx to produce the sound, and of the structure of the throat and mouth to complete the formation of the characteristic sounds. To go further required the development of better instrumentation than the flame watching of Koenig, Toepler, and others. As we shall see, this was provided by the electronics of the new century.

The continued interest in the artificial talking machines such as those of von Kempelen and Kratzenstein (Chapter 2) led to the creation of many models both of the vocal cords and of resonators that would emulate the action of the throat and mouth. A detailed study of such models was given by Sir Richard Paget [125] in 1930 (Fig. 7-37). Some of these models were developed by Rayleigh. In the models shown in Fig. 7-38, the constriction at

FIGURE 7-37. Plasticine voice models on Lord Rayleigh's organ. (From Paget [125].)

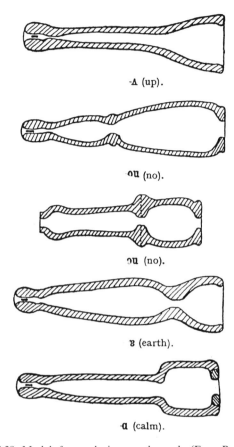

-ʌ (up).

ɔu (no).

ɔu (no).

ɜ (earth).

-ɑ (calm).

FIGURE 7-38. Models for producing vowel sounds. (From Paget [125].)

the left represents the larynx, while the separation of the rest of the tube represents an attempt to simulate two resonance frequencies observed in the particular vowels.

In his book in 1928 [126], Harvey Fletcher reviewed the knowledge of tongue and lip position in the creation of vowel sounds in English (Fig. 7-39). Other speech developments were attempts to simulate the sounds by electrical circuits (Fig. 7-40) [127] and the production of an artificial larynx [128], used by people after a tracheotomy.

It was during this period that the study of speech sounds made the transfer from the study of flames disturbed by the sound and the study of tracings of sound from phonograph equipment to the use of electronically recorded signals. We have mentioned the work of Merritt and Nichols in recording manometric flames (Chapter 6). A final paper on such devices was written by Joseph G. Brown at Stanford in 1911. Near the end of the paper, he wrote: "It is doubtful whether the curves obtained from the

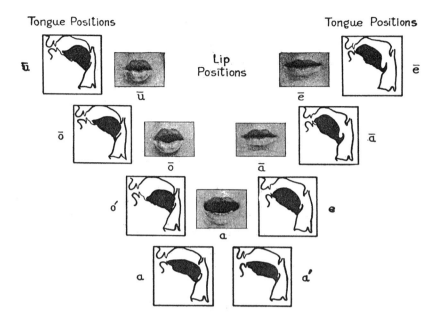

FIGURE 7-39. Vowel sounds. (From Fletcher [126].)

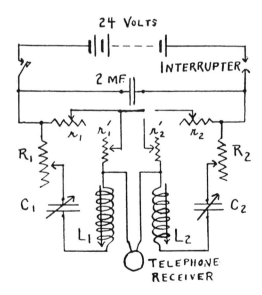

FIGURE 7-40. Electric model of speech. (From J.Q. Stewart [127].)

vibrating flame . . . will serve as well as other means for analyzing sound waves." The flame technique had had its final day [129].

Edison's development of the phonograph stimulated a number of scientists into making use of this instrument to show the vibrations of speech. L. Hermann [130] and Louis Bevier [131] at Rutgers made photographs of phonographic responses, while Edward Wheeler Scripture transferred the vibrations of the phonographic recorder to a rotating drum (Fig. 7-41) [132], thus producing recordings of both speech and music.

In Chapter 3, we made reference to an article on experimental phonetics in *Nature* by John McKendrick. In this same article, he reviewed the ideas of L. Hermann [133] on the nature of vowel sounds. Hermann, a Professor of Physiology at the University of Königsberg, suggested that the oral cavity produced a partial tone that had no relation to the fundamental emitted by the larynx, and called this tone the "formant." McKendrick wrote in his article "A vowel . . . is a special acoustic phenomenon, depending on the intermittent production of a special partial, or 'formant,' or 'characteristique.' The pitch of the 'formant' may vary a little without altering the character of the vowel [134]." Thus the word *formant* entered the language of acoustics.

Among the "other methods" referred to by Joseph Brown was the *phonodeik*, invented by Dayton Miller [135]. The basic structure of this device is shown in Fig. 7-42. The sound is collected through the horn *h*, at the end of which is a thin glass diaphragm *d*. As described by Miller,

behind the diaphragm is a minute steel spindle mounted in jeweled bearings, to which is attached a tiny mirror *m*; one part of the spindle is fashioned into a small pulley; a string of silk fibers . . . is attached to the center of the diaphragm and being wrapped once around the pulley is fastened to a spring tension piece; light from a pin hole is focused by a lens and reflected by the mirror to a moving film *f* in a special camera. [136]

The quality of the recordings from this instrument is demonstrated in Fig. 7-43, which shows the detail of the singing of single notes from (and by) a famous operatic sextet [137].

The next step forward in the examination of speech sounds came with the development of the condenser microphone and the vacuum tube amplifier [138]. The microscope converted the acoustic signal to an electrical one. This was then amplified and the result signal applied to the oscillograph, which was basically a string galvanometer with a mirror. The variations of the position of a spot of light reflected from the mirror were recorded on a film strip rotating on a drum [139].

The availability of this equipment produced a great flood of measurements of various speech sounds [140]. The famous sentence "Joe took father's shoe bench out," must first have been uttered about this time; it soon became a kind of test standard for speech analysis.

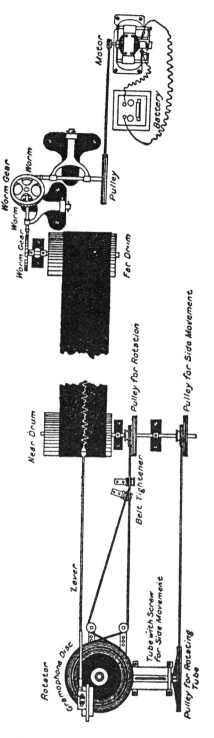

FIGURE 7-41. Scripture's apparatus for tracing talking-machine records [132].

(a)

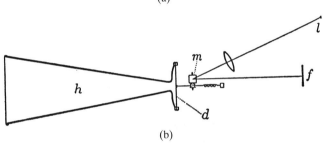

(b)

FIGURE 7-42. (a) The phonodeik used for photgraphing sounds; and (b) principle of the phonodeik. (From D.C. Miller [135].)

This period also saw the first attempts at recognition of various speech sounds. George A. Campbell followed up on a report by Lord Rayleigh [141] on his ability of distinguish between s and f in reproduced speech, Campbell conducted a systematic analysis of human ability to recognize a wide variety of speech sounds over the telephone [142]. This was a problem to which Fletcher and J.C. Steinberg later turned [143]. This method became known as articulation testing.

7.10. Musical Acoustics

In 1924, Wegel and C.R. Moore invented an electrical harmonic analyzer to measure the harmonic content both of speech and music, and from that time on, we have had a clear understanding of the waveform and the harmonic content of both speech and music [144]. Their paper was soon followed by a paper by Harvey Fletcher on the determination of the pitch

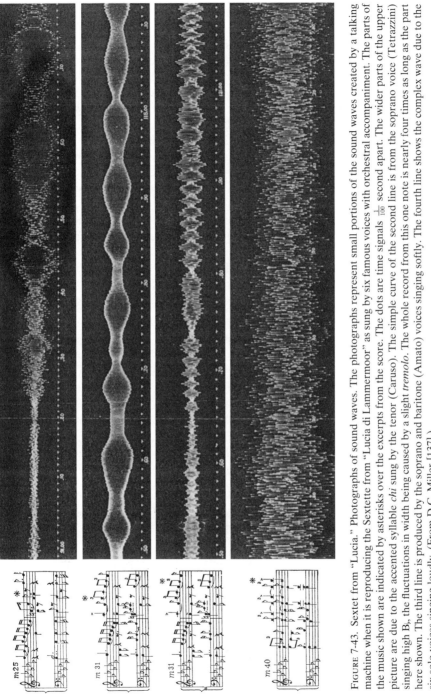

FIGURE 7-43. Sextet from "Lucia." Photographs of sound waves. The photographs represent small portions of the sound waves created by a talking machine when it is reproducing the Sextette from "Lucia di Lammermoor" as sung by six famous voices with orchestral accompaniment. The parts of the music shown are indicated by asterisks over the excerpts from the score. The dots are time signals $\frac{1}{100}$ second apart. The wider parts of the upper picture are due to the accented syllable *chi* sung by the tenor (Caruso). The simple curve of the second line is from the soprano voice (Tetrazzini) singing high B♭, the fluctuations in width being caused by a slight *tremolo*. The whole record from this one note is nearly four times as long as the part here shown. The third line is produced by the soprano and baritone (Amato) voices singing softly. The fourth line shows the complex wave due to the six solo voices singing loudly. (From D.C. Miller [137].)

of a musical tone. He reported the extraordinary result that "even when the fundamental and first seven overtones were eliminated for the vowel /ah/ sung at an ordinary pitch for a baritone, the pitch remained the same [145]." Using a set of electronic oscillators with frequencies at 10-Hz intervals from 100 to 1000, he found that "three consecutive component frequencies were sufficient to give a clear musical tone of definite pitch corresponding to 100." Fletcher interpreted these results as due to nonlinearities in the ear, following the ideas of von Helmholtz discussed in Chapter 6.

The work on violin strings in the nineteenth century was followed by that of the famous Indian physicist Sir Chandrasekhara Venkata Raman (1888–1970) and his colleagues [146], [147]. Raman made the shadow of the string pass through a narrow slit and fall upon a photographic plate that was moved at constant speed, so that a vibration curve was produced of that portion of the string. He also attached a pin on the bow and photographed its shadow (Fig. 7-44) [148]. Since the slope of the curve is proportional to the speed of both string and bow, it is clear that the string and bow move at the same speed until the string breaks away.

Raman also did important studies on the "wolf note" of the violin. In a paper in 1915, G.W. White described this unpleasant feature of stringed instruments of the violin family:

On all stringed instruments of the violin type a certain pitch can be found which it is difficult and often impossible to produce by bowing. . . . At this pitch . . . the bow refuses to "bite" and a soft pure tone is almost impossible to obtain; if the pressure of the bow on the string is increased the tone resulting is usually of an unsteady nature with considerable fluctuations in intensity. [149]

Or, as put by Edward John Payne, "there is a wolf somewhere in all fiddles [150]." Although this note had long been known, little research had been done on it. From his experiments, White concluded that the "fluctuations in intensity are due to beats which accompany the forced vibration impressed

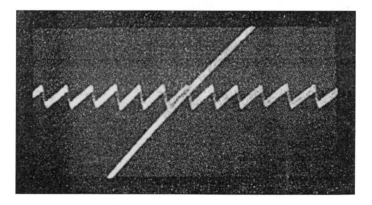

FIGURE 7-44. Speed of point under bow compared with speed of bow. (From Raman [146].)

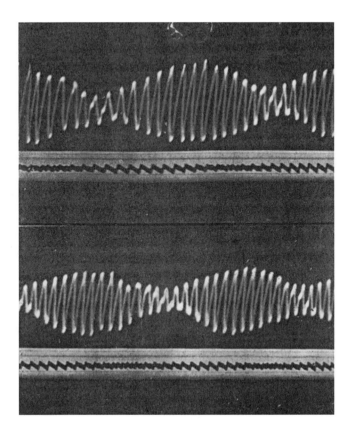

FIGURE 7-45. Simultaneous vibration-curve of belly and string of violoncello at the "wolf-note pitch. (From Raman [151].)

on the resonator and which by reacting on the string may interfere with even bowing."

Raman proposed a different explanation, based on the inability of the string to supply energy fast enough to the resonant mode of the belly of the instrument, so that the string frequency jumps to its octave. The belly resonance then fades out, whereupon the string returns to the original frequency. This behavior is clearly demonstrated in Fig. 7-45, where the two vibration curves are plotted simultaneously [151].

7.11. The Origin of the Decibel

The decibel belongs to all of acoustics, rather than to any particular branch. One might say that the decibel has two family trees, one in experimental

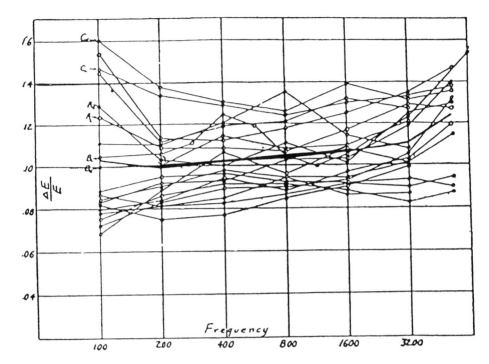

F<small>IGURE</small> 7-46. Knudsen's measurements of the Fechner ratio as a function of frequency. The heavier line represents the average values [152].

psychology, and the other in engineering. In Chapter 6, we discussed its psychological ancestor, the Weber–Fechner law, connecting the response to stimuli in biological systems. An important corollary of this law was the dependence of the response on the logarithm of the stimulus.

In 1923, Vern Oliver Knudsen (1893–1974), working at the University of Chicago and using electronic equipment, measured the Fechner ratio k for 19 different individuals. The results are shown in Fig. 7-46 [152]. While the results show a substantial variation from person to person, the average value of k was close to 0.10 over the range from 100 to 3000 Hz (the heavy curve in the figure). At frequencies above 200 Hz, Knudsen's results were consistent with those of Deenik (points superposed on the Knudsen graph). Knudsen also measured the value of k as a function of the intensity of the sound [153].

It should be noted that the graph in Fig. 7-47, which is a replot of Knudsen's data by Harvey Fletcher, has the intensity plotted in powers of ten, and the "loudness of the tone" in numbers that are clearly $10 \log I_R$, where R is the "intensity ratio [154]." Clearly, the decibel was in the wings, ready for a curtain call.

On the engineering side, we learn from a paper by William H. Martin (1889–1974) in 1924 that the unit "mile of standard cable" had been in use

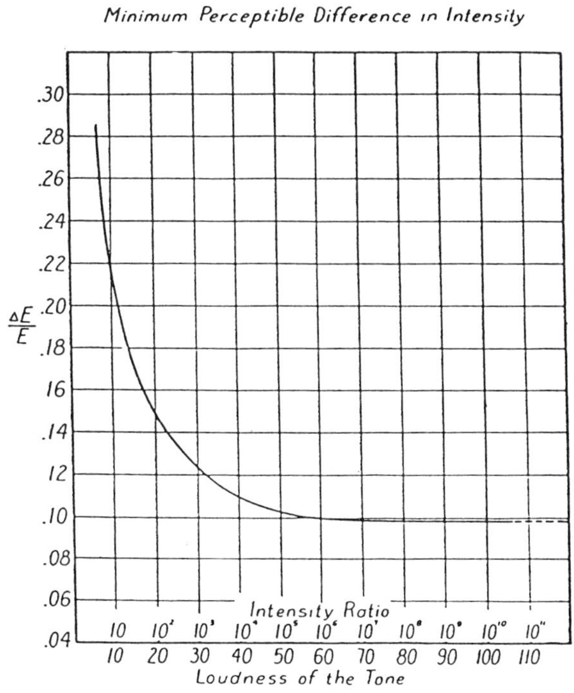

Minimum Perceptible Difference in Intensity

FIGURE 7-47. Fletcher's plot of Knudsen's data on the Fechner ratio as a function of intensity [152].

in telephone engineering in the United States for more than 20 years as a measure of "the transmission efficiency of telephone circuits and apparatus [155]." The effect on the power output of a transmission line upon the insertion of other telephone apparatus or lines into the existing line could be measured by comparing the loss with that resulting from adding a certain amount of "standard telephone cable" at the same point. Thus, a mile of standard cable became the unit of measuring this insertion power loss.

The unit was rather unsatisfactory, since, if the power ratio P_{out}/P_{in} is expressed as $\exp(2ia)$, the value of a for the cable then in use was equal to $(\pi fRC)^{1/2}$, i.e., the unit dependent on the frequency used (f), as well as on the resistance (R) and capacitance (C) of the cable per unit length, and these latter numbers could be changed upon change in the cable. To avoid such difficulties, the engineers at the Bell Laboratories devised the "sensation unit," also called the transmission unit," abbreviated TU [155], [156], such that loss in TU is given by $\log(P_{in}/P_{out})/\log 10^{0.1}$. This unit is, of course, identical with the decibel. But still, there was no name.

From the literature, it is clear that there was substantial debate over the usefulness of the TU relative to a unit based on the natural logarithm. Several international committees discussed the matter, and a series of letters in the British journal *The Electrician* in 1924–1925 debated the relative advantages and disadvantages of the two unites [157]. These included one by Colonel Sir Thomas Fortune Purves (1871–1950) [158], Chief Engineer of the British Post Office, which may have been decisive. He wrote:

The term "bell" has indeed been suggested [for the name of the TU] and it combines a graceful compliment to the inventor of the telephone with the symbols β, ε and *l*, which have become so prominent in transmission theory. Unfortunately, it is an everyday English word. . . . I think my preference [over other suggested names] would be [the bell] abbreviated to the form "bel." [159]

The final result of the work of the international committees was to establish the neper as the unit of loss in terms of the natural logarithm (after Sir John Napier, 1550–1617, the inventor of the natural logarithm), and the bel as the unit of loss in terms of the denary logarithm. The decibel was almost immediately adopted as the operative unit.

Overview

This chapter has exhibited much of the unity of acoustics in the use of the new electronic equipment to make measurements throughout the range of acoustical research topics. One was able to solve old problems in the ultrasonic transmission in fluids and new ones such as those of psychological acoustics and underwater sound. Much more could be expected for the future.

Notes and References

[1] R. Bruce Lindsay, Historical introduction to the reprinting of Rayleigh's *Theory of Sound*, Dover, New York, NY, 1945, p. xxix.
[2] F.V. Hunt, *Electroacoustics*, Harvard University Press, Cambridge, MA, 1954. Reprinted by the Acoustical Society of America, Woodbury, NY, 1982, p. 65.
[3] F.V. Hunt, *Electroacoustics*.
[4] In 1911, in his cork-walled room on the Boulevard Haussman, Marcel Proust was able to listen to performances of the Paris Opera over a special telephone, called the theatrophone. (George D. Painter, *Marcel Proust*, 2nd ed., Random House, New York, NY, 1989, vol. 2, p. 168.)
[5] F.V. Hunt, *Electroacoustics*, pp. 40–41.
[6] S.H. Preece, *Proc. Roy. Soc. (London)* 30, 408–411 (1880).
[7] George Forbes, *Proc. Roy. Soc. (London)* 42, 141–142 (1887).
[8] F. Braun, *Ann. Physik Chemie* 65, 358–360 (1898).
[9] W.S. Tucker and E.T. Paris, *Proc. Roy. Soc. (London)* A221, 389–430 (1921). W.S. Tucker, *J. Roy. Soc. Arts (London)* 71, 121–134 (1923).
[10] H.D. Arnold and I.B. Crandall, *Phys. Rev.* [2] 10, 22–38 (1917).

[11] E.C. Wente, *Phys. Rev.* [2] **10**, 39–63 (1917).

[12] E.C. Wente, *Phys. Rev.* [2] **10**, 39–63 (1917). In one of the most modest claims for an instrument ever made, Wente wrote: "The sensitiveness of the transmitter is not absolutely uniform, but varies only about a hundred per cent between zero and 10,000 cycles." (p. 59). If he had plotted his vertical axis on a logarithmic scale, as we do with decibels, and discarded the portion above 10,000 Hz, the curve would have looked a lot flatter (we know better today how to disguise our data!)

[13] A.G. Webster, *Am. Inst. Electr. Eng. Proc.* **38**, part 1, 889–898 (1919); *Nature*, **110**, 42–45 (1922).

[14] H.S. Osborne, A.G. Webster, and A.E. Kennelly, *Am. Inst. Electr. Eng. Proc.* **38**, part 1, 721–723 (1923). Reprinted in *Benchmark/16*, pp. 85–86. Parts of some of the papers in this section are reprinted in this volume. Webster stood up strongly for the old way of doing things. As Miller remarks (p. 86) "The handwriting was now on the wall, but Webster would not look." See also, H.B. Miller. *J. Acoust. Soc. Am.* **61**, 174–181 (1977). Tragically, Webster killed himself in 1923, leaving, according to a remark from K.K. Darrow, a note that said "Physics has gotten beyond me, and I can't catch up."

[15] C.E. Seashore, *Psychol. Monogr.* **16**, 2–12 (1914).

[16] F. Spandöck, "Loudspeakers and Telephones," in *Technical Aspects of Sound* (E.G. Richardson, ed.), Elsevier, Amsterdam, 1953, pp. 331–382.

[17] C. Cuttriss and J. Redding, US Patent No. 242,816 (filed 28 November 1877) issued 14 June 1881.

[18] E.W. Siemens, German Patent No. 2355 (filed 14 December 1877), granted 30 July 1878; British Patent No. 4685 dated 10 December 1977 (provisional spec. filed 10 December 1877); "sealed" 1 February 1878; complete spec. filed 30 April 1878.

[19] F.V. Hunt, *Electroacoustics*, pp. 58–59.

[20] F. Guthrie, *Proc. Roy. Soc. (London)* **21**, 168–169 (1873); *Phil. Mag.* **46**, 257–266 (1873).

[21] J. Elster and H. Geitel, *Phil. Mag.* **14**, 161–184 (1882).

[22] Edison entered this observation in his notebook in 1881, but made no further use of it. See Lloyd Taylor, *Physics, the Pioneer Science*, Houghton Mifflin, New York, NY, 1941, p. 738.

[23] Percy Dunsheath, *A History of Electrical Engineering*, Pitman, New York, NY, 1962, pp. 266–276. See also J.A. Fleming, *Fifty Years of Electricity*, Wireless Press, London, UK, 1921; J.A. Fleming, *Memories of a Scientific Life*, Wilcox and Follett, London, UK, 1934.

[24] Lee De Forest, *Father of Radio, an Autobiography*, Wilcox and Follett, Chicago, IL, 1950.

[25] E. Armstrong, *Proc. Inst. Radio Engrs.* **3**, 215–248 (1915).

[26] H. Hertz, *Electric Waves, Being Researches on the Propagation of Electric Action with Finite Velocity through Space*. English translation, Macmillan, London, UK, 1893.

[27] C.S. Franklin, *Engineer* **117**, 663–664 (1914); **118**, 57 (1914).

[28] Those acousticians "of a certain age" may remember the use of "regenerative feedback" in short-wave radio receivers in the 1930s; when the amount of feedback became too great, the weak radio signal was overwhelmed by a whistling sound that was, of course, the resonant frequency of the feedback

oscillator. Those of an even greater age might recall the "howling telephone" of the 1920s, a phenomenon that recurs regularly in public address systems.

[29] *Encyclopeedia Britannica*, vol. 12, William Benton, Chicago, IL, 1958, p. 450.

[30] F. Braun, *Ann Physik Chemie* **60**, 552–229 (1897); **63**, 324–328 (1897)

[31] O. Heaviside, *Collected Papers*, vol. 2, Macmillan, London, UK, 1892, p. 371; *Phil. Mag.* **24**, 479–502 (1887).

[32] A.G. Webster, *Proc. Nat. Acad. Sci.* **5**, 275–282 (1919). Professor Webster read a paper at the Dec. 1914 meeting of the American Physical Society and its title appears in *Phys. Rev.* **5**, 177 (1915), but no further details were published.

[33] A. Kennelly and G.W. Pierce, *Proc. Amer. Acad. Arts Sci*, **48**, 111–154 (1912).

[34] A. Kennelly, *Electrical Vibration Instruments*, Macmillan, New York, NY, 1923, Chapter 13. Some discussion of priorities and other references may be found in F.V. Hunt, *Electroacoustics*, p. 66. A long and well-illustrated paper on mobility and impedance analogies was later given by F.A. Firestone, *J. Acoust. Soc. Am.* **28**, 1117–1153 (1956).

[35] D.C. Miller, *Anecdotal History of the Science of Sound*, Macmillan, New York, NY, 1935, Plate XV.

[36] W.C. Sabine, in *The American Architect*, 1900. Reprinted in *Collected Papers in Acoustics*, by W.C. Sabine, Harvard University Press, Cambridge, MA, 1922, and later by Dover, New York, NY, 1954. Tradition has it that Sabine ran downstairs from his study, shouting to his mother, "Mother, it's a hyperbola!" in the spirit of Archimedes famous "Eureka!"

[37] W.C. Sabine, loc. cit. Many years later (in the 1970s), Paul Bamberg at Harvard tried to repeat Sabine's measurements in all the rooms at Harvard mentioned in Ref. 36. When he came to the Fogg Art Museum, however, he found the dimensions were all wrong. Apparently, Sabine's exertions did not rescue everything, and the building had been torn down and a new one erected in its place—no doubt, with better acoustics! (Private communication to the author.)

[38] W.C. Sabine, *Am. Acad. Arts Sci. Proc.* **42**, No. 2, 1906; *Collected Papers in Acoustics*, pp. 69–105, especially pp. 103–104.

[39] W.S. Franklin, *Phys. Rev.* **16**, 372–374 (1903).

[40] W.C. Sabine, *Collected Papers in Acoustics*, p. 23.

[41] W.C. Sabine, *Proc. Amer. Acad. Arts Sci.* **52**, 49–82 (1906).

[42] W.C. Sabine, *The American Architect* **104**, 257 (1913).

[43] A.H. Davis, *Proc. Phys. Soc. (London)* **38**, 234 (1926). *Modern Acoustics*, G. Bell, London, UK, 1934, Plate VII.

[44] P.E. Sabine, *The American Architect*, June 1924. In 1919, Paul Sabine became the Director of Riverbank Laboratories in Geneva, IL, a laboratory that had been founded the year before by Colonel George Fabyan as a research vehicle for Wallace Clement Sabine. The history of this famous acoustical laboratory has recently been published: John W. Kopec, *The Sabines at Riverbank*, Acoustical Society of America, Woodbury, NY, 1997.

[45] Samuel Lifshitz, *Phys. Rev.* **25**, 291–294 (1925). See also Samuel Lifshitz, *J. Russian Phys. Chem. Soc.* **40** (1924).

[46] J.H. Jeans, *The Dynamical Theory of Gases*, 1st ed., Cambridge University Press, Cambridge, UK, 1904, Chapter XVI.

[47] W. Altberg, *Ann. Physik* **23**, 267–276 (1907).

[48] N. Neklepajew (Neklapaev), *Ann. Physik* **35**, 175–181 (1911).

[49] P. Lebedev, *Ann. Physik* **35**, 171–174 (1911). There is a hint in Lebedev's remarks of the molecular relaxation theories that were to follow in the next 20 years. Unfortunately, Labedev was near the end of his short life. He had resigned his university position in protest of government policies. Harrassed by tsarist officials, his health collapsed, and he died the following year, at the age of 46. He was one of the many Russian scientists who would suffer government oppression and even death in this century—first under the tsar, and then under Soviet rule: L.D. Landau, A.D. Sakharov, L.V. Shubnikov, N.I. Vavilov, A.A. Vitt, to name just a few.

[50] A. Einstein, *Sitzber. Berliner Akad.* 380–385 (1920). English translation, *Benchmark*/1, pp. 267–272.

[51] These developments are discussed in Hunt, *Electroacoustics*, pp. 51–53 and in W.G. Cady, *Piezoelectricity*, rev. ed., Dover, New York, NY, 1964, vol. 1, pp. 1–9.

[52] W.G. Cady, *Phys. Rev.* **17**, 531(A) (1921); *Proc. I.R.E.* **10**, 83–114 (1922). See also his *Piezoelectricity*.

[53] G.W. Pierce, *Proc. Am. Acad. Arts Sci.* **59**, 79–106 (1923); **60**, 271–190 (1925). The court battle, which lasted until 1953, is discussed by Hunt, *Electroacoustics*, pp. 53–57.

[54] G.W. Pierce, *Proc. Am. Acad. Arts Sci.* **60**, 271–290 (1925).

[55] D.L. Rich and W.H. Pielemeier, *Phys. Rev.* **25**, 117(A) (1925). The full report of this work did not appear until several years later [W.H. Pielemeier, *Phys. Rev.* **34**, 1184–1203 (1929)]. This paper will be discussed in Chapter 8.

[56] This institution has included such names as Francis Fenlon, John Johnson, John Schilling, Eugen Skudrzyk, John Snowdon, and Jiri Tichy among its faculty, not to mention many distinguished graduates and others, such as Isadore Rudnick, Wesley Nyborg, and Allan Pierce, who spent some time there.

[57] A survey of atmospheric acoustics, given by E.H. Brown and F.F. Hall, Jr, in *Rev. Geophys. Space Phys.* **16**, 47–110 (1978), is especially useful in tracing the development of this field.

[58] L.P. Teisserenc de Bort, *C.R. Acad. Sci. Paris* **134**, 987–989 (1902).

[59] G. van den Borne, *Z. Phys.* **11**, 483–488 (1910). Van den Borne was making more precise the qualitative observations of this phenomenon that date back at least to Samuel Pepys, who wrote of the "Zones of silence" in England during the English–Dutch naval battle in the Straits of Dover: "Up . . . to Whitehall . . . we find the Duke [of York [later James II] at St. James [Park, London] we saw hundreds of people listening at the Gravell pits and to and again in the park to hear the guns . . . yet at Deale and Dover, to last night, they did not hear one word of a fight. . . . This . . . makes room for a great dispute in Philosophy: how we should hear it and not they, the same wind that brought it to us being the same that should bring it to them. But so it is." June 4, 1666 (*The Shorter Pepys*, Bell and Hyman, London, UK, 1985, p. 625).

[60] R. Emde, *Meteorol. Z.* **33**, 351–360 (1916). For a discussion of Emde's results see D.I. Blokhintzev, *Acoustics of a Moving Inhomgeneous Medium* (English translation by R.T. Beyer and D. Minzter), Brown University, Providence, RI, 1952, pp. 41–42.

[61] F.J.W. Whipple, *Nature* **111**, 187 (1923). In a remark that was a forecast of great things to come, Whipple wrote: "Further progress in our knowledge of the temperature of the outer atmosphere and of its motion would be made if Prof. Goddard could send up his rocket. The times of passage of the sound waves from the bursting rockets would give immediate information as to the temperature of the air." It also suggests the great esteem in which Goddard's work on rockets was held.

[62] F.A. Lindemann and G.M.B. Dobson, *Proc. Roy. Soc. (London)* **A102**, 411–437 (1923); **A103**, 339–342 (1923). Lindemann (Lord Cherwell) was Churchill's scientific advisor before and during World War II.

[63] E. Schrödinger, *Phys. Z.* **18**, 445–453 (1917). Schrödinger was an artillery officer in the Austrian army during World War I and was working on acoustic problems of the military. It was an assignment closer to his profession than that of another Austrian of the time whom we might claim for acoustics, the great violinist and composer, Fritz Kreisler, who did nothing very musical during his wartime military duty. He did, however, make an observation on the sound of shells that involved the Doppler shift. See F. Kreisler, *Four Weeks in the Trenches*, Houghton Mifflin, Boston, MA, 1915.

[64] L.V. King, *Phil. Trans. Roy. Soc. (London)* **A218**, 211–293 (1919).

[65] G.I. Taylor, *Phil. Trans. Roy. Soc. (London)* **A215**, 1–26 (1915).

[66] Anyone who flies today is familiar with "clear-air turbulence," that often occurs at 10 km altitude or higher. Such disturbances clearly form the basis for much of Tyndall's "acoustics clouds."

[67] A. Powell, *J. Acoust. Soc. Am.* **98**, 1839–1841 (1995). Rayleigh's contributions to aeroacoustics are set forth in some detail in this article.

[68] Rayleigh, *Nature*, **95**, 66–68 (1915).

[69] M. Wilson, *American Science & Invention*. Bonanza, New York, N.Y., 1954, pp. 108–113.

[70] M. Lasky, *J. Acoust. Soc. Am.* **61**, 283–297. This article is a rich treasury of information and photographs of underwater sound work during the first part of this century.

[71] L.F. Richardson, British patent 11,125, 1912. A description of the early days of underwater sound is given by F.V. Hunt, *Electroacoustics*, pp. 44–53. See also H. Hayes, *Proc. Am. Phil. Soc.* **59**, 1–48 (1920); E. Klein, *J. Acoust. Soc. Am.* **43**, 931–947 (1968); and R.J. Urick, *Sound Propagation in the Sea*, Report of the Defense Advanced Research Project Agency, 1979.

[72] US Patent 1,207,388 (filed Jan. 29, 1913, issued Dec. 5, 1916). Fessenden had done pioneering work in transmitting speech by radio, and had originated the idea of using a continuous carrier frequency to "carry" the audio signals. See F.V. Hunt, *Electroacoustics*, pp. 39–40.

[73] F.V. Hunt, *Electroacoustics*, p. 45.

[74] F.V. Hunt, *Electroacoustics*, pp. 45–52.

[75] French patent No. 505,703 (demandé 17 Sept 1918, délivré 14 May 1920).

[76] H. Hayes, *Proc. Am. Phil. Soc.* **59**, 6–40 (1920).

[77] E. Klein, *J. Acoust. Soc. Amer.* **43**, 935–940 (1968). This article contains an excellent review of these early days of underwater sound work in the US Navy.

[78] F.V. Hunt, *Electroacoustics*, p. 52.

[79] Some description of this period may be found in M. Lasky, *J. Acoust. Soc. Am.* **61**, 283–297 (1977). R.J. Urick, *Principles of Underwater Sound for Engineers*, McGraw-Hill, New York, NY, 1967, pp. 4–6. Urick repeats a story by A.B. Wood that the original meaning of ASDIC was "Anti-Submarine Division," with the "ic" ending added, to resemble "acoustic."

[80] G.W. Pierce, *Proc. Am. Acad.* **59**, 81 (1923); **60**, 271 (1925); **63**, 1 (1928).

[81] W.H. Besant, *A Treatse on Hydrostatics and Hydrodynamics*, Deighton, Bell, Cambridge U.K., 1859, Sec. 158.

[82] O. Reynolds, *Phil. Trans. Roy. Soc. (London)* **174**, 935 (1883).

[83] Rayleigh, *Phil. Mag.* **34**, 94–98 (1917).

[84] A detailed history of echolocation by bats, along with an extensive bibliography, is given by Donald R. Griffin in *Listening in the Dark*, Yale University Press, New Haven, CT, 1958, Chapter 3. See also R. Galambos, *Isis* **34**, 132–140 (1942).

[85] The letters of Spallanzani describing his bat research appear in vol. 3 of *Opere di Lazaro Spallanzani*, Ulrico Heopli, Milan, 1932, 5 vols.

[86] Jurine's work is discussed by Griffin in Ref. [84]. The orignal paper appeared in *J. Phys.* **46**, 145–148 (1798). An English translation appeared in *Phil. Mag.* **1**, 136–140 (1798).

[87] The quotation is taken from Griffin, loc. cit, p. 63. See G. Cuvier, *Leçons d'Anatomie Comparée*, 5 vols., Crochard et Fantin, Paris, 1805, vol. 2, p. 581.

[88] The work of these scientists is described by Griffin, *Listening in the Dark*, pp. 64–65.

[89] S. Dijkgraaf, *Experientia* **5**, 90 (1949). Griffin, *Listening in the Dark*, p. 65.

[90] H. Maxim, *Sci. Amer. Suppl.*, Sept. 7, 1912, pp. 148–150.

[91] G. Hartridge, *J. Physiol.* **54**, 54–57 (1920).

[92] D. Griffin, *Listening in the Dark*, p. 66.

[93] *Benchmark*/8, p. 4. The original paper by Timoshenko appeared in *Phil. Mag.* **41**, 744–746 (1921), but it is also fully reproduced in the Benchmark volume.

[94] A more detailed treatment of Timoshenko's equations may be found in C.L. Dym and I.H. Shames, *Solid Mechanics: A Variational Approach*, McGraw-Hill, New York, NY, 1973, pp. 370–377.

[95] This shape factor continues to appear in the literature. See G.R. Cowper, *J. Appl. Mech.* **33**, 335–340 (1966).

[96] H. Lamb, *Proc. Roy. Soc. (London)* **A93**, 114 (1917).

[97] L. Pochhammer, *Crelle's J. Reine Angew. Math.* **81**, 324–336 (1875).

[98] Rayleigh, *Proc. London Math. Soc.* **20**, 201–225 (1889).

[99] See A.E.H. Love, *On the Mathematical Theory of Elasticity*, Cambridge University Press, London, UK, 1917, Chapter 7. Reprinted by Dover, New York, NY, 1944.

[100] Thorne Lay and Terry V. Wallace, *Modern Global Seismology*, Academic Press, San Diego, CA, 1995, p. 6.

[101] Lay and Wallace, *Modern Global Seismology*, p. 2. See also J.M. Steim, *The very broadband seismograph*. Ph.D. Thesis, Harvard University, 1986.

[102] Samuel Johnson, *A Dictionary of the English Language*, 1756, abridged and reprinted, Barnes and Noble, New York, NY, 1994, p. 492.

[103] Harvey Fletcher, *Speech and Hearing*, D. Van Nostrand, New York, NY, 1929, pp. 99–107.

[104] H. von Helmholtz, *On Sensations of Tone*, Longmans, Green, London, UK, 1875, p. 67.

[105] Harvey Fletcher, *Speech and Hearing*, p. 100. One of the earliest papers on street noise was that by E.E. Free, in *Pop. Sci. Monthly*, **109**, 16–17, 110 (August, 1926).

[106] It is of interest that noise is one of the few words that the acoustics community has given to the rest of science and engineering: all unwanted background signals, whether in acoustics, electricity, radio, or optics, have taken on the name of noise.

[107] R. Murray Schafer, *The Tuning of the World*, A.A. Knopf, New York, NY, 1977, p. 190. Schafer cites an even older lament, from the ancient "Epic of Gilgamesh" (c. 3000 B.C.): "The uproar of mankind is intolerable and sleep is no longer possible by reason of the babel." (Ibid., p. 189.)

[108] All apparently to little avail. The author is mindful of his visit to St. Mark's Cathedral in Venice, where the loud buzz from visiting tourists was periodically interrupted by a sepulchral "Silenzio," spoken over a P.A. system, with little or no resultant change in the noise level.

[109] H.J. Watts, *Brit. J. Psychol.* **7**, 1–43 (1914).

[110] E. ter Kuile, *Pflüg. Arch. Ges. Physiol.* **79**, 146–147, 484–409 (1900).

[111] The naming of the audiogram has been attributed to Carl E. Seashore.

[112] Harvey Fletcher, Ref. [103]; *Phys. Rev.* **15**, 513–576 (1920); *J. Franklin Inst.* **193**, 729 (1922); **196**, 289 (1923); Harvey Fletcher and R.L. Steinberg, *Phys. Rev.* **24**, 306–317 (1924).

[113] Hallowell Davis, *J. Acoust. Soc. Am.* **61**, 264–266 (1977). A short but valuable review of work in psychological and physiological acoustics 1920–1942.

[114] E.P. Fowler and R.L. Wegel, *Trans. Am. Laryngol. Rhinol. Otol. Soc.* **28**, 98–132 (1922).

[115] Hallowell Davis, loc. cit. Davis also asserts that Fowler called "Fletcher's 'sensation units' decibels," but the author has not been able to verify this comment.

[116] T.F. Vance, *Psychol. Monogr.* **16**, 115–149 (1914).

[117] E.G. Shower and R. Biddulph, *J. Acoust. Soc. Am.* **3**, 275–287 (1931).

[118] The graph is taken from S.S. Stevens and H.O. Davis, *Hearing: its Psychology and Physiology*, 1938. Reprinted by the Acoustical Society of America, Woodbury, NY, 1983, p. 85.

[119] G.L. Rich, *J. Expt. Psychol.* **1**, 13–22 (1916); *Am. J. Psychol.* **30**, 121–164 (1919).

[120] H.M. Halverson, *Am. J. Psychol.* **35**, 360–367 (1924).

[121] G.W. Stewart, *Phys. Rev.* (first ser.) **33**, 467–475 (1911); **4**, 252–258 (1914).

[122] A. Mayer, *Phil. Mag.* **2**, 500–507 (1876).

[123] R.L. Wegel and C.E. Lane, *Phys. Rev.* **23**, 266–276 (1924).

[124] *Benchmark*/13, p. 186.

[125] Sir Richard Paget, *Human Speech*, Kegan Paul, Trench, Traubner, London, UK, 1930, Chapters I–V. The book also contains a detailed discussion of the two-resonator theory of speech production, including a mathematical treatment by W.E. Benton (pp. 175–298). Paget was interested in the historical development of language, and discusses the use of sign language and pantomime by primitive peoples. He makes the curious statement that "the significant elements in human speech are the

postures and gestures rather than the sounds. The sounds only serve to indicate the postures and gestures which produced them. We lip-read by ear" (p. 174).

[126] Harvey Fletcher, *Speech and Hearing*, 1st ed., Van Nostrand, New York, NY, 1928, pp. 7–12.

[127] J.Q. Stewart, *Nature* **110**, 311–312 (1922).

[128] J.E. MacKenty, work described in Ref. [126, pp. 12–13]. See also R.R. Riesz, *J Acoust. Soc. Am.* **1**, 273–279 (1929).

[129] James G. Brown, *Phys. Rev.* **33**, 442–446 (1911).

[130] L. Herman, *Pflüger's Arch.* **45**, 42, 44, 347 (1899).

[131] L. Bevier, *Phys. Rev*, **10**, 193–201 (1900).

[132] Edward Wheeler Scripture, *The Elements of Experimental Phonetics*, C. Scribner's, New York, NY, 1904. 1906. The picture is taken from D.C. Miller, *The Science of Musical Sounds*, Macmillan, New York, NY, 1926, p. 77.

[133] L. Hermann, *Arch. Ges. Physiol. (Pflüger's Arch.)* **58**, 264–279 (1894).

[134] J.G. McKendrick, *Nature*, **65**, 182–189 (1901–1902).

[135] D.C. Miller, *Phys. Rev.* **28**, 151(A) (1908); *Science*, **29**, 161–171 (1909).

[136] Dayton C. Miller, *The Science of Musical Sounds*, p. 79. This author has always found the name infelicitous, and that may have hampered the use of the instrument.

[137] Dayton C. Miller, *The Science of Musical Sounds*, Frontispiece. To have obtained the services of Caruso and Tetrazzini must have been quite a coup. The early members of the Acoustical Society of America always seemed to have moved in the highest of musical circles!

[138] I.B. Crandall, *Bell Syst. Tech. J.* **4**, 586–626 (1925). C.F. Sacia, *Bell Syst. Tech. J.* **4**, 657–641 (1925). C.F. Sacia and C.J. Beck, *Bell Syst. Tech. J.* **5**, 393–403 (1926).

[139] Harvey Fletcher, *Speech and Hearing*, 1st ed., p. 29. The replacement of the oscillograph by the cathode-ray oscilloscope took place slightly later.

[140] Harvey Fletcher, *Phys. Rev.* **23**, 428–437 (1924). It is impressive how much of this early work in psychological acoustics was published in the *Physical Review*. It would appear that that journal had a much broader interest in the 1920s than it does today.

[141] Rayleigh, *Phil. Mag.* Ser. 6, **16**, 242–246 (1908).

[142] G.A. Campbell, *Phil. Mag.* Ser. 6, **19**, 152–159 (1910).

[143] H. Fletcher and J.C. Steinberg, *Bell Syst. Tech. J.* **8**, 848–852 (1929).

[144] R.L. Wegel and C.E. Moore, *Bell. Syst. Tech. J.* **32**, 299–323 (1924). Typical waveforms and harmonic spectra of both speech and music, measured by this equipment, are given in Harvey Fletcher, *Speech and Hearing*, 1st ed., Van Nostrand, New York, NY, 1928, pp. 90–95.

[145] Harvey Fletcher, *Phys. Rev.* **23**, 427–437 (1924).

[146] C.V. Raman, *Indian Assn. Cult. Sci. Bull.*, No. 11, p. 43 (1914).

[147] K.C. Kar, *Phys. Rev.* **20**, 148–153 (1922).

[148] Maindra Nath Mitra, *Indian J. Phys.* **1**, 311–328 (1926–1927).

[149] G.W. White, *Cambridge Philos. Soc. Proc.* **18**, 85 (1915).

[150] Cited by A.T. Jones, *Sound*, p. 297.

[151] C.V. Raman, *Phil. Mag.* **32**, 391–395 (1916).

[152] V.O. Knudsen, *Phys. Rev.* **21**, 84–102 (1923).

[153] Knudsen was in effect measuring the differential sensitivity of the ear to frequency and his results compare well with those of T.F. Vance. See Chapter 6 and below.

[154] This graph of Knudsen's data was replotted in a paper by Harvey Fletcher, Ref. [156] below.

[155] W.H. Martin, *Bell Syst. Tech J.* **3**, 400–408 (1924). See also C.W. Smith, *Bell. Syst. Tech. J.* **3**, 409–413 (1924).

[156] Harvey Fletcher, *Bell Syst. Tech. J.* **23**, 145–179 (1924).

[157] A list of references to these papers is included in a note by W.H. Martin, *Bell. Syst. Tech. J.* **8**, 1–2 (1929).

[158] Colonel (later Sir) Thomas F. Purves, C.B.E. (1871–1950), *The Electrician* **94**, 535, 542 (May 8, 1925).

[159] T.F. Purves, *The Electrician* **94**, 535 (May 8, 1925).

8
The Second Quarter of the Twentieth Century

> Madame, with the notes I missed I could have given another concert.
> Remark attributed to the concert pianist Vladimir de Pachman
> after having been congratulated by an enthusiastic listener that
> he had not missed a single note in his performance.

By 1925, the basic elements of electronics were all in place, and it was time for acousticians to use them in attacking acoustical problems with renewed vigor. There was a remarkable growth and diversification in all fields of science, and this was reflected in acoustics by the appearance of new societies, new journals, and new centers of research. Any historian of science has a tiger-like ride in attempting to cover all the significant work and still keep the size of the book within reasonable bounds. Following de Pachmann, this author could write at least another book with the material that has been omitted.

The unity of acoustics at this point can be gleaned by an examination of the Table of Contents of Harvey Fletcher's (1884–1981) book on *Speech and Hearing in Communication* (1st ed., 1929) [1]. Fletcher's book begins with speech, covering the means of its production, and its characteristics of speech, all viewed from an engineering viewpoint, and proceeds through noise and music, the mechanism of hearing and the properties of the ear, masking effects, loudness, speech recognition. The nature of listening spaces, and the electroacoustics of the equipment used in making measurements are also discussed. In so doing, Fletcher covered most of the divisions of modern airborne acoustics, all in a single volume. One might say that the core of modern acoustics is defined by this text.

Another way of looking at the expansion of topics in acoustics is contained in the comment by Ira Hirsh in a review paper in 1980 [2], Hirsh noted the increased number of authors of texts as the century progressed. Thus, Fletcher's book (1929) was followed in 1937, by Stevens and Davis's book on hearing (two authors) [3], and, in 1951, by Stevens *Handbook of Experimental Psychology*, where the field of hearing now required five authors. As for the books on hearing published in the 1970s, Hirsh remarked that "many separate chapters [were] written by as many authors [4]."

8.1. The Acoustical Society of America

The growing sense of identity among acousticians at this time was evidenced by the banding together in 1928 of acousticians from the United States (with a small number from other countries) to form the Acoustical Society of America. This development occurred at about the same time as the appearance of the first edition of Fletcher's book. In a later interview, Harvey Fletcher [5] observed that giving papers at the meetings of the American Physical Society had become less stimulating, because there were so few people there who were interested in what he was doing. Acoustical workers looked for other such workers with whom they could meet and discuss their research.

Much of this early organizational work was carried on by Wallace Waterfall (1990–1974), a young engineer and official of the Celotex Corporation, who had been instrumental in the founding of the American Materials Association. Waterfall began by attempting to organize a group of architectural acousticians. In a conversation with Fletcher, he was urged to broaden the scope of the proposed organization to include physicists and electrical engineers working on acoustical problems. After further consultation with physicists, Floyd Watson (1872–1974) at the University of Illinois and Vern Knudsen at UCLA, Waterfall was able to convene a total of 40 scientists and engineers in a meeting at the Bell Telephone Laboratories, in December 1928 (Fig. 8-1 [6]). The laboratories were then located on West Street in downtown Manhattan. And it was here that the Society was born. Its first official meeting occurred in May, 1929, also at the Bell Telephone Laboratories. The Society began with Fletcher as its President, Knudsen as Vice-President, Waterfall as Secretary, and with the editorial leadership soon given to Watson. The site of the early semiannual meetings of this Society alternated between academic institutions and the laboratories of industry. The growth of the size of the Society that brought about the "hotel period" was far in the future.

This same time period (1928) saw the formation of the American Standards Association, later to be renamed the American National Standards Institute (ANSI). This organization received early support from engineering societies. Standards in acoustics at that time were then developed by the American Society of Mechanical Engineers (ASME). While individual members of the Acoustical Society of America (ASA) played active roles in standards development, the role of the ASA in standards would not become significant until much later (See Chapter 10).

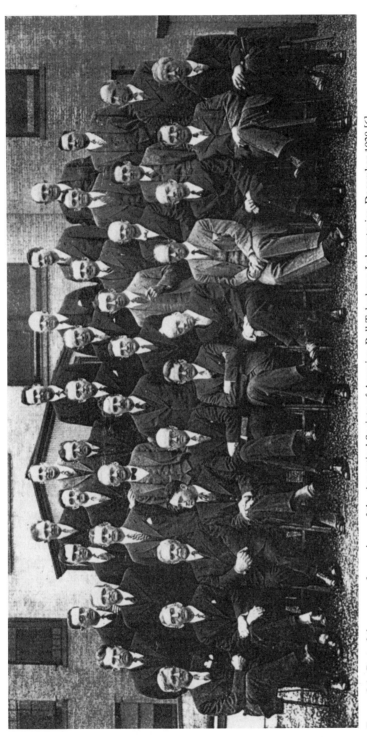

FIGURE 8-1. Part of the group of organizers of the Acoustical Society of America, Bell Telephone Laboratories, December 1928 [6].
Bottom row, from the left: F.A. Saunders, R.V. Parsons, D.C. Miller, W. Waterfall, V.O. Knudsen, H. Fletcher, C.F. Stoddard, J.P. Maxfield, F.R. Watson, F.K. Richtmyer, G.R. Anderson.
Second row, from bottom, from the left: H.A. Erf, H.C. Harrison, J.B. Kelly, R.L. Wegel, H.A. Frederick, N.R. French, C.W. Hewlett, A.T. Jones, I. Wolff, J.B. Taylor.
Third row, from bottom, from the left: L.J. Sivian, E.L. Norton, W.A. MacNair, R.F. Mallina, L. Green, Jr., R.H. Schroeter, H.W. Lamson, C.N. Hickman, D.G. Blattner.
Top row, from the left: W.P. Mason, J.C. Steinberg, V.I. Chrisler, E.J. Schroeter, E.C. Wente, W.C. Jones.

8.2. Ultrasonic Absorption and Dispersion

The improvements of the accuracy of measurement of both sound velocity and sound absorption that was reported on in Chapter 7 led to an explosion of results in the next quarter century. To make an orderly presentation, we shall divide the subject into gases and liquids, theoretical and experimental.

Gases. Theoretical

Experimental work by Pielemeier [7] in 1925 revealed a small but measurable increase in the sound velocity in CO_2 with increase in frequency. David Gordon Bourgin (1900–   ) [8] at the University of Illinois, Karl Herzfeld [9] (1892–1978), and a little later A.J. Rutgers [10] at Amsterdam, developed the ideas of Rayleigh, Jeans, and Einstein into the thermal relaxation theory. In this theory, account is taken of the delay in establishing equilibrium between the energy involved in the translational motions of the molecules of the gases and the energy involved in internal degrees of freedom, such as rotational and vibrational. When the sound wave passes through a gas, the energy of the acoustic wave (at those points where the particle displacement velocity is high) is transferred to the internal degree of freedom through molecular collisions. In the next half-cycle, the energy should be returned from the internal motions to the translational. If the sound frequency is low, there is ample time for the energy to be exchanged, and little energy is lost per cycle. As the frequency increases, however, there is less and less of a complete exchange, and the energy loss per cycle increases. Finally, when the sound frequency becomes very large, little energy gets transferred to the internal degree of freedom, and the energy loss per cycle again becomes small. This energy loss is best measured in terms of the product $2\alpha\lambda$, where α is the absorption coefficient and λ the wavelength. The shape of this curve as a function of frequency is shown in Fig. 8-2(a). This analysis was cleary expressed in the work of Hans Otto Kneser (1901–1985), who worked first at the University of Marburg and then, after World War II, at the Technical University Stuttgart [11].

Along with this absorption of energy, there is an increase in the sound velocity. This dispersion is shown in Fig. 8-2(b) [11]. A connection between dissipation and dispersion was identified for optics by Kronig [12], and later, more generally, by Kramers [13]. Where there is dissipation, there must also be dispersion. A general review of thermal relaxation theory can be found in the 1951 review paper by Jordan J. Markham (1917–   ), Robert T. Beyer (1920–   ) and Robert Bruce Lindsay (1900–1985) [14].

Gases. Experimental

The work on gases in this period begins with that of Pielemeier [15] as was mentioned in Chapter 7. Special assistance was provided by the architectural acoustics community. In a paper in 1929, Paul Sabine described in

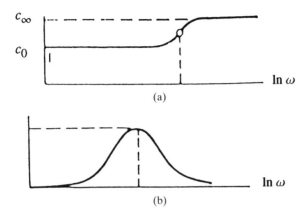

FIGURE 8-2. (a) Variation of absorption coefficient per unit wavelength, 2α, with the log of the frequency. (After Kneser [11].) (b) Variation of c^2/c_0^2 with the log of the frequency [11]. c_0 (c_∞) = sound velocity at very low (very high) sound frequency. (After Kneser [11].)

some detail the way in which a test room could be used to measure the absorption coefficient of the walls of that room. He noted that the total absorption of the room was very sensitive to the relative humidity of the air in the room, at frequencies above 2 kHz [16]. At about the same time, Erwin Meyer (1899–1972) Professor of Physics at the University of Göttingen, used a smaller test room and obtained similar results [17]. Shortly thereafter, Vern O. Knudsen combined his great interest in architectural acoustics with physical acoustics by studying how the measurement of reverberation time in a model chamber could be used to determine the absorption coefficient of the gas in the chamber [18]. As Knudsen wrote, the work of these scientists led

to results which not only are important in connection with problems in architectural acoustics and sound signaling but also are of theoretical interest in connection with the nature of the absorption of sound vibrations in gases. [19]

This bonding between architectural and physical acoustics was further strengthened in 1931, when Kneser spent a year working in Knudsen's laboratory [20]. A diagram of the apparatus used by Knudsen and Kneser is shown in Fig. 8-3. Because of the beam spreading of sound signals in the audio and near-audio range for fluids, accurate measurement of the sound absorption in a beam at low frequencies was virtually impossible, and the reverberation technique has been the principal means for making such measurements in both liquids (below 200 kHz) and gases (in the audio range).

Liquids. Experimental

Sound absorption in liquids is generally much lower than that in gases, so that accurate determination of its values lagged behind the corresponding

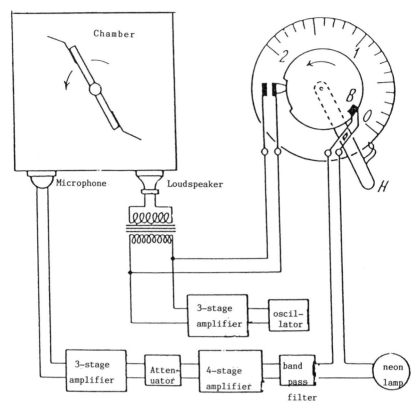

FIGURE 8-3. Knudsen–Kneser apparatus for the measurement of the sound absorption in air. (From Kneser and Knudsen [20].)

work in gases. Much of the early work was flawed because the beam spreading we just mentioned was not accounted for, and often the use of a continuous wave signal caused interference from reflections. Nevertheless, a few sets of reliable measurements appeared before World War II. These were the measurements by P. Biquard [21] in France and P.A. Bazhulin (sometimes transliterated as Bazulin or Bajouline) [22] and also B. Spakovskii (or Spakovskij) [23] in the USSR, using optical techniques. Biquard made measurements in a number of liquids, including water, the alcohols, and other organic liquids, finding the absorption coefficient to be 3 to 100 times that predicted by the classical theories of Stokes and Kirchoff, while Bazhulin made similar measurements, including some in acetic acid and ethyl acetate, where he found the absorption to be as much as a thousand times that predicted by the Stokes–Kirchhoff theories. Spakovskii detected a very small dispersion in the sound velocity of acetic acid, the first such to be measured in a liquid.

Optical techniques for observing ultrasound were pioneered by Peter P. Debye (1884–1966), who noted that, since the sound wave, in its passage through a liquid, altered the density of the medium periodically, and since the index of refraction of a given liquid depends on the density (the law of Clausius and Mosotti [24]), the sound wave could serve as a kind of diffraction grating for a beam of monochromatic light. Since the velocity of light is so much greater than that of sound, the "Grating" was essentially at rest when the light passed through it. Debye made this observation while at MIT on a lecture tour, and the idea was almost immediately confirmed experimentally by Francis Watson Sears (1898–1975) [25], and later by Biquard and Lucas [26]. An elaborate mathematical justification of the effect was worked out in India by C.V. Raman and N.S. Nagendra Nath [27]. Their results showed that the light intensity should be proportional to the square of the Bessel function of the corresponding order (counting the central line as zero). Experimental confirmation of the theory of Raman and Nath was soon provided by Otohiko Nomoto in Japan [28] (Fig. 8-4). Reading the first column in this figure downward and then proceeding to the second, one can see that, as the sound intensity increases, the various orders of Bessel functions appear, disappear, and then reappear in proper fashion.

The international nature of science is well demonstrated here by the number of nations represented by the scientists working on this one problem, even though these same nations were on the verge of World War II. The years of war, of course, interrupted the stream of international collaboration, but it was quickly resumed in 1946. The electronics of radar had produced pulse generators, so that one no longer had to endure the interference of sound reflected from the walls of the measurement chamber. A typical system is shown in Fig. 8-5 [29]. A pulse generator shapes the signal from an rf (radio-frequency) oscillator and the resultant rf pulses are conveyed to a piezoelectric transducer, which generates sound pulses in the sample. By use of a reflector, the signal is returned to the transducer (at a quiescent time) and the resultant electric signal is amplified, and portrayed on an oscilloscope, often in comparison with an exponential signal, so that the decay can be measured directly. The first to make use of this new technique were John R. Pellam and J.K. Galt [30]. They were soon followed by J.M.M. Pinkerton in Britain, whose measurements of sound absorption in distilled water over the temperature range from 0° to 100°C have remained unchallenged for 50 years (Fig. 8-6) [31].

Soon there was a growing amount of experimental data on the sound absorption coefficient. Much of it, however, suffered from a lack of consideration of the spreading of the sound beam as it passed through the medium and few measurements of this period were reliable below a frequency of 15 MHz. Rayleigh [32] had developed an expression for the intensity on the axis of a piston source, which is shown graphically in Fig. 8-7 [33]. It is clear from the graph that energy is lost from the main beam as one moves further

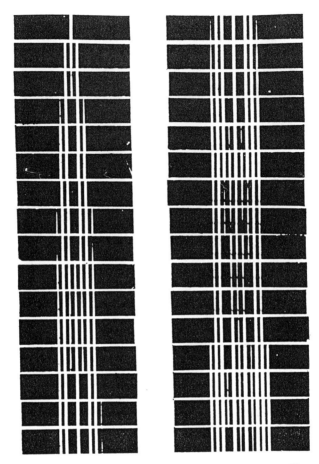

FIGURE 8-4. Debye–Sears effect for a traveling sound wave at various sound intensities. (From Nomoto [28].)

from the source. Various attempts were made to take this loss into account, the most significant one being that due to Arthur Olney Williams, Jr. (1913–1992) [34]. Williams obtained an approximate formula for the correction to be made for the signal received on a circular receiving transducer of the same radius as the source. This is, of course, the exact geometry used when one reflects the transmitted beam back on the receiver. His formula has been given wide use by various experimenters, especially with the simplification of the calculation developed by Seki, Granato, and Truell (1913–1968) [35].

Liquids. Theoretical

While the nature of the extra absorption of sound in gases was well understood by 1950, the status of absorption in liquids was much less certain. One

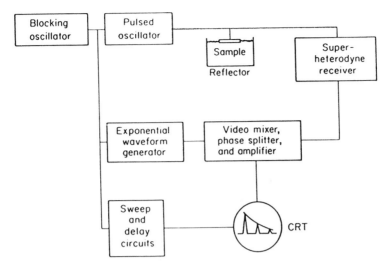

FIGURE 8-5. Typical pulse system for measuring sound absorption in liquids. (From Beyer and Letcher [29].)

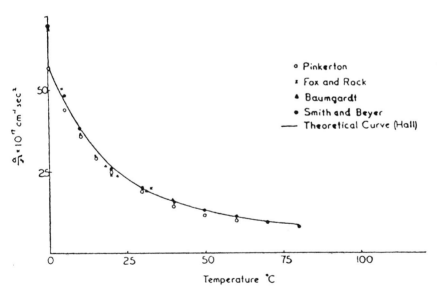

FIGURE 8-6. Temperature dependence of sound absorption coefficient in waiter. (From Markham, Beyer, and Lindsay [31].)

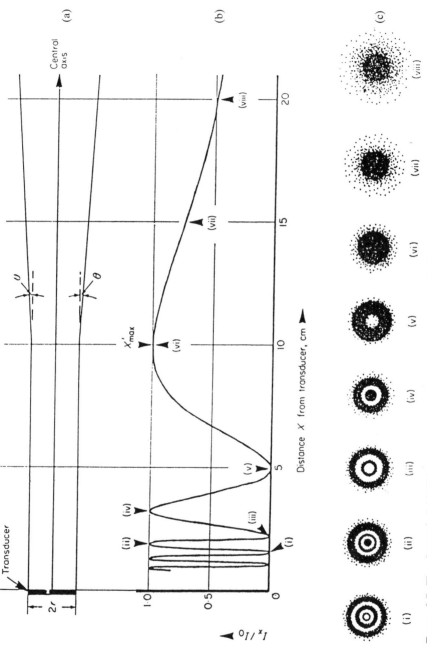

FIGURE 8-7. The ultrasonic field. This example shows the distribution for a 1.5-MHz transducer of radius $r = 1 = 1$ cm. (a) Almost all the ultrasonic energy lies within the limits shown in this diagram. (b) Relative intensity distribution along the central axis of the beam. (c) Ring diagrams showing the energy distribution of beam sections at positions indicated in (b). (From Wells [33].)

began to blame Stokes for his assertion that the volume or bulk viscosity in liquids must be zero (Chapter 2), but no theory was advanced that could account for this quantity until one was proposed in 1948 by Leonard Herbert Hall (1913–   ). Hall, who was a student of Bruce Lindsay at Brown University, hypothesized a structural relaxation in water [36]. In this theory, the molecules were assumed to have two arrangements, one resembling the crystalline state of water and another in which the molecules were close packed. The excess sound absorption was then attributed to a lag in the exchange of energy between the two different structures.

While this theory could account for the absorption in water, and thus provide a value for the bulk viscosity of that liquid, no independent experiment measuring the bulk viscosity has as yet been successfully devised, although a number of attempts have been made [37]. In addition, attempts to apply Hall's theory to other liquids, especially the lower alcohols, have not met with much success. The relaxation frequency for this phenomenon would be in the range 100–1000 GHz, which has not yet been amenable to experiment [38], [39].

8.3. Other Ultrasonic Effects in Fluids

Rao's Rule

There has been considerable attention in ultrasonics to the development of empirical rules relating the sound velocity to the parameters of the medium. The most famous of these is Rao's rule, first presented by M.R. Rao in 1940 [40]. In its simplified form, it states that

$$c^{1/3} V = R,$$

where V is the molar volume (the volume occupied by one gram-molecular weight of the substance) and R is a constant for a given liquid, sometimes called the Rao number [41]. While a great deal of effort has been expended in endeavoring to connect this constant to the molecular structure of organic materials, it served mainly as a rough guideline to the behavior of the sound velocity, but not a sure path for its evaluation [42]. A variant of Rao's rule was later developed by Y. Wada [43].

Acoustic Streaming

We have mentioned the flow of the air in front of a source of sound in air, observed by Faraday and Wheatstone (Chapter 2). This phenomenon was also noted in liquids, beginning in 1926 [44]. An excellent photograph of the effect, due to Leonard Liebermann (1914–   ) [45], is shown in Fig. 8-8. This work led to a detailed theoretical study by Carl Eckart [46]. The connection between acoustic streaming and sound absorption in the me-

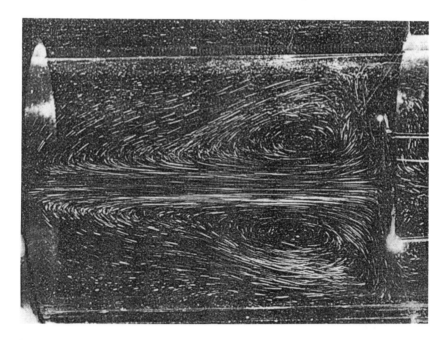

FIGURE 8-8. Photograph of acoustic streaming from a sound source; the motion of the liquid is made visible by a suspension of tiny particles of aluminum. (From Liebermann [45].)

dium was also studied by Liebermann, who attempted to use it as an independent measurement of the second viscosity coefficient. A clarification of this relationship was developed in the 1950s (see Chapter 9).

Another long-standing subject of acoustics in fluids has been the phenomenon of radiation pressure. Rayleigh had discussed this phenomenon of the one-directional force that would be experienced by a detector in the presence of a sound beam [47], and concluded that this radiation pressure would be equal to $(\gamma + 1) \langle E \rangle$ for an ideal gas, where $\langle E \rangle$ is the energy density in the incident sound beam and γ is the ratio of specific heats. In the period covered by this chapter, Leon Brillouin [48] pointed out the tensor character of the pressure in a sound wave, while Pierre Langevin [49] arrived at a value of $\langle E \rangle$ for this pressure. In 1950, Gustav Hertz and H. Mende demonstrated that the difference in the two expressions due to Rayleigh and Langevin came from subtle differences in the method of measurement involving also the difference between Lagrangian and Eulerian coordinates [50].

Low-Temperature Acoustics

Measurements of the sound absorption coefficient in liquefied gases were carried out by Galt in 1948 [51], but the greatest interest in low-temperature

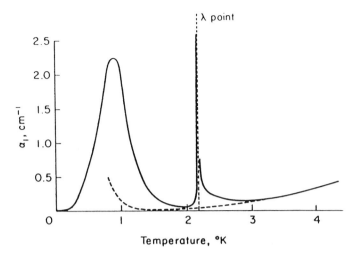

FIGURE 8-9. Attenuation of first sound in liquid helium at 12 MHz; (------) classical viscous absorption. (From Atkins [52].)

acoustics has centered around liquid helium. The variation of the sound absorption coefficient with temperature in liquid helium, taken from the work of Kenneth Atkins, is shown in Fig. 8-9 [52]. The discovery of this complicated behavior came at about the same time as the discovery of the periodic fluctuations of temperature and entropy in liquid helium in the superfluid state (below 2.18 K) [53]. These fluctuations took on the name of second sound, although they are not a sound wave in any traditional sense [54].

An explanation for the sound absorption coefficient for liquid helium was provided in the early 1940s by the theory of Lev Davydovich Landau (1908–1968), who hypothesized the relation between the energy of elementary excitations in the liquid and momentum in the form shown in Fig. 8-10 [55].

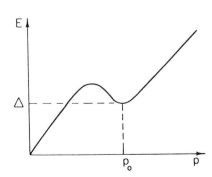

FIGURE 8-10. Dispersion curve for He II. (From Landau [55].)

8.4. Ultrasound in Solids

The classical behavior of sound propagation in solids was well understood by 1925 [56]. This included the relations between sound velocity in an extended solid and elastic constants, the behavior of compressional and shear waves, the interaction of such waves in rods, etc. [57]. The interaction of the various waves in bodies of finite dimensions did pose a limitation on the ability of the observer to determine the elastic constants, but improved techniques, developed by Claus Schaefer and Ludwig Bergmann [58], using the optical diffraction of light by sound waves in transparent solids, greatly facilitated such measurements.

The introduction of pulsed ultrasound after World War II heightened the interest in ultrasonic transmission in solids. Warrne Perry Mason (1900–1986) pursued the subject of electromechanical filters [59] and, together with Herbert J. McSkimin (1915–1981), carried out extensive studies and produced numerous devices, based on these studies [60]. These include a pulse system for measuring velocity and attenuation in a solid sample (Fig. 8-11) [61], a phase comparison method for such measurements (Fig. 8-12) [62], and many others.

The need in wartime electronics for devices to delay one electricalsignal relative to another led to the appearance of ultrasonic delay lines [63]. By passing an electric signal through a wideband piezoelectric transducer, an appropriate acoustic signal could be transmitted through a liquid mercury line or a solid line such as fused quartz. A second piezoelectric transducer then converts the acoustic signal back to the original electric signal. Since the speed of sound in quartz is of the order of $5000\,\mathrm{m\,s^{-1}}$, which is about half a centimeter per microsecond, it is evident that a delay of $100\,\mu s$ could be obtained by a rod of $50\,\mathrm{cm}$ length. Such delay lines were put into frequent use.

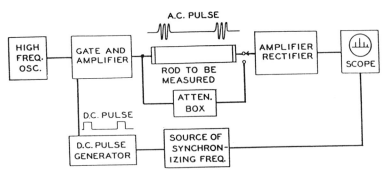

FIGURE 8-11. Pulse system for measuring sound velocity and attenuation in a solid material. (After McSkimin [61].)

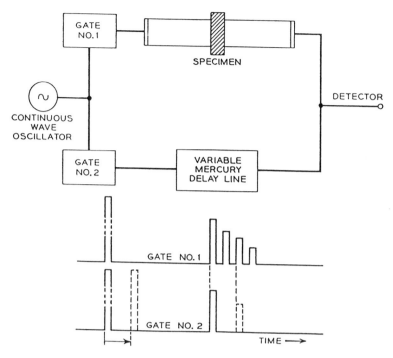

FIGURE 8-12. Phase comparison system for measuring sound velocity and attenuation in a single crystal or other specimen. (After McSkimin [62].)

8.5. Medical Ultrasonics

Work on medical ultrasonics began very slowly in this period. It was demonstrated by H. Freundlich, K. Söllner, and F. Rogowski [64] that the marrow of a bone could be heated without the bone being affected. Extensive tests by A. Dognon and E. and H. Biancani [65] indicated that small samples of liquids could be heated, sometimes by as much as 40° C by an ultrasonic beam. These experiments paved the way for the development of ultrasonic diathermy [66].

The first hints of success in medical diagnosis by means of ultrasound came about 1950. John J. Wild [67] used a pulse–echo system (what was later to be called A-scan) to measure thickness of tissue and also density of tissue. In effect, he was measuring the acoustic impedance ρc (the product of density ρ and the sound velocity c) of the tissue. In another paper, his group found that cerebral tumors could be detected by such a technique [68].

8.6. Other Applications of Ultrasonics

Many applications for ultrasound quickly developed. Since metals are very good conductors of ultrasound, it was early hypothesized that any flaws in a piece of metal would impede such transmission, and thus could be detected. A patent for such a system was obtained by Mühlhäuser in 1931 [69], but the first actual system appears to have been developed by Sokoloff in 1934 [70]. In the first of Sokoloff's methods, the ultrasound passed through a solid (plane-faced) block of metal into a vessel containing a liquid, and the presence of the sound in the liquid was detected by Debye–Sears diffraction. Flaws in the metal interfered with the diffraction patterns (Fig. 8-13). Unfortunately these methods required a comparison passage through a perfect sample, also they did not identify the location of the flaw. However, this early work stimulated a number of investigators in various countries, and devices for what became known as nondestructive testing began to appear [71]. The development of these devices was largely made possible by the availability of pulse generators.

One of the best known of these devices was the reflectoscope, invented by Floyd Alburn Firestone (b. 1898) [72]. Firestone was a professor at the University of Michigan and later the editor of the *Journal of the Acoustical Society of America* (1939–1957). The basis of the device is shown in Fig. 8-14 [73]. An electrical pulse from the generator G is passed to the transducer K and the ultrasonic signal enters the sample W. Upon hitting a flaw F, the signal is returned and picked up by the transducer and the electrical signal is passed along to the vertical plates of an oscilloscope R. The original electrical pulse is also passed to these plates. The horizontal plates of the oscilloscope are driven by the pulses from the generator, so that the screen produces the images shown to the right in the drawing.

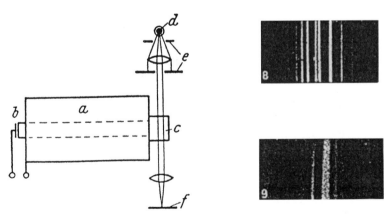

FIGURE 8-13. Testing materials by diffraction of light by ultrasonic waves. (From Sokoloff [70].)

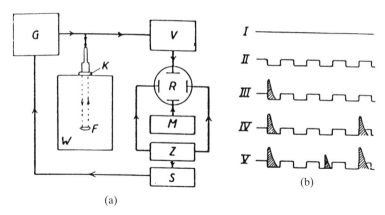

(a)

(b)

FIGURE 8-14. (a) Basic principles of operation of Firestone's ultrasonic reflectoscope. (b) Image appearance on the oscilloscope *R* under the following conditions: I. deflection with the impedance *z* attached; II. deflection with time markers; III. sound pulse emitted by the transducer; IV. emitted pulse and pulse reflected from the base of the sample; and V. emitted pulse, pulse reflected from the base of the sample and pulse reflected from an internal flaw in the sample.

This system eliminated the two problems of the Sokoloff apparatus, and was quickly adopted by industry in testing materials such as automobile tires, railroad rails, and metallic casting of various types. Pohlmann [74] developed various devices to render the sound visible in such test equipment.

One of the basic features of all sound waves is the local vibration of the medium. This vigorous shaking was found to lead to a number of different physical effects. These included the production of stable emulsions of immiscible liquids (Robert W. Wood and Alvin L. Loomis, 1927 [75]), the dispersion of solid particles in a liquid (N. Marinesco, 1933 [76] and B. Claus, 1935 [77]), and, oddly enough, a coagulating effect on particulate matter in the air (O. Brandt, 1937) [78]. The dispersive effects of ultrasound have been used extensively in the cleaning of metal surfaces by exposure to ultrasonic radiation, while the coagulating effects of aerosols have been put to good use in ultrasonic precipitators in industrial smoke stacks and other sources of pollution.

8.7. Nonlinear Acoustics

After a rather dormant period at the beginning of the twentieth century, the subject of nonlinear acoustics became more of interest in the second quarter. Richard Dudley Fay (1891–1964) provided much of this stimulus by a paper in 1931, in which he noted that a sound wave of finite amplitude became distorted as it traveled, thereby enriching itself in its harmonics.

But these harmonics are more rapidly absorbed than is the fundamental, so that he hypothesized the existence of the "almost stable waveform," and found that this waveform took on the shape of a sawtooth [79].

An even more significant paper was that by Eugene Fubini-Ghiron (1913–   ) [80] in 1935. In this paper he was able to solve the Earnshaw equation (Chapter 6) explicitly. He thus demonstrated that the pressure amplitude in a nondissipative fluid was proportional to an infinite series in the harmonics of the original signal; the amplitude of each harmonic in this series was proportional to the Bessel function of corresponding order divided by its argument. In turn this argument was a function of the distance and of the original amplitude of the sound pressure [81]. This was a valuable contribution but it remained virtually unknown until the later 1950s [82]. In 1962, David Theobald Blackstock (1930–   ) [83] pointed out that the original solution of this equation had been provided by Bessel himself when he invented the Bessel functions.

An important experimental advance, although little noted at the time, was a paper by Albert Lauris Thuras (1888–1945), R.T. Jenkins, and H.T. O'Neil [84] at the Bell Telephone Laboratories, which was a study of the output of horn-type loudspeakers carrying intense sound. They found that the level of the second harmonic of the fundamental tone actually increased with distance and with the amplitude of the fundamental pressure, both in complete agreement with the perturbation analysis of the acoustic wave equation. Any further development in this area, however, and to wait until well after World War II.

8.8. Atmospheric Acoustics and Aeroacoustics

In this period the two fields of atmospheric acoustics and aeroacoustics blended, since the concerns of propagation through the atmosphere were found to be inextricably interwined with the problem of generation of turbulence in the same atmosphere. Significant advances in these fields were provided at this time by the Russian school. The contributions of A.M. Obukhov [85] and A.N. Kolmogorov [86] followed the work of G.I. Taylor in Britain on vortices in the atmosphere. Kolmogorov derived his "2/3 law" in 1941, but was anticipated in part by Obukhov. However, because of World War II, the work of Kolmogorov was not immediately appreciated in the West (and of Obukhov even less so), and the law was rederived independently by Western scientists in the late 1940s. A detailed account of these developments may be found in the books by G.K. Batchelor [87] and D.I. Blokhintzev [88].

The "two-thirds law" connected the mean energy E involved in the fluctuations of turbulence with the scale λ of that turbulence, and stated that E was proportional to $\lambda^{2/3}$. Thus the energy of homogeneous and isotropic turbulence is concentrated in large-scale turbulence [89]. These gains in turbulence theory, plus the assumption made by Blokhintzev [90], that only wave numbers in the inertial range had an effect on the propagation of low-frequency and audible sound, opened up the way for applying the theory of Obukhov and Kolmogorov to the calculation of the attenuation of sound by turbulence. Use would be made of this in the 1950s and later (Chapter 10).

Sound propagation in a quiescent (nonturbulent) atmosphere was not neglected. Peter Bergmann (1915–   ) [91] showed that a static density gradient in the atmosphere would produce dispersion in sound propagation and later [92] began the study of scattering in terms of correlation functions.

The Russian school also pioneered the study of the production of sound by airflow of various kinds [93]. Blokhintzev studied the production of sound by vortices, and concluded that airplane propeller noise would be proportional to the sixth power of the velocity, anticipating later work on jet noise done by Lighthill [94]. Another feature of this period was the observation by Yudin that sound generation by vortices might not require the presence of a solid surface (the necessity for which had been the general belief at the time), stating that he was "obliged to suppose that the origin of the vortex noise lies in the variable force acting on the medium during the flow past the body [95]."

8.9. Underwater Sound

Government stimulation of defense-related industries fell off markedly in the period between World War I and World War II, and underwater sound was no exception. In the 1950s, the author heard an official of the Navy Department give an after-dinner speech in which he remarked that the Navy had not lost any ground in the interval between the two wars. At the end of World War I the navy's submarines could do 4 knots underwater, and in 1941, at the onset of World War II, the navy had submarines that could still do 4 knots underwater! The onset of the war, however, caused an enormous increase in military research and development, including underwater sound devices, and a corresponding increase in our knowledge of the entire subject of underwater sound. As the electronic equipment became more sophisticated, it was possible to shape the beam of sound projected from the transducer so that the signal could be more sharply directed (Fig. 8-15) [96].

In order to get longer ranges on the sound-producing equipment—which began to be called SONAR (SOund NAvigation and Ranging [97])—more and more powerful sound sources were used, but it was found that there were limitations. It was already known that the production of air bubbles in the vicinity of high-speed propellers—the phenomenon known as cavitation—could severely damage those propellers. Now it was found that high-

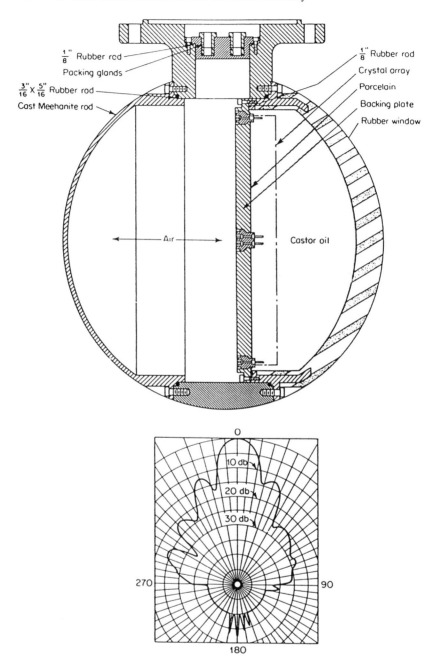

FIGURE 8-15. Construction and beam pattern of a piezoelectric projector comprising a crystal array cemented to a backing plate. (From Urick [96].)

intensity sound could also produce such bubbles—cavitation—and that these severely limited the intensity of the resultant signal [98].

Another limitation was the absorption of sound by sea water. As indicated above, measurements of sound absorption in liquids were sketchy and often inaccurate in the period before 1940 (Chapter 8). Hence, when E.B. Stephenson made seawater measurements of sound absorption at frequencies of 30 kHz and lower in the 1930s [99], the fact that he obtained much larger values for the ratio of $2\alpha/f^2$ (α = absorption coefficient, and f = frequency) than had been obtained in laboratory measurements by Biquard [100], the difference might still be blamed on experimental errors from poor measuring devices. During World War II, measurements were carried out at the University of California Division of War Research and elsewhere. Some of these were published by Leonard Liebermann in 1948 [101]. His measurements were made in the fresh water of a large reservoir and in the sea water of San Diego Bay. The fresh water measurements indicated that the sound absorption was several times that predicted by Stokes [102] and Kirchhoff [103], but the surprise result was that the sea water absorption was much greater than that of fresh water at frequencies below 1 MHz, i.e., in the sonar frequency range, thus supporting Stephenson's findings. A plot of the various measurements reported at that time is shown in Fig. 8-16 [104]. At first, this excess absorption was attributed to the presence of NaCl [105], but later work by Robert W. Leonard (1910–1967), Paul C. Combs, and Leslie R. Skidmore [106] indicated that the dominant agent was $MgSO_4$.

The increased use of underwater signaling resulted in an accumulation of data on the sound velocity in the ocean. It was well known that the sound velocity increased with temperature (about $5\,\text{m}\,\text{s}^{-1}$ per degree C in the vicinity of $0°C$), and also increased very slightly with pressure. Detailed records indicated that the temperature of the ocean fluctuated with depth in the surface layers, depending on the weather conditions, but fell off gradually to the value corresponding to the temperature at which the density of sea water is a maximum (near $0°C$). Thus the influence of temperature on sound velocity predominates in the near-surface region, but at great depths the pressure dependence becomes important. Typical distributions of the sound velocity with depth are shown in Fig. 8-17 [107]. This figure shows the faster speed of sound in the near-surface layers, due to warmer water, and faster speeds again at great depths due to the effect of hydrostatic pressure on the sound speed.

In 1943, Maurice Ewing (1906–1974) suggested that this dependence of the sound velocity on the depth should lead to a two-dimensional decay of the sound wave, with most of the energy staying in a pancake-like channel centered on the sound velocity minimum. Experimental studies proved the validity of this surmise, and the subject of SOFAR was born (SOund Fixing And Ranging). Some of results of Ewing's calculations are shown in Fig. 8-18) [108].

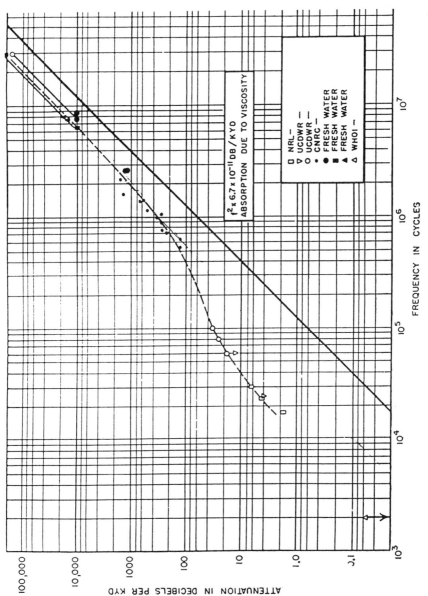

FIGURE 8-16. Dependence of attenuation coefficient on frequency: NRL, 13 and 16; UCDWR, 14; UCDWR, 18; CNRC, 19; Fresh Water, 20; Fresh Water, 22; Fresh Water, 23; WHOI, Chapter 9 of *Physics of Sound in the Sea* [104]. The points and their references are identified in Ref. [104].

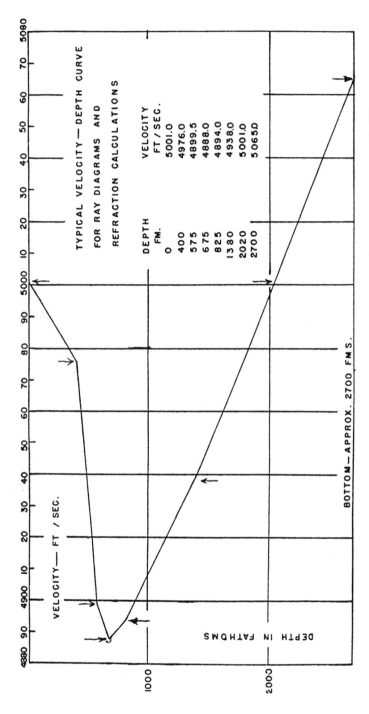

Figure 8-17. Sound velocity–depth curve from *Atlantis* hydrographic data. (From Ewing and Worzel [107].)

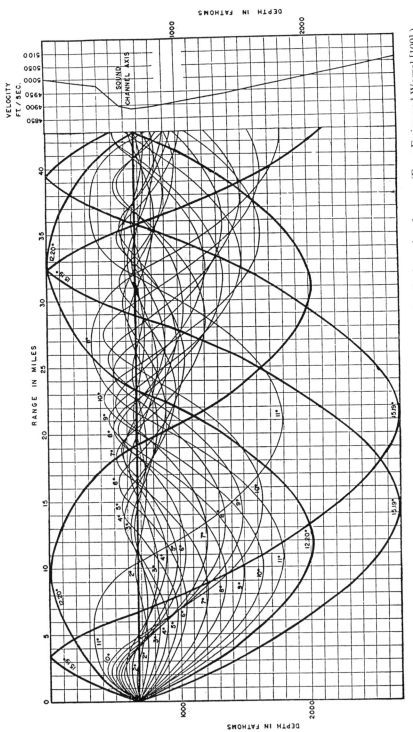

FIGURE 8-18. Ray diagram for typical Atlantic Ocean sound channel—sound channel and refracted surface ray. (From Ewing and Worzel [109].)

A few years later, the Russian acousticians Leonid Maksimovich Brekhovskikh (1917–   ) and Lazar Davydovich Rozenberg (1908–1968)—two of the ablest Russian acousticians of the post-war period—published papers indicating that they had also discovered the sound-channel phenomenon during the war. Since each group was working under conditions of war-time secrecy, it remains difficult to determine the original discoverer [109].

Attempts at following the sound rays through an ocean of varying temperatures called for great and greater computational skills. Ray tracing became a naval practice during World War II and blossomed into a major industry in the post-war period. In the simplest form of ray tracing, the ocean is divided into a finite set of horizontal layers. In each layer, the sound velocity is assumed to vary linearly with depth. Knowing the angle of incidence for the ray entering the first layer, one can compute its path and determine the angle at which the ray departed the layer. This value can then be used as the incident angle for the next layer, and so on. A sample of these layers is shown in Fig. 8-19 [110].

Detailed studies of sound propagation in shallow water were also of great interest. Chaim Leib Pekeris (1908–   ), a Lithuanian-born scientist, carried out wartime research in underwater sound at Columbia University in New York, and later became a professor at the Weizmann Institute in Tel Aviv in 1950. He wrote seminal papers on ray tracing in shallow water, employing the theory of normal modes [111]. Similar studies were carried out in the Soviet Union by Brekhovskikh [112].

The intensive study of the ocean in this period revealed the fact that the ocean is quite inhomogeneous. (A somewhat fanciful picture of such inhomogeneities was given by Blackstock in Fig. 8-20 [113].) A range of objects, from tiny dust particles and bubbles to the myriads of aquatic life, as well as the roughness of the surface, especially in stormy weather, and the irregularities of the bottom, all contribute to the scattering of any sound signal traveling through the medium. The totality of such scattering is known as reverberation, and is a counterpart to the phenomena involved in room acoustics [114].

The reception of sonar signals has always been interfered with by various noises in the ocean. Systematic studies of these noises began during World War II. An important summary of this work was given by Knudsen, Alford, and Emling [115], in which experimental results were given for noises due to water motion, marine life, and ship and man-made noise.

8.10. Bioacoustics

In Chapter 7, we introduced the subject of echolocation by bats, and in the previous section we made mention of various animal noises in the sea. This latter is part of an enormous subject—the study of sounds made by all sorts

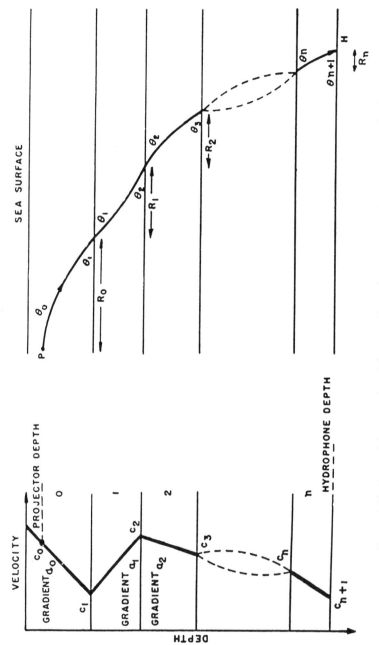

FIGURE 8-19. Ray path in succession of linear gradients. (From *Physics of Sound in the Sea* [110].)

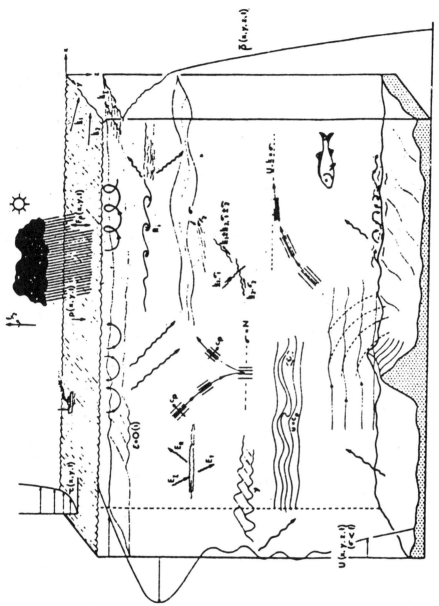

FIGURE 8-20. The inhomogeneities of the ocean. (From Blackstock [113].)

of animal life, and the ability to hear among the same population. The constraints on this book (its size and the author's knowledge) will limit our coverage to the subject of echolocation by bats, whales, and dolphins. This study of the sounds and hearing of animals in general was pioneered by Réné-Guy Busnel [116] and became widespread after the mid-century and continues today [117].

Echolocation by Bats

The study of echolocation by bats began at Harvard in 1938, when Donald Redfield Griffin (1915–   ), a graduate student in zoology, became acquainted with the work on ultrasonic production and reception by G.W. Pierce in the Physics Department. Griffin's career led him first to Cornell, then back to Harvard, where he served as Chair of the Biology Department, and still later to the Rockefeller Institute in New York. Joining in the research was Robert Galambos (1914–   ), who obtained both Ph.D. and M.D. degrees at Harvard, and later became Professor of Physiology and Psychology at Yale [118]. In their first publication [119], Pierce and Griffin noted that, when the bat was held in front of an ultrasonic receiver and was struggling to escape, it emitted short pulses of ultrasound at about 48 kHz. The duration of the pulses was 50–100 ms. Some early examples of pulses emitted by bats are shown in Fig. 8-21 [120].

In subsequent researches, Griffin and Galambos verified that the ultrasonic pulses being emitted by the bat was necessary for its navigation. With its mouth sealed shut, the bat could not navigate [122]. Further, the pulses were directed forward, in a relatively narrow beam, with the same sort of directivity as that possessed by sonar signals [123].

Galambos, who had previously been working on the physiological mechanisms of hearing, became especially interested in the hearing of bats and published a series of important papers on cochlear potentials and neuromechanisms of bats [124].

8.11. Structural Acoustics and Vibration

This period saw a significant increase in the study of nonlinearities in vibration. The work of G. Duffing [125] in 1918 was supplemented by that of Manfred Rauscher. Whereas Duffing began his iteration method of approximation of nonlinear vibrations with the linear vibration, Rauscher made his approximation start with the free nonlinear vibration, thus developing a more rapid convergence [126].

The nonlinear vibrations of strings whose tensions vary appreciably from their initial values because of the displacements was studied by George Carrier (1918–   ) [127]. As Kalnins remarks [128], Carrier showed that the results [of calculation of perturbation coeffeicients] could be regrouped in

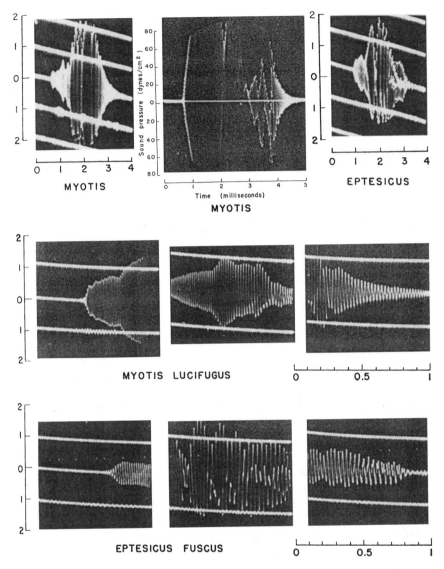

FIGURE 8-21. Cathode ray oscillograph records of pulses of sound from *Myotis lucifugus* and *Eptesicus fuscus*. The veritcal scales are sound pressure (dyn cm^{-2}), the horizontal scales are milliseconds. The record in the center of the upper row is an unusually intense pulse recorded at 5 to 10 cm from the bat's mouth; the others were recorded at 30 cm. Top center record form Ref. [120]; other not previously published. (From Griffin [120].)

a different form so as to be presentable in terms of elliptic functions. In this way, he was able to show graphically the dependence of the period on the change in the tension. Other significant approximation methods were developed at this time, reported in the books by Kryloff and Bogoliuboff [129] and by Andronov and Chaikin [130].

An important contribution to the theory of vibrations of shells was made in 1949 by R.N. Arnold and G.B. Warburton, when they solved in detail the complete boundary-value problem of the free vibration of a finite cylindrical shell [131]. This work, which showed excellent agreement with experiment was the forerunner of other contributions to free-vibration analysis (Chapter 9).

8.12. Noise and its Control

The beginnings of the study of noise as a subdiscipline of acoustics essentially concide with the formation of the Acoustical Society of America. We can divide the noise field into four topics of interest: sources and transmission of noise, its measurement, its effect on man and animals, and, of course, methods of its control.

The short list of noise sources given by Fletcher in his book (Chapter 7) could be supplemented by a long list—airplane noise, motor car horns, pneumatic drills, lawnmowers, and, by the end of the period covered in this chapter, electric guitars and "boomboxes [132]." Sound level meters made their appearance at this time [133], and journal articles and books began to feature tables of noise levels from different sources (Fig. 8-22) [134], [135].

The effect of sound on man was also considered. Donald A. Laird and his colleagues [136] studied the physiological and psychological reactions of humans to various noises. Subsequent studies focused on the levels that produced annoyance and those that resulted in physical damage to hearing.

As the quarter century passed, more and more detailed information on noise sources was made available [137]–[139]. It is of particular interest to compare Knudsen's chart of sound levels, given in 1949 [141], with those presented in Refs. [134], [135] some 20 years earlier. Knudsen cites an even older reference to noises (without dB levels):

Festina calidus mulis gerulisque redemptor;
Torquet nunc lapidem, nunc ingens machina tignum.
Tristia robustus luctantur funera plaustris;
Hac rabiosa fugit canis, hac lutulenta ruit sus:
I nunc, et versus tecum meditare canoros [140].
[Horace (65–8 B.C.). The translation is given in Ref. [141].]

Investigators became sensitive to the fact that there were differences between continued loud noises and isolated spikes of noise. At the same time, more and more attention was directed toward techniques for lowering

NOISE LEVELS OUT OF DOORS DUE TO VARIOUS NOISE SOURCES					
SURVEY OF NEW YORK CITY NOISE ABATEMENT COMMISSION		NOISE LEVEL	OTHER SURVEYS		
DISTANCE FROM SOURCE	SOURCE OR DESCRIPTION OF NOISE		SOURCE OR DESCRIPTION OF NOISE	SURVEY NO.	
FEET		DB			
		130	THRESHOLD OF PAINFUL SOUND	4	
		120			
2	HAMMER BLOWS ON STEEL PLATE-SOUND ALMOST PAINFUL (INDOOR TEST)		AIRPLANE; MOTOR 1600 R.P.M.; 18 FT. FROM PROPELLER	5	
		110	AERO ENGINE UNSILENCED - 10 FT	4	
		100			
35	RIVETER				
15-20	ELEVATED ELECTRIC TRAIN ON OPEN STRUCTURE	90	PNEUMATIC DRILL - 10 FT.	4	
			NOISIEST SPOT AT NIAGARA FALLS	2	
			HEAVY TRAFFIC WITH	7	
15-75	VERY HEAVY STREET TRAFFIC WITH ELEVATED LINE	80	ELEVATED LINE, CHICAGO		
15-50	AVERAGE MOTOR TRUCK		VERY NOISY STREET N.Y. OR CHICAGO	1	
15-75	BUSY STREET TRAFFIC	70	VERY BUSY TRAFFIC, LONDON	4	
15-50	AVERAGE AUTOMOBILE				
3	ORDINARY CONVERSATION				
15-300	RATHER QUIET RESIDENTIAL STREET, AFTERNOON	60	AVERAGE SHOPPING ST., CHICAGO	6	
			BUSY TRAFFIC, LONDON	4	
15-50	QUIET AUTOMOBILE MINIMUM NOISE LEVELS ON STREET:	50	QUIET AUTOMOBILE, LONDON	4	
	IN ENTIRE CITY	MIN. AVERAGE		QUIET ST. BEHIND REGENT ST, LONDON	4
15-500	DAY TIME	MIN. INSTANTANEOUS			
50-500	IN MID-CITY				
50-500	NIGHT	MIN. INSTANTANEOUS	40		
		30	QUIET ST., EVENING, NO TRAFFIC	4	
			SUBURBAN LONDON		
		20	QUIET GARDEN, LONDON	4	
			AVERAGE WHISPER - 4 FT.	3	
		10	QUIET WHISPER - 5 FT.	4	
			RUSTLE OF LEAVES IN GENTLE BREEZE	3	
		0	THRESHOLD OF HEARING		

(a)

FIGURE 8-22. Early noise level tables [134], [135].

sound levels. This almost immediately involved architectural acoustics, since the design of new homes and apartment houses could be modified to produce a quieter environment. In his 1949 paper, Knudsen noted that improvements could be made at both ends; houses could be made quieter by the use of insulating materials, but the sound sources, especially the automobile, could also be quieted. In citing the need in large cities of reducing the level of traffic noise by 50 dB by the time the noise entered the interior of a room, Knudsen remarked:

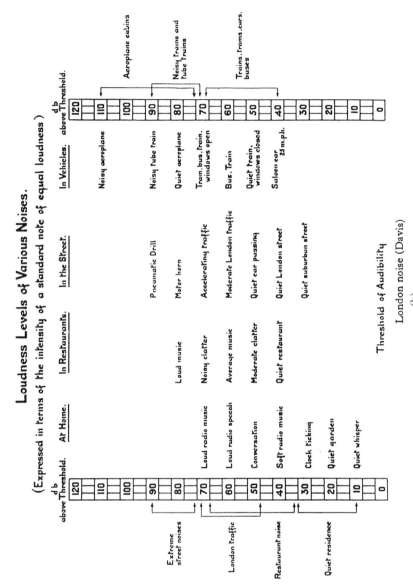

FIGURE 8-22. *Continued*

This [sound level reduction by insulation and interior absorption] would require costly construction, with few or no windows. . . . Practically, it would be much more feasible to establish traffic regulations that would reduce existing traffic noise by about 10 dB. A moderate amount of acoustic designing and gadgetry, including some suitable sound filters, would suffice to reduce this traffic noise by at least 10 dB, ... With such a reduction . . . the problem of constructing rooms so that the noise levels in these rooms would not exceed the acceptable values . . . would be greatly simplified [140].

But control of noise was not enough. Many of the same scientists who studied the problem of noise also studied the effects of vibration, first of heavy machinery, and later, of hand-held devices, such as jackhammers, on man, and vibration isolation became of major concern for those designing the installation and operation of heavy equipment in factories. Such studies led to the development of a new field, that of bioresponse to vibration.

8.13. Architectural Acoustics

While Sabine had made clear our understanding of the role of reverbera- tion time in room acoustics at the turn of the century, there was still much more to be learned. In 1949, Knudsen noted that, while the approval by the audience for musical sounds depends on "the musical taste and the disposi- tion of the listener [140]," a room can be evaluated for speech in a quanti- tative way by borrowing from the telephone engineers, who had developed an *articulation index*. The percentage articulation for a telephone was the percentage of disconnected speech sounds that were heard correctly. Be- cause a perfect articulation score is never achieved, Knudsen used the relation.

$$\text{percentage articulation} = 0.96 k_l k_r k_n k_s,$$

where the various k's include the losses due to inadequate loudness [l], reverberation [r], noise [n], and room shape [s]. Making use of his own speech data and the subjective results for music by Watson [138] and Lifshitz [143], Knudsen arrived at the optimal curves for music and speech shown in Fig. 8-23 [144].

In 1944, two outstanding members of the MIT faculty, Philip H. Morse (1903–1985) and Richard Bolt (1911–   ) reviewed the state of knowledge of room acoustics [145]. Morse, well known for his authoritative texts on vibration and sound and theoretical physics [146], as well as for work in quantum mechanics, pioneered in the application of wave theory to the problems of architectural acoustics, while Bolt worked in room acoustics and normal mode theory, and was later a ground-breaker in the develop- ment of international cooperation in acoustics, which led to the formation of the International Commission on Acoustics, of which Bolt was the first presiding officer. At the time it seemed almost as if this application of wave

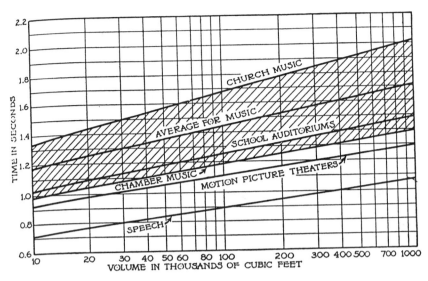

FIGURE 8-23. Optimum reverberation time at 512 cycles for different types of rooms as a function of room volume. (From Knudsen and Harris [144].)

theory would solve everything in room acoustics. Unfortunately, as Thomas Northwood noted later, "sound fields in most rooms are far too complicated to be handled by wave acoustics [147]."

The book by Knudsen and Cyril Harris (1917–   ) [148] first published in 1950, well sums up the state of architectural acoustics at mid-century. Main concerns were the development and measurement of sound-absorptive materials, the control of air-borne noise and noise in ventillating systems, and sound amplification systems. One is struck by the slight attention paid at this time to the psychological problems of the listener [149].

8.14. Physiological Acoustics

A new star appeared on the acoustics horizon in the 1930s—Georg von Békésy. Born in Budapest in 1899, he studied at the University of Bern, Switzerland, and received his doctorate in physics at the University of Budapest. He then worked for both the Hungarian Telephone Co. and the University of Budapest. While his first publication was on the nonlinear properties of the iron in telephone receivers (he was interested in determining the quality of reception) his second paper turned to the study of the ear itself and, in particular, to oscillations of the basilar membrane. And from that time on, his interest in the ear never waned; more than 100 papers on that subject would appear bearing his name, and, in 1962, he was awarded the Nobel prize in physiology and medicine for his work on the ear.

His career was interrupted by World War II, but upon its end he moved, first to Stockholm, then to Harvard, and finally to the University of Hawaii. At each institution, papers came forth, delving into the details of operation of the human ear [150]. His work was characterized by ingenious use of apparatus to probe into the ear, and by the development of mechanical models to simulate the ear's actions.

In discussing the pattern of vibrations in the cochlea (Chapter 11 of Ref. [150]) von Békésy made the following remarks:

For more than a century the vibratory pattern of the cochlear partition has been the central problem of audition. . . . Because for a century no numerical values concerning the mechanical properties of the cochlear partition were available, there were no restrictions on the imagination and probably every possible solution of the problem was proposed. It seemed to the writer that the only way to solve the problem was to open up the cochlea and observe the action during the presentation of a tone [151].

Easier said than done, but perhaps not for von Békésy!

von Békésy first constructed a mechanical model of the cochlea (a typical one is shown in Fig. 8-24) [152]. He also determined the static properties of the basilar membrane, finding that the stiffness of the membrane varied by a factor of 200 from one end to the other. Following his own advice, von Békésy proceeded to open up the cochlea of a cadaver and observe the motions of the cochlear structure [153]. Figure 8-25 shows von Békésy's results for the outward stimulation of the stapes [154]. Further analysis (Fig.

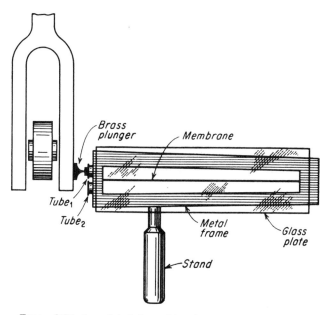

FIGURE 8-24. A model of the cochlea. (From von Békésy [152].)

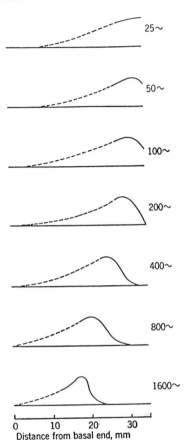

FIGURE 8-25. von Bekésy's measurements of amplitude of motion in the cochlea, for various frequencies. The solid lines of the curves represent the measurements and the broken lines of an extrapolation based on further observations. (Wever and Lawrence after von Békésy [154].)

8-26) indicated that the peak of the amplitude moved toward the stapes as the frequency increased [155], while the role of the basilar membrane in the creation of eddies is shown in Fig. 8-27 [156].

von Bekésy's work served to confirm von Helmholtz's "place theory" for cochlear reception. In his book of 1953, Fletcher observed that after 1928 "any [cochlear] theory had to be consistent with the data of von Békésy [157]." That remark is borne out by the number of theories of the cochlea

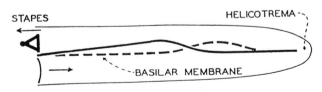

FIGURE 8-26. Diagram of a traveling wave set up on the basilar membrane by outward movement of the stapes. (Stevens and Davis [155] after the original work of von Békésy.)

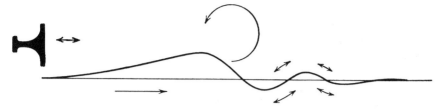

FIGURE 8-27. Vibration pattern of the basilar membrane and eddy effects in the surrounding fluid. (From von Békésy [156].)

developed in this period, all using the work of von Békésy as a starting point. Otto F. Ranke (1931) considered the cochlea as a tube with compliant walls, J.A. Reboul (1937) advanced the idea that the different frequencies varied in their stimulation of the basilar membrane with the low frequencies moving the membrane as a whole, while Jozef Zwislocki (1946), L.A. de Rosa (1947), L.C. Peterson and B.P. Bogert (1950), and W.H. Huggins (also 1950) added other refinements [158].

von Békésy was perhaps the first to observe eddy currents in the fluid of the cochlea. He studied these by way of another mechanical model, shown in Fig. 8-28 [159]. This idea was picked up by Zwislocki, who recognized the importance of hydrodynamics in understanding the problem and

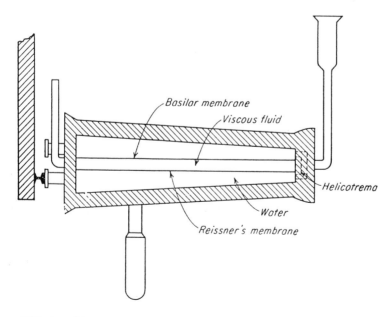

FIGURE 8-28. A cochlear model in which Reissner's membrane is represented. (From von Békésy [159].)

reasoned that the existence of such fluid motions would inevitably lead to nonlinearities. This had an important consequence; since von Helmholtz's time, it had been assumed that the inner ear was a linear system, and that any nonlinearities (such as those responsible for Tartini tones) were provided by the bones of the middle ear. Now the source of the nonlinearity could be found in the cochlea itself.

This nonlinear behavior of the cochlea was also observed in the work of Ernest Glen Wever (1902–1991), Charles W. Bray, and Merle Lawrence (1915–   ) [160]. In a series of experiments on the stimulation of animal ears by acoustic tones, they were able to record the electrical response of the cochlea to single tones and combinations of tones, and left no doubt that the combination tones reported by von Helmholtz found their origin in the cochlea itself.

In 1930 Wever and Bray had reported an electrical response of the ear of cats to auditory stimulation [161]. In this work, they thought that sound frequencies—as high as several thousand hertz—were directly transmitted over the auditory nerve to the brain, thus supporting the "telephone theory of hearing." This "Wever–Bray effect," as it became known, received severe criticism from E.D. Adrian, Detlev W. Bronk, and G. Philips [162], especially since it was well known at the time that nerve fibers elsewhere in the body could not transmit nerve impulses at frequencies above 1000 Hz. In 1935, however, Hallowell Davis [163] demonstrated, again on cats, that the Wever–Bray effect has two components. These components can be described as

(1) the cochlear microphonics, a specific ac response of the cochlea, syncrhronized with the acoustic signal, and (2) the action potentials of the cochlear nerve that, at low frequencies, are superposed on the microphonics [164].

In an attempt to determine the effects of sound on single fibers of the auditory nerve, Robert Galambos and Hallowell Davis reported on the response of such fibers to acoustic stimuli applied to the intact ear [165]. Their initial conclusion was that the spikes that they recorded were from single-fiber activity. Later, after some criticism and rethinking, they accepted the explanation that their recording electrodes may have been in contact with cells of the cochlear nucleus rather than with the cochlear nerve [166].

These ideas led to the conclusion that a single fiber could not handle impulse rates of thousands of impulses per second that could be experienced in sound waves. In his book of 1949, E.G. Wever wrote

We are forced to the conclusion that the high rates [of several thousand per second] are a result of many fibers acting in concert. This conception of auditory nerve action constitutes what has been called the volley principle [167].

An example of the volley principle and the effect of intensity was given by Wever (Fig. 8-29) [168].

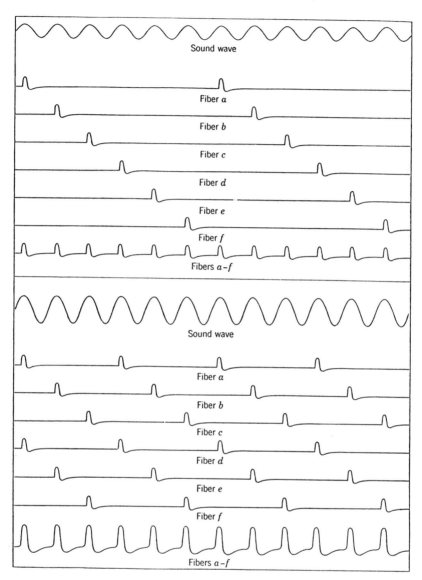

FIGURE 8-29. Intensity representation in the volley principle. Above, the pattern of discharge in a group of nerve fibers at one intensity; and below, this pattern at a higher intensity. The frequency is represented in both patterns, but the discharge is larger in magnitude in the second. (From E.G. Wever [168].)

8.15. Psychological Acoustics

By the 1930s, the minimum auditory thresholds had been well established. Figure 8-30 recapitulates the work of many observers [169]. The variation is perhaps due to the inadequate absorption of the walls of the room in which the measurements were made. Sivian and White give a picture of their "sound stage" (Fig. 8-31), which was probably the closest approach to an "anechoic chamber" as had been made by that time (1933). At the same time Sivian and White also found another peculiarity in the difference between the minimum audible pressure at the eardrum and the minimum audible field. In 1952, just after the period covered by this chapter, William Munson and Francis Wiener [170] followed up on this difference and found an average difference of 6 dB between the pressures required for detection when the source was an earphone and when it was a loudspeaker. The problem remained unresolved at that time. The work of Sivian and White also suggested that the limiting value of the minimum audible sound pressure was probably due to the various noises made by the human body—physiological noise [171].

The improved equipment of the period made it possible to develop audiometers for the measurement of hearing and hearing loss in individuals, and much information was gained about partial deafness. Figure 8-32 shows the results of measurements by C.C. Bunch [172] of a particular form

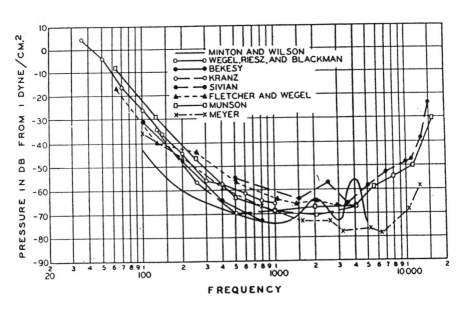

FIGURE 8-30. The minimum audible pressure at the eardrum as determined by various experimenters. (After Sivian and White [169].)

FIGURE 8-31. Sound stage and sound source, with observer in position. (From Sivian and White [169].)

of hearing loss characterstic of older people: the average loss of a number of older individuals relative to the normal hearing of people in the 20–30 year range. This dropoff in hearing at the higher frequencies is now known by the term presbyacusis [173].

Quantification of Psychoacoustics

The introduction of mathematical analysis into psychoacoustics through Fechner's law was the beginning of an ultimately successful "storming of the heights" by the quantifiers. Earl Schubert writes [174]

Boring (1942) remarked about Fechner's struggle with quantification of sensations that it is difficult for moderns to understand the depth of the philosophical schism that separates those convinced of the folly of the assignment of numbers to sensations and those who saw it as a fruitful endeavor. But at least for the last two decades [Schubert is writing in 1978] moderns in audition have been largely spared that struggle by the clear exposition of S.S. Stevens.

The area of psychoacoustics that includes such quantification—"the identification of dimensions central to auditory perception" [175]—is labeled

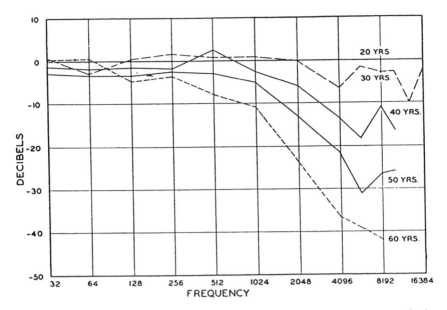

FIGURE 8-32. The audiograms of people at different age levels. The ordinate records the hearing loss in decibels, relative to the hearing of people whose ages lie between 20 and 30 years (zero ordinate). (After Bunch [172].)

classical psychoacoustics by Schubert. This includes the relation of loudness to signal amplitude and pitch to signal frequency, and was the center of attention of psychoacousticians in this period.

Loudness

The need for a unit for describing the subjective sensation of loudness became apparent in the 1930s. After several attempts at quantification by different Bell Laboratories researchers, H. Fletcher and W.A. Munson [176] proposed a *loudness level*, "obtained by adjusting the intensity level of a reference tone until it sounds equally loud as judged by a typical listener." The reference level was taken to be a pure tone at 1000 Hz. The curves that ultimately were produced by such measurements are shown in Fig. 8-33 [177]. In his book on hearing, Stevens observed "There is some possibility that the German word *phon* will come to be the accepted name for the loudness level [178]." And so it did.

However, Stevens was more directly responsible for two other units to be used for measurement of subjective quantities in hearing: the *sone* [179] for loudness, and the *mel* [180] for pitch. In the article just cited, Stevens pointed out that, while physicists regard pitch and frequency as the same ("Pitch is specified by the period or frequency of vibration," wrote one physicist in his textbook on sound [181]), experiments have shown that the

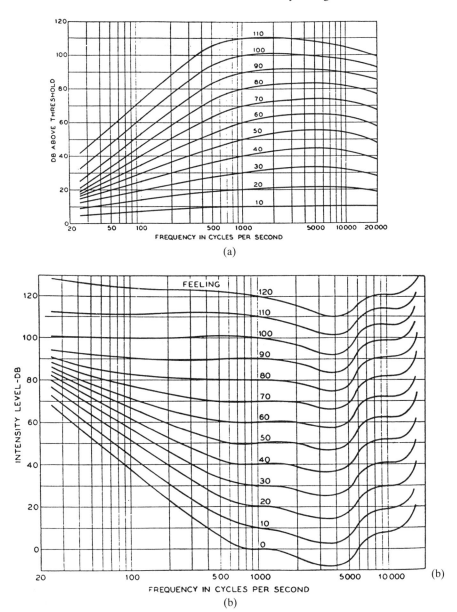

FIGURE 8-33. (a) Loudness level contours versus sensation level; and (b) loudness level contours versus intensity level. (From Fletcher [177].)

pitch of a particular frequency can be varied by increasing the sound intensity—increasing the pitch at high frequencies and lowering it at lower ones [182].

The sone, as the unit of subjective loudness, is closely related to the phon. In fact, they are connected by the relation $\log_{10} S = 0.0301P - 1.204$, where S is the loudness in sones and P the loudness level in phons. This law reflects Stevens preference for a power, rather than a logarithmic law [183].

Pitch

The definition of the mel is somewhat more complex. The pitch of a 1000-Hz tone at 40 dB above threshold is defined to be 1000 mels. A tone whose pitch is judged by a listener to be double that of the 1000-Hz tone is said to have a pitch of 2000 mels. Stevens and his colleagues were also able to conclude that the difference limens (thresholds) or DL for pitch are of equal subjective magnitude across the spectrum, while the DL's of loudness increased with increasing sound intensity [180], [182].

The nature and perception of pitch continued to attract researchers. It was known from earlier work of Fletcher [184] that the pitch of a complex tone, that consisted of harmonics of a certain tone but was lacking the fundamental, would still be identified by the listener as that of the missing fundamental. This was attributed to the production of the difference frequency between the second and third harmonics, formed within the ear itself—the so-called subjective tones of Tartini et al. This work led to a vigorous discussion between supporters of place theory in which each segment of the basilar membrane vibrates in resonance at a particular frequency (von Helmholtz, Fletcher, Stevens, and Davis) and those who saw that all parts of the cochlea respond to all frequencies, transmitting the sound waveform to the auditory nerve, so that "periodicity rather than spectral location could be the basis for pitch" [185] (Seebeck, Wever, and Bray).

J.F. Schouten gave support to the periodicity theory by his papers [186] on the perception of subjective tones and to the phenomenon of residue pitch—the residue being "the pitch that remains if one cancels out the difference-tone distortion that appears at the frequency of the fundamental, and which has a different perceptual quality [because of phase differences] than the fundamental sinusoid of the same pitch [187]." At about the same time, George A. Miller and Walter G. Taylor [187] demonstrated that subjects could match the pitch of either a sinusoidal wave or a square wave to that of the interruption rate of white noise, giving further support of the periodicity theory.

It might be noted that most of the attention paid historically to pitch and other hearing phenomena was to sounds of relatively long duration. The increased attention paid to temporal aspects of pitch in the 1940s led to a number of researches on signals of short duration. Earlier studies by P. Kucharski [189] and W. Bürck, P. Kotowski, and H. Lichte [190] were followed up by that of Doughty and Garner in 1947 [191]. In the latter paper, the view was taken that there is a certain duration threshold for pitch perception, dividing the responses into "tone pitch" and "click pitch." In a subsequent paper [192], these same two authors studied the duration thresholds for each of these phenomena.

Masking and Time Resolution

In the measurements on masking by Wegel and Lane (Chapter 7), peaks were noted at multiples of the masking tone. This was attributed to the presence of distortion products (combination tones) at these frequencies. The technique used in measurement was the so-called best-beat method [193], and was later employed by Fletcher [194] and von Békésy [195] for measuring nonlinear distortion in the ear. This difficulty of measurement was noted by James P. Egan and Harold W. Hake [196] and avoided in their experiments by using a band of noise "narrow enough to preserve the general shape of the tonal masking pattern but broad enough to minimize the effects of beats and difference tones [197]."

This formed the beginning of the study of critical bandwidth. In 1940, Fletcher used white noise (equal energy at all frequencies) for masking purposes [198]. He then reduced the band width of the noise until he reached a point at which the original sinusoidal signal, which is centered in the noise land, became perceptible. This was labeled the critical bandwidth. It amounted to about 50 Hz at a signal frequency of 1000 Hz.

Early measurements on the ability of the ear to separate two tone bursts separated in time were reported by Bürck, Kotowski, and Lichte. They arrived at a value of 10 ms for this minimum time [199]. By the use of pairs of clicks, F.J.J. Butendijk and A. Meesters found the time interval to be somewhat shorter [200], while Hans Wallach, Edwin B. Newman, and Mark R. Rosenzweig found that the minimum value depends somewhat on whether one measured decreasing or increasing time intervals during the experiment [201].

Localization and the Role of Binaural Hearing

The next research step on localization after Rayleigh occurred in 1936 when "Smitty" Stevens and E.B. Newman took to an elevated chair above the roof of a building at Harvard (Fig. 8-34) to measure localization in a roughly field-free medium—the open air of Cambridge [202]. They quickly found considerable uncertainty between sounds directly in front and directly behind the listener. When the front-back ambiguity was eliminated

FIGURE 8-34. Localization measurements from an outdoor "anechoic space" on top of Harvard. Note the absence of vertical reflecting surfaces at the level of the observer. (From Stevens and Newman [202].)

from consideration, the signals showed a systematic behavior, with the number of location errors rising with frequency, up to about 3000 Hz, and descending thereafter. In later times, David M. Green noted that the usual explanation of this is the duplex theory, which we shall discuss in Chapter 9 [203].

An oft-discovered effect, variously known as the Haas effect, the precedence effect, and the principle of the first wavefront, became a significant study at this time. Put briefly, the effect "states that within a certain time frame, the earlier-arriving signal (wavefront) will dominate over a later-arriving signal (e.g., echo) in determining what we hear [204]." The effect was reported in 1949 by H. Wallach, E.B. Newman, and M.R. Rozenzweig [205] and also by Helmut Haas [206], one of Erwin Meyer's students at Göttingen, but the trail can be traced all the way back to Joseph Henry [207].

Studies on binaural hearing suggested that the absolute sound thresholds, when measured binaurally, were about 3 dB lower than those for monaural listening [208].

8.16 Speech

Not only did the improved electronics of this period move forward the experimental study of speech, but they also contributed significantly to our understanding of the speech production process. The knowledge that radio signals consisted of a carrier wave of high frequency, modulated by low-frequency (audio) signals suggested that speech was in fact a similar combi-

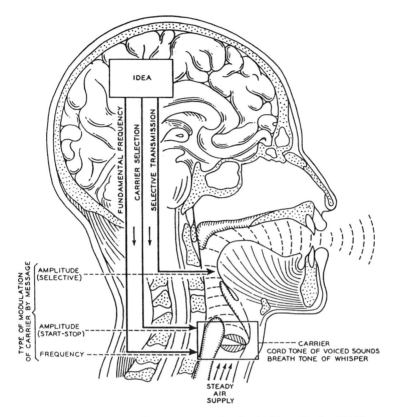

FIGURE 8-35. The vocal system as a carrier circuit. (From Dudley [210].)

nation of a basic carrier of the sound energy plus a shaping or modulating of that carrier by the vocal tract, the tongue, and the lips [209]. A pioneering effort in understanding this parallel and in applying it to speech synthesis was made by Homer Dudley in 1939 [210]. As suggested in Fig. 8-35, the brain operates the air supply and vocal cords to produce the carrier signal in a particular frequency range. It then controls the position of the tongue and the shape of the mouth and lips, so as to modulate the original sound. With this in mind, Dudley was able to develop an electrical system (Fig. 8-36) to simulate the human system and produce an artificial speech. This instrument, the Voder (Voice Operation DEmonstratoR), was first exhibited at the New York World's Fair, 1939. This research at the Bell Laboratories formed part of the laboratories mission in using *visible speech* patterns as an aid in teaching the deaf. As such, it was carrying on a mission dear to the hearts of both Melville and Alexander Bell (recall Chapter 5).

As we saw in the previous chapter, by 1925 it was possible to view the complicated waveform of speech sounds, especially if they were sustained

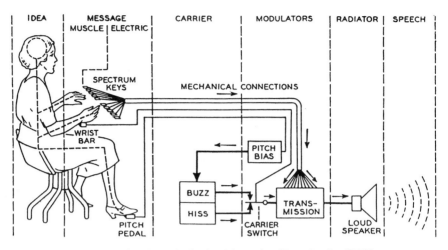

FIGURE 8-36. Schematic circuit of the voder. (From Dudley [210].)

for an interval of time. The problem of recording in detail the short-term sounds that are involved in ordinary speech was still unsolved. In 1930, R.K. Potter [211] developed the basis for the sound spectrograph, shown schematically in Fig. 8-37. The sounds could be recorded on a magnetic tape mounted on a cylinder. The tape can be played again and again, thus transforming a transient into a stationary signal, and the attached variable narrow bandpass filter could be adjusted for different frequencies. This

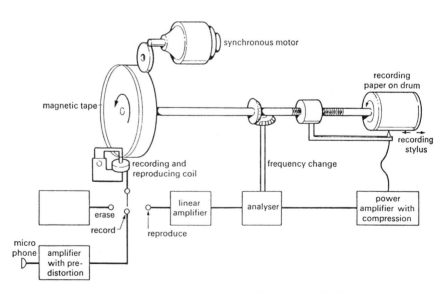

FIGURE 8-37. Sound spectrograph. (After Potter [211].)

FIGURE 8-38. Oscillogram: "Joe took father's shoe bench out." (From Fletcher [214].)

device was later used by Potter and others to determine the spectrum of speech sounds [212], [213]. An example of the phrase "Joe took father's shoe bench out" is plotted in Fig. 8-38 [214].

Speech comes to us from the vocal tract, and the physiology and physics of the system from lungs to the nasal passages, and ultimately, to the brain, have long been the center of study. Figure 8-39 shows a stylized form of these passages [215]. The air flow from the lungs is interrupted by the vocal cords, in the process known as *phonation*. As early as 1840, Johannes Müller [216] had conceived of speech production as the combination of the source (lungs plus vocal cords) and a filter (the various passages above the larynx, now referred to as the *supralaryngeal filter* [217]. These ideas were applied in this period by Chiba and Kajiyama [218], who took x-rays of the tract to determine the appropriate dimensions, and calculated the appropriate resonances for several vowels.

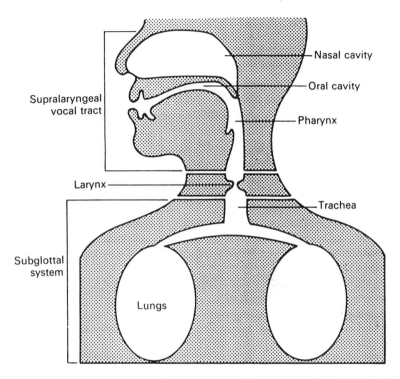

Supralaryngeal vocal tract

Nasal cavity

Oral cavity

Pharynx

Larynx

Trachea

Subglottal system

Lungs

FIGURE 8-39. The three physiological components of human speech production. (From Liebermann and Blumstein [217].)

The term *formant* had existed in music theory since early in the century [219] but it was not until the period around 1950 that it came into common use for speech analysis [220]. In 1950, H.K. Dunn wrote

It has been known for more than two hundred years that the different vowels have associated with them different frequency regions in which the sound is more intense than elsewhere in the spectrum. The name "formant" has been applied to these regions, and will be used in this paper. [221]

It was soon determined that there were several basic formant frequencies. Figure 8-40 (from Ref. [221]) shows the spectrograms (see below) of a male speaker in pronouncing a variety of vowels. The dark areas are the resonant bands of the formants. Two are visible in most cases, and a third is sometimes observed.

Speech scientists in the 1920s began the study of articulation testing. The classic paper on the subject at the time was due to Fletcher and J.C. Steinberg [222]. In 1947, N.R. French and Steinberg introduced the *articulation index (AI)*, whose magnitude "is taken to vary between zero and unity, the former applying to when the received speech is completely unintelligible, the latter to the condition of best intelligibility [223]." An exten-

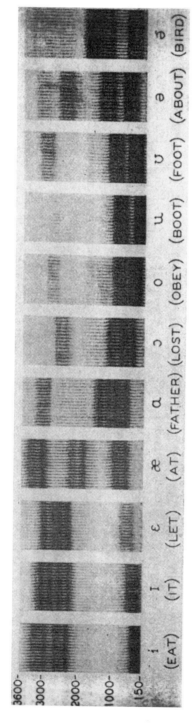

FIGURE 8-40. Sections from the spectrograms of a series of vowels, all from the same male speaker. A frequency scale in cycles per second is given at the left. The analyzing bandwidth used was 300 cycles per second. (From Dunn [221].)

sive description of methods of articulation testing was given shortly thereafter by James P. Egan [224]. These methods were used as a predictor of intelligibility for telephone systems. In particular, Beranek applied the methods of Fletcher and Steinberg to the design of speech communication systems [225].

The wartime use of signaling equipment under a wide variety of conditions, including signal distortion and noise, prompted a number of studies of these effects on speech intelligibility [226]. The loss of speech intelligibility by masking with a variety of extraneous sounds, including sine waves, square waves, and pulses, was studied by S.S. Stevens, Joseph Miller, and Ida Truscott [227] and, shortly thereafter, two important papers by Licklider and colleagues studied the effects of clipped [228] and interrupted [229] speech.

8.17 Music

A number of books on the physics and engineering aspects of music appeared in the period around 1950 [230]–[232]. In looking over their contents, one is struck with the fact that most of the subject matter (especially the texts by Barthomolew and Wood) have been covered in this history, not so much in the music sections, but throughout the text—in wave motion and vibrations, electroacoustics, architectural acoustics, speech and hearing, and so forth, demonstrating once again the fundamental unity of acoustics. The chief gaps have been those of detailed studies of musical instruments, and of the phenomenon of vibrato. The study of musical instruments is an important part of the book by Olson, and of a later book by Arthur Benade [233], while vibrato had been given a detailed analysis in a collection of papers edited by Carl E. Seashore [234].

One of the developments of this period was the use of mechanical and electrical equivalent circuits for different instruments. Figure 8-41 shows a schematic drawing of a struck string and its mechanical equivalent network [235], while Fig. 8-42 shows a schematic of an organ pipe along with electrical and acoustical equivalent networks [236].

The study of the vibrations of the violin came back into fashion in the 1930s with the work of Hermann Backhaus (1885–1958). Using a mechanical bowing device, he vibrated the instrument at its principal resonances, and, using capacitor pickup, he was able to determine nodal lines at various frequencies (Fig. 8-43) [237]. This type of research was continued in Germany by one of his students, Hermann F. Meinel (1904–1977) [238], who later worked with Erwin Meyer; and in the United States by Frederick A. Saunders (1875–1963), who published numerous studies of violins constructed by the great masters [239].

The significance of transient signals also received much attention [240]. The ability to making tape recordings now made possible the playing of music backward. As noted by Charles A. Taylor,

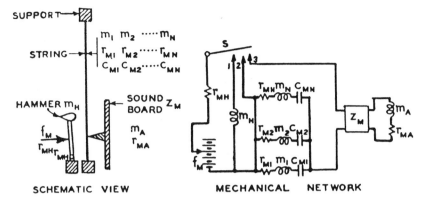

FIGURE 8-41. Schematic view and the mechanical network of a struck-string instrument. In the mechanical network: f_M = driving force; τ_{MH} = mechanical resistance of the generator; m_H = mass of the hammer; $m_1, c_{M1}, \tau_{M1}, m_2, c_{M2}, \tau_{M2}, \ldots, m_N, c_{MN}, \tau_{MN}$ = lumped masses, compliances, and mechanical resistance representing the string; m_A and τ_{MA} = mass and mechanical resistances of the air load; and z_M = quadripole representing the sound bopard which couples the string to the air. (From Olson [235].)

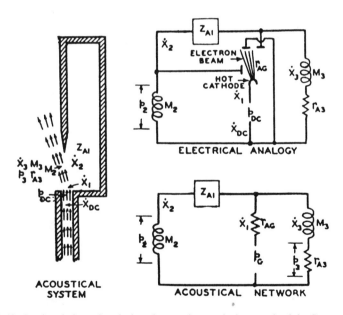

FIGURE 8-42. Sectional view, electrical analogy and acoustical network of the flue organ pipe. In the electrical analogy and acoustical network, p_{DC} = pressure in the direct-current air supply; X_{DC} = the direct volume current flow; τ_{AG} = acoustical resistance of the equivalent acoustical generator p_G = pressure of the equivalent acoustical generator; z_{A1} = acoustical impedance of the pipe; M_2 = inertance of the pipe opening; M_3 and τ_{A3} = inertance and acoustical resistance of the air load; p_2 = pressure drop across M_2; and X_1, X_2, and X_3 = alternating-volume currents. p_3 = pressure drop across τ_{a3} or the output pressure. (From Olson [236].)

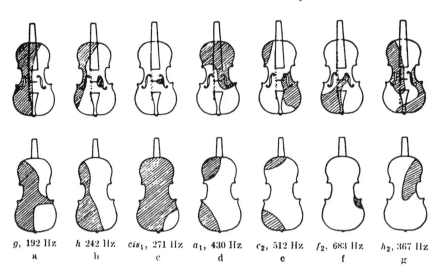

g, 192 Hz h 242 Hz cis_1, 271 Hz a_1, 430 Hz c_2, 512 Hz f_2, 683 Hz h_2, 367 Hz

a   b   c   d   e   f   g

FIGURE 8-43. Nodal lines on a violin. (From Backhaus [237].)

It is clear that any continuous sinusoidal components present will be unaffected by the reversal—that is, the harmonic content will remain the same. Any transient effects, however, will obviously occur in abnormal places, and hence will become obvious. [241]

This technique was first exploited by Erwin Meyer and G. Buchmann [242], and later by W.H. George [243]. George was earlier responsible for confirmation of Kaufmann's theory of the transient action of the hammer on the strings of a piano (Chapter 7) [244].

Vibrato is the variation of the pitch of a sung note above and below the basic pitch at a regular rate, "a vibrant, palpitating thrill in the voice; a pulsating quality which carries waves of feeling on it . . . giving a deviation from a perfectly steady tone [245]." The rate of these fluctuations is about $6-7 S^{-1}$. The availability of recording equipment in the 1920s made it possible to study the vibrato in some detail [246]. Figure 8-44 shows the vibrato for a note sung by Caruso from the aria "Favorita" (Spirit so Fair) [247]. A similar vibrato can be produced on stringed instruments [248].

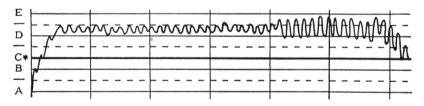

FIGURE 8-44. A vibrato-crescendo by Caruso. Victor Record No. 6005-A. (From M. Metfessl [247].)

Overview

This chapter has followed the research developments of the immediate post-World War II period. In 1950, one might have thought that acoustical science was proceeding at full tilt. But, as the next chapter will show, it was only the beginning. Enormous strides forward would be made in technology, in our understanding of phenomena of speech and hearing, fostered by rapid advances in various forms of signal processing. As so often has been the case, the past was only prologue.

Notes and References

[1] Harvey Fletcher, *Speech and Hearing*, Van Nostrand, New York, NY, 1929. *Speech and Hearing in Communication*, 2nd ed., Van Nostrand, New York, NY, 1953.

[2] I.J. Hirsh, *J. Acoust Soc. Am.* **68**, 46–52 (1980).

[3] Stanley Smith Stevens and Hallowell Davis, *Speech and Hearing*, Harvard University Press, Cambridge, MA, 1938. Reprinted by the Acoustical Society of America, Woodbury, NY, 1983. Stevens and Davis were two of the most significant investigators in physiogical and psychological acoustics in the mid-twentieth century. A graduate of Stanford, "Smitty" Stevens (1906–1973) took his Ph.D. at Harvard in 1933 and remained at that institution for the rest of his career, serving as Professor and Director of its Psychophysics Laboratory. Davis (1896–1992) did both his undergraduate and graduate studies at Harvard, receiving an M.D. in 1922. He remained at Harvard until after World War II, when he became Director of the Central Institute for the Deaf in St. Louis, and Research Professor of Otolaryngology at Washington University.

[4] I.J. Hirsh, loc. cit. One might remark that the *Shock and Vibration Handbook* (Cyril Harris, ed.), McGraw-Hill, New York, NY, 1986, had at least 60 authors, while the forthcoming *Encyclopedia of Acoustics* (Malcolm Crocker, ed.), has an untold number.

[5] Harvey Fletcher, *Oral History*, Niels Bohr Library, American Institute of Physics, College Park, MD.

[6] F.A. Firestone, *J. Acoust. Soc. Am.* **26**, 882 (1954).

[7] W. Pielemeier, *Phys. Rev.* **34**, 1184–1203 (1929).

[8] D.G. Bourgin, *Nature*, **122**, 133 (1928); *Phys. Rev.* **34**, 521–526 (1929); **42**, 721–730 (1932); **50**, 355–361 (1936); *Phil. Mag.* **7**, 821–840 (1929); *J. Acoust. Soc. Am.* **4**, 108–111 (1932); **5**, 57–59 (1934).

[9] K.F. Herzfeld and F.O. Rice, *Phys. Rev.* **31**, 691–695 (1928).

[10] A.J. Rutgers, *Ann. Physik*, **16**, 350–359 (1933).

[11] H.O. Kneser, *Handbuch der Physik* (S. Flügge, ed.), Springer-Verlag, Berlin, 1961, vol. XI/1.

[12] R. Kronig, *J. Opt. Soc. Am.* **12**, 547–557 (1926).

[13] H.A. Kramers, *Physik Z.* **30**, 522–523 (1929).

[14] J.J. Markham, R.T. Beyer, and R.B. Lindsay, *Rev. Mod. Phys.* **23**, 353–411 (1951).

[15] W. Pielemeier, *Phys. Rev.* **34**, 1184–1203 (1929); **35**, 1417 (A) (1930); **36**, 1005–1007 (1931).

[16] Paul E. Sabine, *J. Franklin Inst.* **207**, 341–368 (1929).

[17] E. Meyer, *Z. Techn. Physik* **7**, 253–259 (1930).

[18] V.O. Knudsen, *J. Acoust. Soc. Am.* **3**, 126–138 (1931). Knudsen, an early president of the ASA, was later Dean of the Graduate School and then Chancellor of UCLA.

[19] V.O. Knudsen, *J. Acoust. Soc. Am.* **3**, 126 (1931).

[20] H.O. Kneser and V.O. Knudsen, *Ann. Physik* **21**, 682–696 (1934).

[21] P. Biquard, *C. R. Acad. Sci. Paris* **188**, 1230–1234 (1929); **193**, 226–229 (1931); **196**, 257–259 (1933); **197**, 309–311 (1933).

[22] P. Bazhulin, *Phys. Z. Sowjetunion* **8**, 354–358 (1935).

[23] B. Spakovskij, *C. R. Acad. Sci. Leningrad*, **3**, 588 (1934).

[24] H.A. Lorentz, *Ann. Physik* **9**, 641–665 (1880); L. Lorenz, *Ann. Physik* **11**, 70–103 (1880). The relation between this law and the earlier law of Clausius–Mosostti is discussed in P. Debye *Polar Molecules*, reprinted by Dover, New York, NY, c. 1950, pp. 11–14.

[25] P.P. Debye and F.W. Sears, *Proc. Nat. Acad. Sci. Washington* **18**, 410–414 (1932). The effect is known as the Debye–Sears effect.

[26] P. Biquard and R. Lucas, *C. R. Acad. Sci. Paris* **194**, 2132–2134 (1932); **195**, 121–123 (1932); *Rev. Acoustique* **3**, 198 (1934).

[27] C.V. Raman and N.S. Nagendra Nath, *Proc. Indian Acad. Sci.* **2**, 406–411, 413–420 (1935); **3**, 75–78, 119–125, 459–465 (1936).

[28] O. Nomoto, *Proc. Phys. Math. Soc. Japan* **22**, 314–319 (1940).

[29] R.T. Beyer and S.V. Letcher, *Physical Ultrasonics*, Academic Press, New York, NY, 1969, p. 81.

[30] J.R. Pellam and J.K. Galt, *J. Chem. Phys.*, **14**, 608–614 (1946).

[31] J.M.M. Pinkerton, *Nature*, **160**, 128–129 (1947). The graph is taken from J.J. Markham, R.T. Beyer, and R.B. Lindsay, *Rev. Mod. Phys.*, **23**, 353–411 (1951). Other data on the graph are from E. Baumgardt, *C. R. Acad. Sci. Paris* **202**, 203–206 (1936); F.E. Fox and G.D. Rock *Phys. Rev.* **70**, 68–73 (1946); M.C. Smith and R.T. Beyer, *J. Acoust. Soc. Am.* **20**, 608–610 (1948).

[32] Rayleigh, *Theory of Sound*, vol. 2, p. 162. Reprinted by Dover, New York, NY, 1946.

[33] P.N.T. Wells, *Physical Principles of Ultrasonic Diagnosis*, Academic Press, London, UK, 1969, p. 54.

[34] A.O. Williams, Jr., *J. Acoust. Soc. Am.* **23**, 1–6 (1951).

[35] H. Seki, A.V. Granato, and R. Truell, *J. Acoust. Soc. Am.* **28**, 230–238 (1951).

[36] L. Hall, *Phys. Rev.* **73**, 775–781 (1948).

[37] A general discussion of Hall's work and that of others on structural relaxation is given by A.B. Bhatia, *Ultrasonic Absorption*, Oxford University Press, Oxford, UK, 1967. Reprinted by Dover, New York, NY, 1985, Chapter 10.

[38] A group of distinguished scientists under the leadership of L. Rosenhead, discussed the issues of first and second viscosity at a conference in Britain in 1954, but with little in the way of success [L. Rosenhead, *Proc. Roy. Soc. (London)* **226A**, 1–69 (1954)].

[39] T.A. Litovitz and E. Carnevale, *J. Appl. Phys.* **26**, 816–820 (1955).

[40] M.R. Rao, *Indian J. Phys.* **14**, 109–115 (1940).

[41] The quantity R was called the "molecular sound velocity" by R.T. Lageman and W.S. Dunbar, *J. Phys. Chem.* **49**, 429–436 (1945), although it is clearly not a velocity.

[42] A review of the work on Rao's rule in this period is found in W. Schaaffs, *Ergeb. Exakt. Naturw.* **25**, 109 (1951).

[43] Y. Wada, *J. Phys. Soc. Japan* **4**, 280–282 (1949).

[44] A. Meissner, *Z. Tekhn. Physik* **7**, 585–592 (1926).

[45] L.N. Liebermann, *Phys. Rev.* **75**, 1415–1422 (1949).

[46] C. Eckart, *Phys. Rev.* **73**, 68–76 (1948).

[47] Rayleigh, *Phil. Mag.* **3**, 338–346 (1902). A later summary of this work appears in R.T. Beyer, *J. Acoust. Soc. Am.* **98**, 3021–3024 (1995).

[48] L. Brillouin, *Ann. Physik* (X) **4**, 528–586 (1925).

[49] Langevin lectured on this subject in 1923, and much of his material was reproduced in the papers by P. Biquard, *Rev. Acoustique* **1**, 93–109 (1932); **2**, 315–335 (1933).

[50] G. Hertz and H. Mende, *Z. Physik* **114**, 354–367 (1939). A review of this work appears in R.T. Beyer, *Am. J. Phys.* **18**, 25–29 (1950).

[51] J.K. Galt, *J. Chem. Phys.* **16**, 505–507 (1948).

[52] K.R. Atkins, *Liquid Helium*, Cambridge University Press, London, UK, 1959, p. 132.

[53] V.P. Peshkov, *J. Phys. USSR* **8**, 381. (1944). A discussion of these phenomena can be found in R.T. Beyer and S.V. Letcher, *Physical Ultrasonics*, Academic Press, New York, NY, 1968. A detailed study of liquid helium is given in J. Wilks, *The Properties of Liquid and Solid Helium*, Oxford University Press, London, UK, 1967.

[54] If acoustics had restricted itself to sound in the narrow sense of 'things that can be heard," we would have lost at least half the field. Fortunately, acoustics has always followed the path of inclusion.

[55] L.D. Landau, *J. Phys. USSR* **5**, 71–90 (1941); **11**, 91–92 (1944).

[56] H. Lamb, *Dynamical Theory of Sound*, E. Arnold, London, UK, 1925, p. 125.

[57] L. Bergmann, *Ultrasonics*, G. Bell, London, UK, 1938, 161ff. This is the English translation of the first German edition (1937). The book has had numerous later editions, of which the third (1942) and the sixth (1954) are the most significant. Unfortunately, the later editions were not translated into English.

[58] The many papers of Schaefer and Bergmann from the 1930s are listed in the various editions of Ref. [57].

[59] See W.P. Mason, *Electrical Transducers and Wave Filters*, Van Nostrand, New York, NY, 1942; 2nd ed., 1948. See also W.P. Mason, *Physical Acoustics and the Properties of Solids*, Van Nostrand, New York, NY, 1958. When Mason retired from the Bell Telephone Laboratories, he held the record for the most patents (190) issued to a single employee of that very patent-conscious organization.

[60] W.P. Mason, *Physical Acoustics and the Properties of Solids*, Chapter 4; W.P. Mason, W.O. Baker, H.J. McSkimin, and J.H. Heiss, *Phys. Rev.* **75**, 936–946 (1949); H.J. McSkimin, *J. Acoust. Soc. Am.* **22**, 413–418 (1950).

[61] W.P. Mason, *Physical Acoustics and the Properties of Solids*, p. 101.

[62] W.P. Mason, *Physical Acoustics and the Properties of Solids*, p. 102.

[63] A.G. Emslie and R.L. McConnell, *Radar System Engineering*, McGraw-Hill, New York, NY, 1946, Vol. 1, Chapter 16. Dvaid L. Arenberg, *J. Acoust. Soc. Am.* **20**, 1–26 (1948).

[64] H. Freundlich, K. Söllner, and F. Rogowski, *Klin. Wochenschrift* **11**, 1512 (1932).

[65] A. Dognon and E. and H. Biancani, *Ultrasons et Biologie*, Gauthier-Villars, Paris, 1937.

[66] Further work in this area was reported by E.J. Baldes, P.A. Nelson, and J.F. Herrick, *J. Acoust. Soc. Am.* **22**, 682 (A) (1950). P.A. Nelson, J.F. Herrick, and F.H. Krusen, *Arch. Phys. Med.* **31**, 6, 687–695 (1950).

[67] J.J. Wild, *Surgery* **23**, 27, 183–188 (1950).

[68] L.A. French, J.J. Wild, and D. Neal, *Cancer* **3**, 705–706 (1950).

[69] O. Mühlhäuser, German patent No. 569598, 1931.

[70] S. Sokoloff, *Phys. Z.* **36**, 142–144 (1935); *Tech. Phys. USSR* **2**, 522 (1934).

[71] A survey of the state of material testing in 1954 is given by L. Bergmann, *Der Ultraschall*, 6th ed., S. Hirzel, Stuttgart, 1954, pp. 709–761.

[72] Floyd A. Firestone, American patent Nos. 2,280,226 (1942); 2,592,134, 2,592,135, and 2,601,779 (1945); *J. Acoust. Soc. Am.* **17**, 287–299 (1946). Floyd A. Firestone and Julian R. Frederick, *J. Acoust. Soc. Am.* **18**, 200–211 (1946).

[73] L. Bergmann, *Der Ultraschall*, p. 731.

[74] R. Pohlman, *Die Technik* **3**, 465–470 (1948). See also L. Bergmann, *Der Ultraschall*, pp. 754–760.

[75] R.W. Wood and A.L. Loomis, *Phys. Rev.* **29**, 373 (A) (1927); *Phil. Mag.* **4**, 417–436 (1927).

[76] Neda Marinesco, *C. R. Acad. Sci. Paris* **196**, 346–348 (1933).

[77] B. Claus, *Z. Techn. Phys.* **16**, 80–82 (1935). Others who worked in this very international field included S.N. Rschevkin and E.P. Ostrowsky, *Acta Physicochem. URSS* **1**, 741 (1935) and Naoyasu Sata and Seiito Watanabe, *Kolloid-Z.* **73**, 50–57 (1935).

[78] O. Brandt, *Z. Phys. Chem. Unterr.* **50**, 1 (1937).

[79] R.D. Fay, *J. Acoust. Soc. Am.* **3**, 222–241 (1931). This paper is discussed in considerable detail in R.B. Lindsay, *Mechanical Radiation*, McGraw-Hill, New York, NY, 1960, pp. 279–285.

[80] E. Fubini-Ghiron, *Alta Freq.* **4**, 532–581 (1935). A nearly complete translation of this paper appears in *Benchmark/19*, pp. 118–177. Fubini-Ghiron later shortened his name to Fubini.

[81] The mathematical formulation is

$$\frac{u}{u_0} = 2\sum_{n=1}^{\infty} J_n \left[\frac{J_m(nx/l)}{nx/l} \right] \sin n(\omega t - kx),$$

where u is the particle velocity, n the number of the harmonic, l the so-called discontinuity distance, and k the wave number: $1/l = (B/A + 1)Mk$, where A, B are coefficients in the expansion of pressure changes in terms of density changes (see Chapter 7). M = acoustical Mach number = u_0/c.

[82] W. Keck and R.T. Beyer, *Phys. Fluids* **3**, 346–352 (1980).

[83] D.T. Blackstock, *J. Acoust. Soc. Am.* **34**, 9–30 (1962).

[84] A.L. Thuras, R.T. Jenkins, and H.T. O'Neil, *J. Acoust. Soc. Am.* **6**, 173–180 (1935).

[85] A.M. Obukhov, *Dokl. Akad. Nauk Georgian SSR Nos.* **4–5**, 453 (1941).

[86] A.N. Kolmogorov, *Dokl. Akad. Nauk SSSR* **30**, 301 (1941).

[87] G.K. Batchelor, *The Theory of Homogeneous Turbulence*, Cambridge University Press, Cambridge, UK, 1953, Introduction and Chapter 5.

[88] D.I. Blokhintzev, *Acoustics of a Moving Inhomogeneous Medium*, Gos. Izd. Tekhn. Teoret. Lit-Ry. Moscow, 1946. (English translation by R.T. Beyer and D. Mintzer, Brown University, Providence, RI, 1952.)

[89] D.I. Blokhintzev, *Acoustics of a Moving Inhomogeneous Medium*, p. 48ff.

[90] D.I. Blokhintzev, *J. Acoust. Soc. Am.* **18**, 332–334 (1946).

[91] P.G. Bergmann, *J. Acoust. Soc. Am.* **17**, 329–333 (1946).

[92] P.G. Bergmann, *Phys. Rev.* **70**, 486–492 (1946).

[93] D.I. Blokhintzev, *Acoustics of a Moving Inhomogeneous Medium*, Chapter 4. An exhaustive study of Russian work in aeroacoustics has been provided by Alan Powell, *J. Acoust. Soc. Am.* **92**, 41–56 (1992), which includes an extensive bibliography.

[94] J.M. Lighthill, *Proc. Roy. Soc. (London)* **A211**, 524–547 (1952); **A222**, 1–32 (1954).

[95] E.Y. Yudin, *Z. Tekhn. Fiz.* **14**, 561–567 (1944). A discussion of Yudin's work may be found in Alan Powell, *J. Acoust. Soc. Am.* **36**, 177–195 (1964). In this paper, Powell recalls watching a locomotive blow off steam and being impressed by the fact that "each time that a particularly large eddy formed on the edge of the turbulent steam jet he heard a definite impulsive sound. This focused attention on the idea that the origin of aerodynamic sound might be attributed to the *process of formation of eddies or vortices* Out of such small and ordinary observations, great ideas still come!

[96] R.J. Urick, *Principles of Underwater Sound for Engineers*, McGraw-Hill, New York, NY, 1947, p. 62.

[97] The name SONAR is attributed to Professor F.V. Hunt at Harvard.

[98] F.G. Blake, *Harvard Univ. Acoust. Res. Lab. Tech. Mem.* **9** (1949).

[99] E.B. Stephenson, US Naval Res. Lab. Rept S-1204 (1935); S-1546, 1939.

[100] P. Biquard and R. Lucas, *C.R. Acad. Sci. Paris* **193**, 226–229 (1931).

[101] Leonard Liebermann, *J. Acoust. Soc. Am.* **20**, 868–873 (1948).

[102] G.G. Stokes, *Trans. Cambridge Philos. Soc.* **8**, 287 (1845).

[103] G. Kirchhoff, *Ann. Physik* **134**, 177 (1868).

[104] *Physics of Sound in the Sea*, Summary technical report of Division 6, NDRC, vol. 8, Washington, DC, 1946. Reprinted by the Department of the Navy, HDQ Naval Material Command, Washington, DC, 20360, 1969, page 105. Graph references include: (13) Carl Eckart, NDRC 6.1-sr30-1532, Report U-236, Project NS-140, UCDWR, July 6, 1944; (14) Report U-307, Nobs-2074, Sonar Data Division, CUDWR, Mar. 16, 1945; (16) E.B. Stephenson, Report S-1466, NRL, Aug. 12, 1938; (18) F.A. Everest and H.T. O'Neil, NDRC C4-sr30-494 UCDWR, revised July 30, 1942; (19) G.J. Thiessen, OSRD Liaison Office III-I-830, Report PS-162, CNRC, June 10, 1943; (20) F.E. Fox and G. Rock, *J. Acoust. Soc. Am.* **12**, 505–510 (1941); (22) G.W. Willard, *J. Acoust. Soc. Am.* **12**, 438–448 (1940); (23) P. Biquard and R. Lucas, Ref. [100] above. This volume contains a wealth of information about the state of underwater sound just after World War II, written by a number of outstanding scientists who had been connected with the classified wartime work.

[105] L. Liebermann, Ref. [101, p. 873].

[106] R.W. Leonard, P.C. Combs, and L.R. Skidmore, *J. Acoust. Soc. Am.* **21**, 63 (A) (1949).

[107] Maurice Ewing and J. Lamar Worzel, in *Propagation of Sound in the Ocean*, The Geological Society of America, Waverly Press, Baltimore, MD, 1948, Memoir 27, Part III, pp. 1–35.

[108] Maurice Ewing and J. Lamar Worzel, *Propagation of Sound in the Ocean*, p. 19.

[109] L.M. Brekhovskikh, *Dokl. Akad. Nauk SSSR* **62**, 469 (1948); **69**, 157 (1949). L.D. Rosenberg, *Dokl. Akad. Nauk. SSSR* **69**, 174 (1949). These papers were translated into English by the author of this text for the Research Analysis Group at Brown University, Providence, RI. He still has a copy of the translations, since they were among the first papers that he ever translated from the Russian.

[110] *Physics of Sound in the Sea*, p. 57. The concept of a layered medium was later popularized by W. Maurice Ewing, Wenceslas S. Jardetzky, and Frank Press (1924–   ) in their book, *Elastic Waves in Layered Media*, McGraw-Hill, New York, NY, 1957, and by Brekhovskikh in his book, *Waves in Layered Media*, Akad. Nauk. Moscow, 1957. English translation by David Liebermann, Academic Press, New York, NY, 1960; 2nd ed., Moscow. English translation by R.T. Beyer, Academic Press, New York, NY, 1980.

[111] C.L. Pekeris, *J. Appl. Phys.* **17**, 678–684, 1108–1124 (1946); *J. Acoust. Soc. Am.* **18**, 295–315 (1946); *Proc. I.R.E.* **35**, 453 (1947); in *Propagation of Sound in the Ocean*, Memoir 27, The Geological Society of America, Waverly Press, Baltimore, MD, 1948.

[112] L.M. Brekhovskikh, *Waves in Layered Media*, Chapter 5.

[113] From a paper given by D.T. Blackstock (private communication).

[114] The mathematical analysis of this subject is well treated in *Physics of Sound in the Sea*, Chapters 12–15, and by R.J. Urick, op. cit., Chapter 8.

[115] V.O. Knudsen, R.S. Alford, and J.W. Emling, *J. Mar. Res.* **7**, 410–429 (1948).

[116] R.-G. Busnel, ed., *Acoustic Behavior of Animals*, Elsevier, New York, NY, 1963.

[117] *Symposium on Marine Bioacoustics*, Macmillan, New York, NY, 1966, vol. 2, 1967.

[118] A summary of the work of Griffin, Pierce, and Galambos may be found in D.R. Griffin, *Listening in the Dark*, Yale University Press, New Haven, CT, 1958, especially Chapter 3. Another view of this early work is given by A.D. Grinnell in the dedication (to D.R. Griffin) of *Animal Sonar Systems* (R.-G. Busnel and J.F. Fish, eds.), Plenum Press, New York, NY, 1979, pp. xix–xxiv. The author of this current history (R.T.B.) recalls with fondness his joining with Don Griffin in a bat-hunting expedition in the attic of a rural schoolhouse in Ithaca, New York, in 1950. The yield was slight; we caught one bat.

[119] G.W. Pierce and D.R. Griffin, *J. Mammal.* **19**, 454–455 (1938).

[120] D.R. Griffin, *Listening in the Dark*, opposite, p. 90.

[121] D.R. Griffin, *J. Acoust. Soc. Am.* **22**, 247–255 (1950).

[122] R. Galambos and D.R. Griffin, *Anat. Rec.* **78**, 96 (1940); *J. Exp. Zool.* **86**, 481–506 (1941); **89**, 475–490 (1942). See also D.R. Griffin, *Sci. Amer.* (March 1954).

[123] D.R. Griffin, *Listening in the Dark*, pp. 69–80.

[124] R. Galambos, *J. Acoust. Soc. Am.* **14**, 41–49 (1941); *Physiolog. Rev.* **34**, 497–528 (1542); R. Galambos and H. Davis, *J. Neurophys.* **6**, 39–58 (1943).

[125] G. Duffing, *Erzwungene Schwingungen veränderlicher Eigenfrequenz*, Braunschweig, 1918.

[126] M. Rauscher, *J. Appl. Mech.* **20**, 237 (1938). A good discussion of this and the next few references is given in *Shock and Vibration Handbook* (Cyril M. Harris, ed.), 3rd ed., McGraw-Hill, New York, NY, 1988, pp. 4-12–4-19.

[127] G.F. Carrier, *Quart. Appl. Math.* **3**, 157–165 (1945).

[128] *Benchmark*/8, p. 293.

[129] N. Kryloff and N. Bogoliuboff, *Introduction to Nonlinear Mechanics*, Princeton University Press, Princeton, NJ, 1949.

[130] A.A. Andronow and C.E. Chaikin (Khaikin), *Theory of Oscillations*, Princeton University Press, Princeton, NJ, 1949. In one of the saddest remarks that can be made, Khaikin, in the preface to the second Russian edition of this book, paid belated and poignant tribute to his coauthor:

> The writer of this Preface is the only one of the three authors of this book who is still alive. Aleksandr Adol'fovich Vitt, who took part in the writing of the first edition of this book equally with the other two authors, but who by an unfortunate mistake was not included on the title page as one of the authors, died in 1937. (A.A. Andronov, A.A. Vitt, and S.E. Khaikin, *Theory of Oscillations*, Pergamon Press, Oxford, UK, 1968, Introduction.)

> In other words, Vitt had been arrested during the great purges in the Soviet Union in 1937 (thus becoming an nonperson), and was executed in prison during the same year, but nothing could be said about it at the time.

[131] R.N. Arnold and G.B. Warburton, *Proc. Roy. Soc. (London)* **197A**, 138–156 (1949).

[132] It has been said that half of the acousticians of the world spend their time making louder sounds (noises?) while the other half try to quiet them down. As the old saying goes, we work both sides of the street.

[133] Harvey Fletcher, *Speech and Hearing*, pp. 102–105. See also E.E. Free, *J. Acoust. Soc. Am.* **2**, 18–29 (1930).

[134] Rogers H. Galt, *J. Acoust. Soc. Am.* **2**, 30–58 (1930); Rexford S. Tucker, *J. Acoust. Soc. Am.* **2**, 39–64 (1930); John S. Parkinson, *J. Acoust, Soc. Am.* **2**, 65–74 (1930). It is of interest to note that the first papers at the third meeting of the Acoustical Society of America (which became the first five papers in vol. 2 of the Journal) were all devoted to noise and its measurement.

[135] A.H. Davis, *Modern Acoustics*, G. Bell, London, UK, 1934, p. 271.

[136] Donald A. Laird and Kenneth Coye, *J. Acoust. Soc. Am.* **1**, 158–163 (1929); Donald A. Laird, *J. Acoust. Soc. Am.* **1**, 256–262 (1930) (this paper contains over 30 references to research on the effects of noise); E.L. Smith and D.A. Laird, *J. Acoust. Soc. Am.* **2**, 94–98 (1930); W.G. King and D.A. Laird, *J. Acoust. Soc. Am.* **2**, 99–102 (1930). See also A.H. Davis, *Modern Acoustics*, pp. 260–261.

[137] V.O. Knudsen, *J. Acoust. Soc. Am.* **1**, 56–78 (1929). This was only the second technical paper to appear in the *Journal of the Acoustical Society of America*, and is believed to be the first paper anywhere outside of the Bell system in which the word "decibel" was used.

[138] F.R. Watson, *Acoustics of Buildings*, Wiley, New York, NY, 1930; *J. Franklin Inst.* (July 1924).

[139] D.F. Seacord, *J. Acoust. Soc. Am.* **12**, 183–187 (1940).

[140] V.O. Knudsen, *J. Acoust. Soc. Am.* **21**, 296–301 (1949).

[141] Horace, *Epist*, ii, 2. The translation cited by Knudsen runs as follows:

> The hot tempered contractor is hurrying about with his carriers and mules;
> A mighty machine turns here a stone, lifts there a wooden beam.
> Mournful funerals contend with heavy wagons (to see which makes more
>   noise);
> A mad bitch flees over that way, a filthy sow wallows around here.
> But now, along with you; I am resolved to meditate on my songs.

[142] V.O. Knudsen, Ref. [140, p. 297].

[143] S. Lifshitz, *Phys. Rev.* **27**, 618–621 (1925).

[144] V.O. Knudsen and C.M. Harris, *Acoustic Designing in Architecture*, Wiley, New York, NY, 1950, p. 173.

[145] P.H. Morse and R.H. Bolt, *Rev. Mod. Phys.* **16**, 69–150 (1944).

[146] Philip H. Morse, *Vibration and Sound*, McGraw-Hill, New York, NY, 1936; 2nd ed., 1948. Reprint of 2nd edition by the Acoustical Society of America, Woodbury, NY, 1981. Also Philip H. Morse and Herman Feshbach, *Methods of Theoretical Physics*, (2 vols.), McGraw-Hill, New York, NY, 1953; Philip H. Morse and K. Uno Ingard, *Theoretical Acoustics*, McGraw-Hill, New York, NY, 1968. Bolt was the director of the Acoustics Laboratory at MIT after World War II and was later a cofounder of the consulting firm of Bolt, Beranek, and Newman.

[147] *Benchmark*/10, p. 117.

[148] V.O. Knudsen and C.M. Harris, *Acoustic Designing in Architecture*, 1950. Reprinted with corrections and a new Foreword by Cyril Harris, Acoustical Society of America, Woodbury, NY, 1978. The acoustics community owes a debt of gratitude to Cyril Harris, first, for co-authoring the book, then for making the text available to that Society for reprinting, and finally, for having inspired the beginning of the reprint series of the Acoustical Society, which now totals more than 20 volumes of famous acoustics texts.

[149] It is amusing to contemplate that, at the same time, the underwater sound community was concentrating its attention on the psychological problems of underwater sound detection, and largely ignoring the problems of signal processing. In 1950, the two communities seem to have passed one another, traveling in opposite directions! (See Chapter 9.)

[150] The remarkable volume, *Experiments in Hearing*, by Georg von Békésy, McGraw-Hill, New York, NY, 1960 (reprinted by the Acoustical Society of America, Woodbury, NY, 1989), contains the most significant portions of von Békésy's publications up to 1960 with the German-language articles translated by E.G. Wever, and all edited by Wever for uniformity of style.

[151] G. von Békésy, *Experiments in Hearing*, p. 403.

[152] G. von Békésy, *Experiments in Hearing*, p. 407.

[153] It was shown in a later period (Chapter 9) that the behavior of the cochlea of a cadaver is somewhat different from one in a living animal.

[154] G. von Békésy, *Acta Otolaryngol.* **32**, 60–84 (1944). The drawing is reproduced in Wever and Lawrence, *Psychological Acoustics*, p. 253.

[155] G. von Békésy, *Phys. Z.* **34**, 577–582 (1933). The drawing is reproduced in Stevens and Davis, *Hearing*, p. 279.

[156] G. von Békésy, *Experiments in Hearing*, p. 422.

[157] Quoted by J. Tonndorf, in *Benchmark*/15, p. 102. The second edition of Fletcher's book, *Speech and Hearing* contains a wealth of detail on the work of Fletcher that was based on von Békésy's findings.

[158] The information here is too extensive to fit into this book. The specific references are O.F. Ranke, *Gleichrichte Resonanzteorie*, Munich, 1932; J.A. Reboul, *Le Phénomène de Wever et Bray*, Montpelier, 1937; *J. Zwislocki, Experientia* **2**, 415–1917 (1946); L.A. De Rosa, *J. Acoust Soc. Am.* **19**, 621–628 (1947).

[159] G. von Békésy, *Experiments in Hearing*, p. 423.

[160] E.G. Wever, C.W. Bray, and M. Lawrence, *J. Acoust. Soc. Am.* **11**, 427–433 (1940); *J. Exper. Psychol.* **27**, 469–496 (1940). See also E.G. Wever and M. Lawrence, *Physiological Acoustics*, Princeton University Press, Princeton, NJ, 1954, pp. 133–176.

[161] E.G. Wever and C.W. Bray, *Proc. Nat. Acad. Sci.* **16**, 344–350 (1930).

[162] E.D. Adrian, D.W. Bronk, and G. Philips, *J. Physiol.* **73**, 2P–3P (1931). See also E.D. Adrian, *J. Physiol.* **71**, xxviii–xxix (1931).

[163] H. Davis, *J. Acoust. Soc. Am.* **6**, 205–215 (1935).

[164] *Benchmark*/15, p. 154.

[165] R. Galambos and H. Davis, *J. Neurophysiol.* **6**, 39–57 (1943).

[166] R. Galambos and H. Davis, *Science*, **108**, 513. (1948). The problem, however, was revived in the 1960s (see Chapter 9).

[167] E.G. Wever, *Theory of Hearing*, Wiley, New York, NY, 1949, p. 165.

[168] E.G. Wever, *Theory of Hearing*, p. 173.

[169] The graph is taken from L.J. Sivian and S.D. White, *J. Acoust. Soc. Am.* **4**, 288–321 (1933). The individual results are those of John P. Minton and J. Gordon Wilson, *Proc. Nat. Acad. Sci.* **9**, 269–278 (1923); R.L. Wegel, R.R. Riesz, and R.B. Blackman, *J. Acoust. Soc. Am.* **4**, 6 (A) (1932); G. von Békésy, *Ann. Physik* **13**, 111–136 (1932); F.W. Franz, *Phys. Rev.* **21**, 573–584 (1923); L.J. Sivian, previously unpublished, Bell Telephone Laboratories, Sept. 1928; H. Fletcher and R.L. Wegel, *Phys. Rev.* **19**, 553–565 (1922); W. Munson, previously unpublished, Bell Telephone Laboratories, July 1932; E. Meyer, *Z. f. Hals, Nasen u. Ohrenheilkunde*, p. 443 (1930).

[170] W.A. Munson and F.M. Wiener, *J. Acoust. Soc. Am.* **24**, 498–501 (1952).

[171] Schubert attributes these noises to "blood flow, breathing, implicit muscle activity," plus the noise level of the first-order auditory (eighth nerve) neurons" (*Benchmark*/13, p. 9). As W.L. Brogden and G.A. Miller put it, "a human being is a relatively noisy organism." *J. Acoust. Soc. Am.* **19**, 620–623 (1947).

[172] C.C. Bunch, *Arch. Otolaryngol.* **9**, 625–636 (1929). The drawing is reproduced in Stevens and Davis, *Hearing*, p. 68.

[173] From the Greek "presby," meaning elder and "acusis," meaning hearing.

[174] *Benchmark*/13, p. 60.

[175] *Benchmark*/13, p. 61.

[176] H. Fletcher and W.A. Munson, *J. Acoust. Soc. Am.* **5**, 82–108 (1933).

[177] H. Fletcher, *Speech and Hearing in Communication*, 2nd ed., Van Nostrand, New York, NY, 1953, p. 188.

[178] S.S. Stevens and H. Davis, *Hearing: Its Psychology and Physiology*, p. 111.

[179] S.S. Stevens, *Psychol. Rev.* **43**, 405–416 (1936).

[180] S.S. Stevens, J. Volkmann, and E.B. Newman, *J. Acoust. Soc. Am.* **8**, 185–190 (1937).

[181] E.H. Barton, *A Textbook on Sound*, Macmillan, London, UK, 1914, p. 9.

[182] S.S. Stevens, J. Volkmann, and E.B. Newman, *J. Acoust. Soc. Am.* **8**, 185–190 (1937).

[183] In 1961, Stevens wrote a paper entitled "To honor Fechner and repeal his law," *Science* **133**, 80–86 (1961). Stevens method of determining loudness is spelled out in detail in David M. Green, *An Introduction to Hearing*, Lawrence Erlbaum, Hillsdale, NJ, 1976, p. 291. This is essentially the international standard ISO R 532, Method A.

[184] H. Fletcher, *Phys. Rev.* **23**, 427–437 (1924).

[185] *Benchmark*/13, p. 126.

[186] J.F. Schouten, *Proc. Kon. Ned. Akad. Wetensch. Amsterdam* **41**, 1086–1093 (1938); **43**, 356–365 (1940).

[187] E.D. Schubert, loc. cit.

[188] G.A. Miller and W.G. Taylor, *J. Acoust. Soc. Am.* **20**, 171–182 (1948).

[189] P. Kucharski, *Comp. Rend. Soc. Biol.* **104**, 1249–1252 (1930).

[190] W. Bürck, P. Kotowski, and H. Lichte, *Elektr. Nachr. Tec.* **12**, 355–367 (1935).

[191] J.M. Doughty and W.R. Garner, *J. Exp. Psychol.* **37**, 351–365 (1947).

[192] J.M. Doughty and W.R. Garner, *J. Exp. Psychol.* **38**, 478–494 (1948).

[193] "a pure tone stimulus will result in a region of maximal displacement along the basilar membrane according to the place principle.... [If] a second tone is added whose frequency is slightly different ... [and] if the frequency difference is small, then the two resulting excitation patterns along the cochlear partition will overlap considerably, so that the two stimuli will be indistinguishable.... [If the] "two tones are equal in level then the resulting beats will alternate between maxima that are twice the level of the original tones and minima that are inaudible due to complete out-of-phase cancellation. Such beats are aptly called 'best beats'" (Stanley A. Gelfand, *Hearing*, Marcel Dekker, New York, NY, 1981, p. 281.)

[194] H. Fletcher, *J. Acoust. Soc. Am.* **1**, 211–343 (1930).

[195] G. von Békésy, *Experiments in Hearing*, p. 334.

[196] H.P. Egan and H.W. Hake, *J. Acoust. Soc. Am.* **22**, 622–630 (1950).

[197] *Benchmark*/13, p. 187.

[198] H. Fletcher, *Rev. Mod. Phys.* **12**, 47–65 (1940).

[199] W. Bürck, P. Kotowski, and H. Lichte, loc. cit. It was later noted that the frequency difference they used was too small, and that a wider frequency separation could produce a smaller value for this time interval. See Chapter 9.

[200] F.J.J. Buytendijk and A. Meesters, *Commentat. Pontif. Acad. Sci.* **6**, 557–576 (1942).

[201] H. Wallach, E.B. Newman, and M.R. Rosenzweig, *Am. J. Psychol.* **62**, 315–336 (1949). As Schubert remarked in his *Benchmark* volume (Ref. [197], p. 258]). "Judgments of click separation are not at all simple."

[202] S.S. Stevens and E.B. Newman, *Am. J. Psychol.* **48**, 297–306 (1936).

[203] David M. Green, *An Introduction to Hearing*, p. 296.

[204] Stanley A. Gelfand, *Hearing*, p. 308. An historical review of the subject with many interesting comments was given by Mark R. Gardner, *J. Acoust. Soc. Am.* **43**, 1243–1248 (1968).

[205] H. Wallach, E.B. Newman, and M.R. Rosenzweig, *Am. J. Psychol.* **62**, 315–336 (1949).

[206] H. Haas, Doctoral dissertation, University of Göttingen. An English translation of this appeared in Library Com. 363, Dept. Sci. Indust. Rest., Garston, Watford, England, 1949. A German article was later published in *Acustica* **1**, 49–58 (1951), and an English translation of this appears in *J. Audiol. Eng. Soc.* **20**, 146–159 (1972) and also in *Benchmark*/13 pp. 164–177.

[207] In 1907, Alexander Graham Bell wrote "The quality or 'timbre' of the human voice, I believe, is due in very minor degree to the vocal cords, and in a much greater degree to the shape of the passage through which the vibrating column of air is passed." (A.G. Bell, *The Mechanism of Speech*, Volta Bureau, Washington, DC, 1907, p. 23.)

[208] Stevens and Davis, *Hearing*, p. 52.

[209] These studies include the reviews of I.J. Hirsh, *Psychol. Bull.* **45**, 193–206 (1948) and J.C.R. Licklider in *Handbook of Experimental Psychology* (S.S. Stevens, ed.), Wiley, New York, NY, 1951, p. 1030. See also I. Pollack, *J. Acoust. Soc. Am.* **20**, 52–57 (1948).

[210] Homer Dudley, *Bell Syst. Tech. J.* **19**, 495–515 (1940). See also H. Dudley. R.R. Riesz, and S.A. Watkins, *J. Franklin Inst.* **227**, 739–764 (1939); H. Dudley, *Bell Labs. Record* **17**, 122–126 (1939b).

[211] R.K. Potter, *Proc. I.R.E.* **18**, 581, 1–11 (1930).

[212] Detailed information is given in R.K. Potter, Kopp, and H. Green, *Visible Speech*, Van Nostrand, New York, NY, 1947.

[213] The July 1946 issue of *JASA* begins with six articles on the subject of visible speech, all from the Bell Laboratories: Ralph K. Potter, *J. Acoust. Soc. Am.* **18**, 1–3 (1946); J.C. Steinberg and N.R. French, *J. Acoust. Soc. Am.* **18**, 4–18 (1946); W. Koenig, H.K. Dunn, and L.Y. Lacy, *J. Acoust. Soc. Am.* **18**, 19–49 (1946); R.R. Riesz and L. Schott, *J. Acoust. Soc. Am.* **18**, 50–61 (1946); Homer Dudley and Otto O. Gruenz, Jr., *J. Acoust. Soc. Am.* **18**, 62–73 (1946); G.A. Kopp and H.C. Green, *J. Acoust. Soc. Am.* **18**, 74–89 (1946). See also R.K. Potter, G.A. Kopp, and H.C. Green, *Visible Speech*, Van Nostrand, New York, NY, 1947, and R.K. Potter and J.C. Steinberg, *J. Acoust. Soc. Am.* **22**, 807–820 (1950). A new and enduring technique had been developed.

[214] By all odds the most popular phrase in speech testing. It and the companion phrase "She was waiting at my lawn," were probably both introduced by Harvey Fletcher. They appear in the second edition of his *Speech and Hearing in Communication*, p. 25, but are not in the first edition. Fletcher observed that "they [this phrase, along with "she was waiting on my lawn"] contain all the fundamental sounds in the English language that contribute appreciably toward the loudness of speech."

[215] Philip Liebermann and Sheila E. Blumstein, *Speech Physiology, Speech Perception, and Acoustic Phonetics*, Cambridge University Press, Cambridge, UK, 1988, p. 4.

[216] Johannes Müller, *Elements of Physiology*. Translated from the German, Taylor and Walton, London, 1842, pp. 1002ff.

[217] Philip Liebermann and Sheila E. Blumstein, *Speech Physiology*, pp. 13–14.

[218] T. Chiba and J. Kajiyama, *The Vowel: Its Nature and Structure*, Tokyo-Kaisekan, Tokyo, 1941.

[219] G. Oscar Russell, *The Vowel*, Van Nostrand, New York, NY, 1929.

[220] Harvey Fletcher did not mention the word *formant* in his 1929 volume, although he did describe its nature, but he used it in the 1953 edition.

[221] H.K. Dunn, *J. Acoust. Soc. Am.* **22**, 740–753 (1950).

[222] H. Fletcher and J.C. Steinberg, *Bell Syst. Tech. J.* **8**, 848–852 (1929). In *Benchmark/11*, p. 13, the volume editor, Mones Hawley, noted that Fletcher and Steinberg entitled their paper "Articulation Test Methods," but shied away from using this term for speech intelligibility ("The authors say it is better applied to the determination of speaking abilities." At that time, speech therapists had used the term to denote one's ability to form and use speech sounds, and even the term "articulation index," which soon became a standby in speech intelligibility testing, had been used in a different sense by K.S. Wood in measuring the speech defects of children. (K.S. Wood, *J. Speech Hear. Dis.* **11**, 266–275 (1946).)

[223] N.R. French and J.C. Steinberg, *J. Acoust. Soc. Am.* **19**, 90–119 (1947).

[224] J.P. Egan, *Laryngoscope* **58**, 955–991 (1948).

[225] L.L. Beranek, *Proc I.R.E.* **35**, 880–890 (1947).

[226] In the Preface to *Benchmark/11*, its editor, Mones Hawley, remarked "My interest in speech intelligibility first became vital about two miles above Anzio, Italy, on 16 February 1944, when my airplane was destroyed because of misunderstandings on the interphone. The other members of my crew were killed."

[227] S.S. Stevens, Joseph Miller, and Ida Truscott, *J. Acoust. Soc. Am.* **18**, 418–424 (1946).

[228] J.C.R. Licklider and I. Pollack, *J. Acoust. Soc. Am.* **20**, 42–51 (1948).

[229] G.A. Miller and J.C.R. Licklider, *J. Acoust Soc. Am.* **22**, 167–173 (1950).

[230] Wilmer T, Bartholomew, *Acoustics of Music*, Prentice Hall, New York, NY, 1942.

[231] Alexander Wood, *The Physics of Music*, The Sherwood Press, Cleveland, OH, 1944.

[232] Harry F. Olson, *Musical Engineering*, McGraw-Hill, New York, NY, 1952.

[233] Arthur Benade, *Horns, Strings and Harmony*, Anchor Books, Garden City, NY, 1960.

[234] Carl E. Seashore, ed., *Studies in the Psychology of Music*, vol. 1: *The Vibrato*, Iowa University Press, Iowa City, IA, 1932.

[235] Harry F. Olson, *Musical Engineering*, p. 127.

[236] Harry F. Olson, *Musical Engineering*, p. 137.

[237] H. Backhaus, *Z. Phys.* **62**, 142–166 (1930); **72**, 218–225 (1931). A list of Backhaus's other publications on musical acoustics may be found in *Benchmark/6*, p. 19. Another valuable historical reprinting of violin research papers, together with a historical review of the field since the sixteenth century may be found in *Research Papers in Violin Acoustics 1975–1993* (C.M. Hutchins, ed., V. Benade, assoc. ed.), Acoustical Society of America, Woodbury, NY, 1996.

[238] H.F. Meinel, *Elek. Nachr. Tech.* **14**(4), 119–134 (1937).

[239] A list of Saunders' publications in musical acoustics may be found in *Benchmark/6*, p. 19.

[240] Charles A. Taylor, *The Physics of Musical Sounds*, Crane, Russak, New York, NY, 1965, Chapter 6.

[241] C.A. Taylor, *The Physics of Musical Sounds*, p. 88.

[242] E. Meyer and G. Buchmann, *Sitzber. Preuss. Akad. Wiss. Phys.-Math.* **32**, 535 (1931).

[243] W.H. George, *Acustica* **4**, 225 (1954).

[244] W.H. George and H.E. Beckett, *Proc. Roy. Soc. (London)* **114A**, 111–136; **116A**, 115–138 (1927).

[245] Milton Metfessel in *The Vibrato* (Carl E. Seashore, ed.), University of Iowa Press, Iowa City, IA, 1932, pp. 14–15.

[246] J. Kwalwasser, *Psychol. Monogr*, **36**, No. 1, 84–108 (1926).

[247] Milton Metfessl, *The Vibrato*, p. 62.

[248] Louis Cheslock, *An Introductory Study of the Violin Vibrato*, Res. Studies in Music, Peabody Conservatory, Baltimore, MD, No. 1, 1931.

9
The Third Quarter: 1950–1975

> Today the world suffers from an overpopulation of sounds. There is so much acoustic information that little of it can emerge with clarity.
>
> R. Murray Schafer [1]

Academic research in America was enormously stimulated by the rapid advances in electronics that had taken place during World War II, and made available after the end of that war. At the same time, the financial support of research, first by the Office of Naval Research, and then by the National Science Foundation, and the National Institutes of Health, as well as the great expansion and proliferation of colleges and universities as a result of government support of the education of war veterans, all allowed an increase in the number and size of research undertakings. Meanwhile, the major industrial establishments, having participated in wartime research and development, recognized the significance of support of fundamental research as they converted from wartime to Cold War conditions. The field of acoustics, like all fields of science, benefitted from this growth. And in Europe, so badly battered by the war, the Marshall plan and other assistance programs enabled a gradual return of the universities to their former levels of research and eventually beyond them.

At a conference on education in acoustics in 1964, R. Bruce Lindsay portrayed the field of acoustics in terms of a wheel, whose hub was "mechanical radiation in all material media" (Fig. 9-1). As one moves out from the center of the wheel, one encounters the major subdivisions of acoustics, and then the disciplines of which these subdivisions were members, and finally, the four basic areas of Arts, Earth Sciences, Engineering Science, and Life Sciences [2].

This wheel, called Lindsay's wheel, has often been copied, modified, and cited in the literature. From the viewpoint of this text, it serves as an excellent summary of the vast scope of the field of acoustics, and the interlocking nature of its various parts. While it has been said of acoustics that it is such an interdisciplinary science that it is difficult, if not impossible,

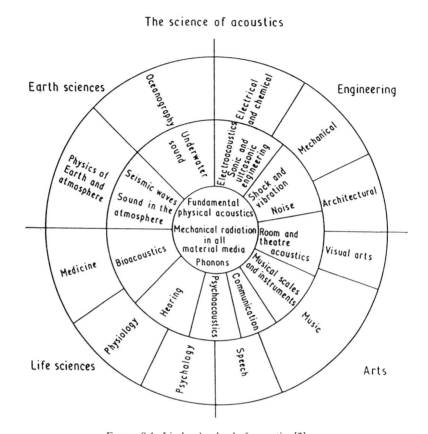

FIGURE 9-1. Lindsay's wheel of acoustics [2].

to isolate the discipline itself, the wheel is an enduring reminder of its essential unity.

9.1. Societies and Journals

Just as the development of physics in the 1920s led to a splitting off of various groups from the American Physical Society, such as the Acoustical Society of America, so the continued growth of research in acoustics in the United States led to the formation of more specialized associations. The Audio Engineering Society began in 1948. The American Speech and Hearing Association (ASHA) in 1925, the different specialty groups of the Institute of Electrical and Electronic Engineering (IEEE) in the 1950s, the Institute of Noise Control Engineering (INCE) in 1971, and Ultrasonics in Medicine in 1951, and most of these societies have sponsored their own journals.

Abroad, each nation had its own group of acousticians, who gradually coalesced into some more or less formal organization. Thus, the Acoustical Society of Japan (ASJ) was established in 1937, the British Institute of Acoustics, formed as a merger of three different societies, the Group of French-Language Acousticians (GALF), the Group of Acousticians of Latin America (GALA), the German Acoustical Society (DAGA), and the Russian Acoustical Society only a few years back. To keep some of this on an organized basis, and to encourage international cooperation, the International Commission on Acoustics (ICA) was established in 1951 as a branch of the International Union of Pure and Applied Physics (IUPAP), holding an international congress every 3 years since that time.

The European acousticians joined together in 1971 to form the Federation of European Acoustical Societies (FASE) and, still later, the European Acoustics Association (EAA), which is an association of the national societies. On the international scene, various new journals have appeared, including *Acustica* (Switzerland, 1951), *Akusticheskii Zhurnal* (Moscow, 1956), *Journal of Sound and Vibration* (England, 1964). *Akusticheskii Zhurnal* has been translated into English from its beginning, first as *Soviet Physics Acoustics*, published by the AIP, and more recently, appearing under the title *Acoustical Physics*. And, in this ever-changing world, *Acustica* has merged with the publication of the EAA, to form *ACTA Acustica*.

One way of classifying this vast outpouring of information on acoustics has been through the sytem known as PACS (Physics and Astronomy Classification Scheme), supported by the American Institute of Physics and used by the *Journal of the Acoustical Society of America* (*JASA*). PACS currently lists over 300 separate acoustical topics. While the original PACS system applied only to topics in physics and astronomy, the classification scheme that had been developed by *JASA* over the years has been appended to the main list, so that all the acoustical topics in the life sciences are also covered [3].

9.2. Physical Acoustics

The period 1950–1975 was the occasion for a burst of research in physical acoustics, leading in many different directions. The great increase in research funding, the improvement of electronic devices, and advances in such interfacing fields as electronics and condensed matter physics [4] combined to make this time one of the most exciting in the history of the field.

Scattering

One might have thought that the subject of the scattering of sound from obstacles was an exhausted field of research by 1950. Clebsch had begun the study in 1863, Rayleigh had written much on it, Karl Herzfeld studied the scattering from a small elastic sphere in 1930 [5], while the text by Philip Morse [6] seemed to sum things up. But this topic was far from finished. In underwater sound, a sequence of papers appeared on the scattering of solid spheres in water [7], while different methods were pursued for scattering, both from impenetrable and penetrable objects.

In introducing a new technique in 1969, P.C. Waterman [8] noted that there were three methods widely used for calculating scattering and diffraction at that time—"The separation-of-variables, variational techniques, and the direct numerical solution of integral equations [9]." Waterman introduced an "outgoing spherical (or circular cylinder) partial waves . . . as a basis" for obtaining "the transition matrix T," describing scattering for general incidence on a smooth object of arbitrary shape [10].

Waterman's method, first used only for the scattering of acoustic waves, has also been generalized to electromagnetic waves by Waterman [11], and to elastic waves in solids by Vasundara Varatharajulu and Yih-Hsing Pao [12], and by Waterman [13]. The *T*-matrix became one of the standard methods of studying all types of scattering.

Creeping Waves

The subject of the diffraction of a plane or spherical wave about a regular object, such as a sphere or cylinder, has also continued to be widely studied. A new development occurred in 1954 when Walter Franz demonstrated mathematically the conditions under which a surface wave "creeps" around the cylinder or sphere while continuing to reradiate the sound energy [14]. Figure 9-2 shows a schematic drawing of such a creeping wave, and a schlieren photograph of an actual one [15].

Transducers

One of the first fruits of this expansion in fundamental research in acoustics was the great increase in the number of measurements of the absorption of ultrasound in fluids. The perfection of pulse systems in electronics [16] plus the increased understanding of the mechanisms involved in piezoelectric transduction [17] made it possible for any research laboratory to carry out experimental measurements on ultrasonic absorption. At the same time, new materials became available as transducers. At the end of World War II, natural quartz and other natural crystals [especially KDP (potassium dihydrogen phosphate), ADP (ammonium dihydrogen phosphate), and Rochelle salt] were the standard materials [18], but the electrically polarized ceramic barium titanate soon appeared on the market. These transducers could be molded in any shape and produced cheaply in any amount, and barium titanate, and its many modifications, soon became the transducers of choice [19].

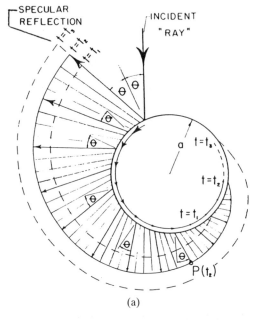

(a)

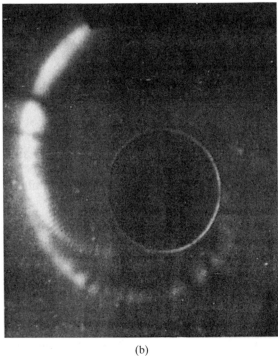

(b)

FIGURE 9-2. Creeping waves. (a) Geometric construction describing the circumferential property of waves radiated from circular cylinders. (b) Schlieren photograph of the wavefronts resulting from incidence of a pulse over a range of angles from 15° to 90°. (From Neubauer [15].)

In this same period, Warren Mason began the editing of what was to be a monumental series of volumes on physical acoustics [20]. Volume IA appeared in 1964, and by 1984, the number of volumes had reached 17. No corner of the field of physical acoustics has been left unreviewed by Mason. Like the list of his patents and his own books, this series is one more lasting tribute to this great acoustician [21].

The study of the solid state from the viewpoint of modern physics has often proceeded quite separately from its study on the side of modern acoustics. The work of Igor Y. Tamm (1895–1971) and Yakov I. Frenkel (1894–1954) in the Soviet Union in the 1930s on the concept of quantized acoustical energy in the form of phonons seemed to bear little relation to beams of ultrasound [22]. But one place in which the two studies have come together into one, for example, has been on the subject of vapor-deposited thin-film piezoelectric transducers [23].

Beginning in 1957, when Baranskii demonstrated that phonons could be generated at the surface of a bar of quartz without the use of a separate transducer [24], through the presentation of a theory of the generation of such phonons by Bömmel and Dransfeld [25], to the production of thin-film piezoelectric transducers by de Klerk and Kelly [26], there has been steady progress toward the union of solid-state theory and practical acoustical solid-state devices. The use of cadmium sulfide films as transducers then became a commonplace.

Ultrasonics in Fluids

The study of ultrasonic propagation in monatomic gases, where the phenomenon of excitation of the internal motions of the atoms could not occur, provided a basis for detailed verification of the classical (Stokes–Kirchhoff) theory of absorption, Martin Greenspan, working at the National Bureau of Standards (now the NIST—National Institute of Science and Technology) improved the techniques of measurement of sound velocity in gases and showed that the absorption and dispersion curves for all the monatomic gases could be plotted on a single curve, using dimensionless variables of c_0/c and $\alpha\lambda_0/2\pi$, plotted against a reduced Reynolds number $r = p/\omega\mu$, where the main components were the ratio of ambient pressure p to the frequency. This latter ratio also made it possible to make measurements at effectively high frequency in gases by going to lower and lower pressures (Fig. 9-3) [27]. Similar measurements were made by Erwin Meyer, and his student Gerhard Sessler (1931–   ), later professor at the University of Darmstadt and a major contributor to the study of electret transducers [28]. They found that the value of $\alpha\lambda_0/2\pi$ approached a constant value as the pressure was reduced to nearly zero. Thus a final answer was obtained to the problem posed by Hauksbee in the seventeenth century. If you could get sound into a very low-pressure gas, it would still propagate with the absorption coefficient approaching a constant at a fixed frequency.

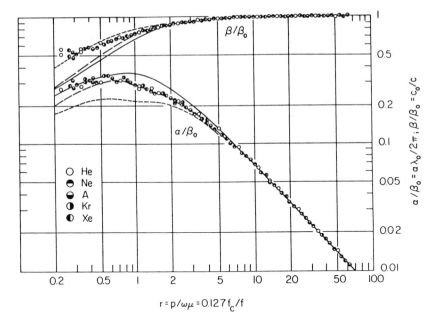

FIGURE 9-3. Results of measurements made at the National Bureau of Standards on five monatomic gases. (From M. Greenspan [27].)

Different theoretical approaches in the study of relaxation processes were discussed at length in the review by Markham, Beyer, and Lindsay [29]. This was updated in a long review by Hans-Jörg Bauer in 1964 [30], especially in the area of irreversible thermodynamics. This latter technique was first used in 1937 by Leonid I. Mandelshtam (1879–1944) and Mikhail A. Leontovich (1903–1981) [31].

The work of Liebermann and Eckart on acoustic streaming (a subject that goes back to Faraday and Rayleigh) in the 1940s was followed in the next decade by the studies of Markham [32] and Wesley LeMars Nyborg [33], who established the connection between streaming and sound absorption, and made it clear that there could be no streaming without absorption, and that measurement of streaming was another way of measuring the total absorption. This was soon confirmed experimentally by John Lamb and Joseph Piercy [34] who developed a method of measuring sound absorption by measurement of the streaming.

The difficulties in establishing the quantitative correctness of the relaxation theory of sound absorption in fluids stemmed from the fact that the two parameters embedded in the theory (the size of the energy jump involved in the relaxation process and the relaxation time τ) could not be measured independently if the sound measurements occurred at too low a frequency, which was generally the case. The form of the expression for the product of the sound absorption α and the sound wavelength λ is given by the expression

$$\alpha\lambda = \frac{A\omega\tau}{1 + \omega^2\tau^2},$$

where the constant A involves the energy jump. At low frequencies ($\omega^2\tau^2 \ll 1$), $\alpha\lambda = A\omega\tau$, so that measurement of $\alpha\lambda$ does not separate A from τ. However, if the frequency of measurement is sufficiently high, so that the maximum of the $\alpha\lambda$ curve can be achieved (see Fig. 8-2(a)), the frequency at which this occurs (which is called the relaxation frequency $\omega_r = 1/\tau$) gives us the value of the relaxation time, while the amplitude of the maximum, $A/2$, enables us to find A. A similar separation of the parameters can be achieved from measurements of the sound velocity at high and low frequency, but in general the sound absorption measurement is easier.

The search was therefore on for measurement of these "relaxation peaks." Such measurements have been accurately made in gases by 1940 [35], but the relaxation frequencies for liquids were generally too high for such measurements until 1949, when John Lamb and J.M. Pinkerton presented such measurements for acetic acid [36]. Thereafter, Lamb [37] and others discovered relaxation peaks in many organic liquids (Fig. 9-4) [38].

While the relaxation frequencies of a number of liquids could be measured by these techniques, many other liquids proved to be resistant. Higher ultrasonic frequencies were necessary. Back in 1922, Leon Brillouin had studied optical scattering in liquids, in particular, the elastic scattering known as Rayleigh scattering. In traditional Rayleigh scattering, the scat-

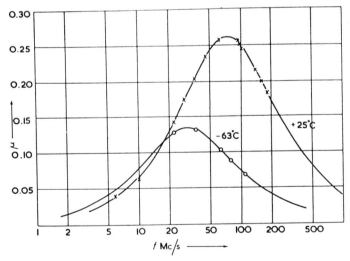

FIGURE 9-4. Relaxation of the total vibrational specific heat in carbon disulfide. The curves are theoretical and the points experimental. (From Andreae, Heasell, and Lamb [38].)

tered light is of the same frequency as that of the initial beam [39]. But Brillouin hypothesized the appearance of scattered light at frequencies above and below that of the initial beam. In modern terms, when a photon of energy h (frequency, f) enters a liquid it will interact with the sea of phonons that constitutes the thermal energy of the liquid. If it absorbs one such phonon (energy hf, f being the frequency of the phonon), it will increase its energy. It can also emit one such phonon, thus losing energy. In general, then,

$$h\nu' = h\nu \pm hf,$$

where ν' is the new frequency of the photon, f is the frequency of the phonon, and h is Planck's constant. If the frequency distribution of the scattered light is measured, $f (= \nu' - \nu)$ can be measured. By measuring the width of the scattered peaks, the sound absorption coefficient at the sound frequency f can also be measured [40]. This made it possible to measure both c and α at frequencies up to 10 GHz [41]. Early measurements by this technique had been made by the Indian physicists C.V. Raman and C.S. Venkateswaren in 1938 and later by Venkateswaren alone [42], but the frequency widths of the light sources used at that time were too great to allow any dispersion. At this point, the laser came to the rescue. The frequency sharpness of the laser beam made quantitative results possible. Studies of Brillouin scattering in benzene have been published by Fleury and Chiao [43], and also by Mash, Starunov, Tiganov, and Fabelinskii [44] (Fig. 9-5).

We noted in Chapter 8 that the increase in sound absorption in sea water (above the fresh water value) was attributed by Leonard to the presence of magnesium sulfate $(MgSO)_4$ in the ocean water. This discovery led to a number of papers by Ernest Yeager and his students at Western Reserve University, and by Konrad Tamm and Gunther Kurtze at Göttingen [45]. Not only did the absorption curve (of α versus $\log f$) from $MgSO_4$ solutions have a relaxation peak (at about 140 kHz), but similar peaks could also be observed in a wide variety of bivalent salt solutions (Fig. 9-6). It was also noted that, by the addition of a monovalent salt, the sound absorption could be reduced to values below that of distilled water [46], [47]. The theory of these complications was developed by Manfred Eigen, a colleague of Tamm and Kurtze at Göttingen [48]. This theory involved a multistep dissociation of the salt and the attached water molecules (Fig. 9-7). This work of Eigen opened up a new field of chemical study—the kinetics of fast reactions— and for his theoretical contributions to this subject, Eigen shared the 1967 Nobel prize in chemistry.

Low-Temperature Acoustics

The study of sound wave propagation in liquid helium had increased in intensity in the late 1940s, as described in Chapter 8. The velocity of propa-

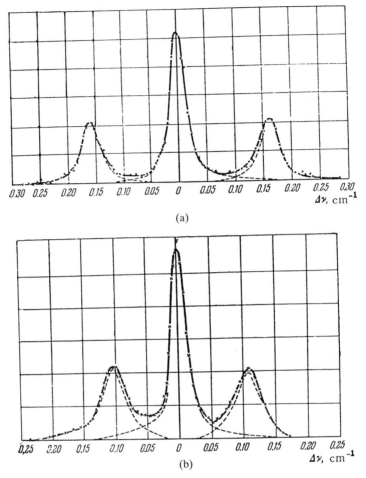

FIGURE 9-5. Intensity distribution in the fine structure of the 6328° line in Mandelshtam–Brillouin scattering in (a) benzene and (b) carbon tetrachloride. (From I.M. Fabelinskii [44].)

gation of such sound was measured accurately. In addition, so-called second sound was discovered [49]—the propagation of waves consisting of periodic oscillations of temperature and entropy (while the pressure and density remain substantially constant). Third sound—a surface wave of the super-fluid component of liquid helium in a thin liquid film—was suggested by Kenneth Atkins in 1958 [50] and subsequently observed by his group at the University of Pennsylvania [51].

Fourth sound was predicted by John Pellam as early as 1948 [52]—a pressure wave in the superfluid component of liquid helium in a porous solid (the so-called "superleak") in which the normal component cannot

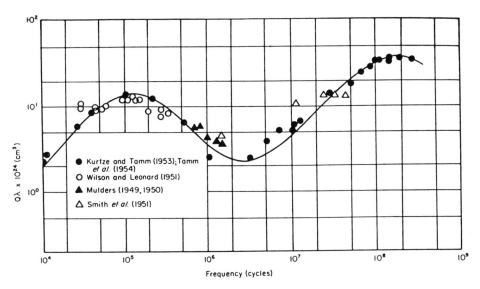

FIGURE 9-6. Absorption cross section per wavelength ($Q\lambda$) for solutions of $MgSO_4$ at $\sim 20°$. (From Stuehr and Yeager [45] and from K. Tamm and G. Kurtze [45].)

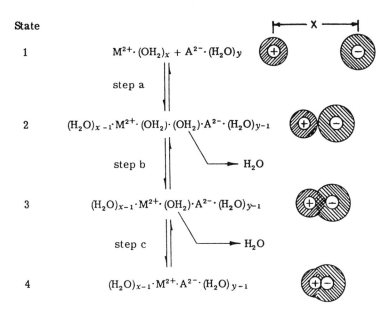

FIGURE 9-7. Ionic interactions and dissociation processes in aqueous solutions of $MgSO_4$. (From J. Stuehr and E. Yeager [45].)

move. However, its experimental verification did not come until 1962, in work by K.A. Shapiro and Isadore Rudnick [53].

There is a "fifth sound," thermal wave, similar to second sound, but one which occurs in a porous solid. It was observed at both UCLA [54] and Pennsylvania State University [55] in 1979. While this discovery is out of the bounds of our time interval (1950–1975), it is appropriate to include it here, thus completing the identification of the five sounds. A graph of the sound velocities, and a table identifying all of the sounds, provided by Rudnick in a 1980 review paper is also provided here (Fig. 9-8) [56].

To these five sounds, we must add one more. In his model of a Fermi liquid [57], Landau treated the interaction between the particles as a perturbation, giving a shift of energy levels and transitions of the particles from one level to another. Because of these interactions, the quantum mechanical description must be in terms of "quasiparticles." Such a model holds only when the quasiparticle lifetimes are sufficiently long that the energy levels are well defined. This limits its validity to temperatures below $0.3\,K$.

In this temperature range, the Landau model predicts a new model of sound propagation—zero sound. As the temperature decreases, the time between the collisions of quasiparticles becomes longer, increasing, according to the Landau theory as T^{-2}. At high ultrasonic frequencies, a point is reached for which $\omega\tau \gg 1$, where τ is the mean thermal phonon lifetime. The resulting zero sound is the analog of sound propagation in rarefied gases (see above), where the molecular mean free path exceeds the wavelength. The existence of zero sound was established experimentally in 1963 by B.D. Keen, P.W. Matthews, and J. Wilks [58]. A somewhat similar phenomenon was predicted to occur in quartz by Humphrey Maris [59]. At low frequencies, where $\omega\tau \ll 1$, sound propagates at its customary velocity. However, at high frequency, the speed of propagation is different. Maris calculated this velocity difference and called the phenomenon zero sound also. In 1970, he and Joseph Blinick (1940–   ) observed the effect experimentally [60].

Another major advance in low-temperature acoustics was the development of the scanning acoustic microscope by Calvin Quate at Stanford University [61]. This device (Fig. 9-9) focused a high-frequency ultrasonic pulse by means of a shaped sapphire rod. The signal reflected from the sample was then retrieved electrically. A sample photograph is shown in Fig. 9-10 [62].

9.3. Ultrasonics in Solids

Prior to 1950, acoustical research in solids was largely limited to measurements of velocity and absorption and a study of how the ultrasonic beams traveled in the medium. But the story after 1950 was a wholly different one.

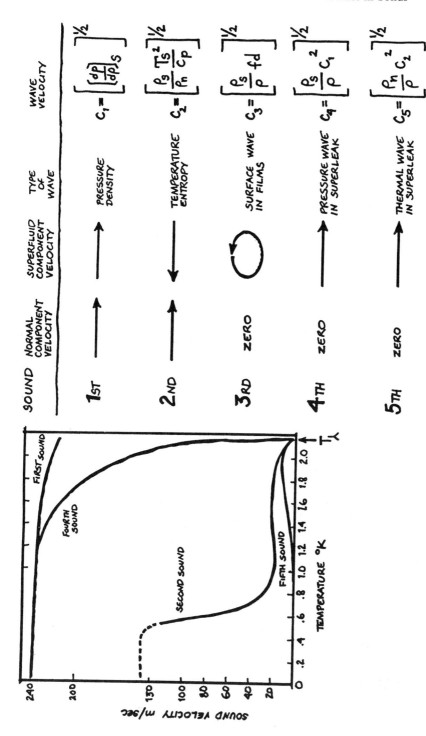

FIGURE 9-8. Characteristics of the five sounds of liquid helium. (From I. Rudnick [56].)

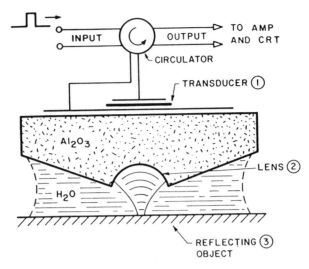

FIGURE 9-9. The ultrasonic microscope. The geometry of the acoustic transducer and lens used in the reflection mode. (From Quate [61].)

The exploitation of the concept of the phonon as the basis of thermal energy in solids and another way of describing Debye waves led to the identification of many processes in solids that were of an acoustical character [63]. We shall describe a few of them.

A. Akhiezer (1911–   ) had theorized, as early as 1939, that the varying strains of a sound wave would disturb the thermal equilibrium of the

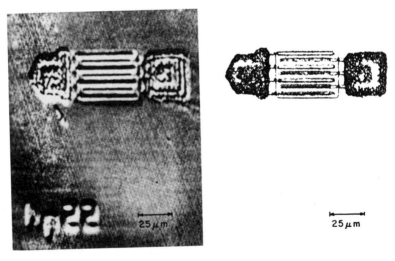

FIGURE 9-10. Acoustic (left) and optic (right) reflection micrographs of a small bipolar transistor. (From Lemon and Quate [62].)

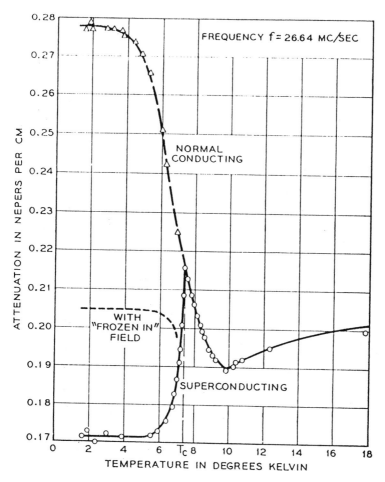

FREQUENCY f = 26.64 MC/SEC

FIGURE 9-11. Comparison of measured attenuation of a lead single-crystal with that calculated from free electron theory. (From H. Bömmel [66].)

phonons, and their return to equilibrium would be a relaxation process, resulting in absorption [64]. This theory was taken up, 20 years later, and studied in great detail by Bömmel and Dransfeld and others [65]. Bömmel and Dransfeld generalized the theory to include a finite relaxation time for the thermal phonons. The effect is known as Akhiezer loss or phonon viscosity.

In 1954, Bömmel first observed a sharp decrease in the acoustic absorption of a metal when it became a superconductor (Fig. 9-11). On the other hand, if a magnetic field is applied so as to keep the metal a normal conductor, the absorption rose to a maximum as the temperature fell toward absolute zero [66]. These results suggested strongly that electrons were largely responsible for sound absorption in metals at low temperatures. A theory for absorption in the normal state was developed by Mason, involving lattice–electron interactions [67].

Another form of absorption in solids is that due to dislocations in the lattice structure, occurring as lines or loops in the lattice and pinned by impurities at certain points. The sound wave passing through the crystal tends to "bow out" these dislocation lines and even causes the lines to break away from their original pinning points. The successful theory of this phenomenon was developed by Andrew Granato and Kurt Lücke in 1956 [68].

The theory of sound absorption in the superconductor was more far-reaching. In the Bardeen–Cooper–Schrieffer (BCS) theory of superconductivity [69], it is the formation of pairs of electrons (the so-called Cooper pairs) that gives rise to superconductivity. As the temperature falls below the critical temperature, the number of these pairs increases and the number of electrons in the normal state decreases. Application [70] of the theory to the acoustical problem yielded the relation

$$\frac{\alpha_s}{\alpha_n} = 2\left(1 + \frac{e\Delta}{kT}\right)^{-1}$$

for the ratio of the absorption in the superconducting state α_s to that in the normal state (α_n) at a given temperature. The quantity Δ is the famous energy gap of the BCS theory. This relation was verified in 1957 by Robert Morse (1920–   ) and Henry Bohm (1929–   ) [71].

A large number of magnetic interactions with sound at low temperature were observed in this period, including acoustic nuclear magnetic resonance, when the sound energy is absorbed by exciting components of the spin system of the solid to higher magnetic levels, acoustic electron paramagnetic resonance, and interactions between ultrasound and the spin wave of a ferromagnetic system [72].

Another phenomenon of great interest has been the magnetoacoustic effect: the appearance of giant oscillations in the sound absorption in single crystals at low temperatures as an applied magnetic field is varied [73] (Fig. 9-12).

9.4. Aeroacoustics

In the 1950s, Sir James Lighthill (1924–1998), building on the framework of earlier research by Rayleigh, performed an analysis of the sounds generated in a fluid by turbulence that would have far-reaching effects in the study of noise generated by jet engines and in the field of nonlinear acoustics [74]. In Rayleigh's analysis of scattering from small-scale inhomogeneities, he had written an equation with the D'Alembertian [75] of the pressure in an inhomogeneous medium on the left-hand side of his equation, and terms

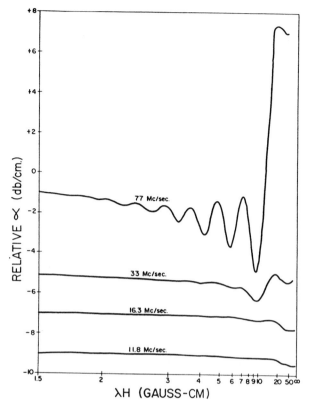

FIGURE 9-12. Giant absorption oscillations in a single crystal of gold as a function of the applied magnetic field. (From Morse [73].)

dependent on the relative inhomogeneities in the sound velocity $\Delta c/c_0$ and the density $\Delta \rho/\rho_0$ as well as other nonlinear terms in the pressure and its derivatives on the right. Since he was interested only in the inhomogeneities, Rayleigh discarded the nonlinear terms, and went on to develop his scattering theory. Lighthill, on the other hand, was interested precisely in the nonlinear aspects and so discarded the inhomogeneity terms and kept the nonlinear ones.

The principal result of Lighthill's analysis was the identification of quadrupole sound sources in the inhomogeneities of turbulence. The fluctuating stresses produce a force per unit volume equal to their inward stress. They therefore act like a dipole field. However, the sound radiated is not to be computed from the total dipole strength per unit volume, since this is in fact zero at any instant of time. Rather, the sound will come from the next higher order of terms, i.e., it will be due to the equivalent quadrupole field, a field that is the limiting case of four simple sources which obey the inverse square law of radiation.

This research led to a large number of papers, by Lighthill, J.E. Ffowcs-Williams, David G. Crichton, and M.S. Howe, among others, that illuminated our understanding of turbulence, sound production, and jet noise [76]. We shall encounter the results of their work in the noise section of this chapter.

The work of Lighthill has also served as the starting point for major developments in nonlinear acoustics, which we shall discuss in the next section.

9.5. More Nonlinear Acoustics

As mentioned in Chapter 8, the next great advance in nonlinear acoustics occurred after a long quiet period. In the 1950s, several observers, including Francis E. Fox, R.B. Lindsay, V.A. Krasilnikov (1912–   ), and their students, reported an increase in sound absorption in liquids when the intensity of the sound level was increased [77]. The experimental results clearly indicated that a sound wave that was originally sinusoidal in shape gradually transformed into a nearly sawtooth shape. Such a signal was rich in harmonics, and since the sound absorption coefficient in most fluids increases with the square of the frequency, the absorption had to increase rapidly. Toward the end of the 1950s, the work of Fubini was rediscovered [78], but the most significant theoretical advance was provided first by J. Mendousse, and then by Rem Viktorovich Khoklov and S.I. Soluyan [79]. The basis of this advance was the viewing of the sound wave from the viewpoint of an observer traveling with the wave. As Mendousse put it,

one imagines an observer moving with the wave velocity c, riding the wave, so to speak; and one assumes that only slow changes occur in the state of the medium near this observer (there would be no change at all for a nondissipated, nondistorted wave). In a suitable system of coordinates, some of the partial derivatives are then very small and can be neglected in many places where they occur. [80]

By carrying out this analysis, with appropriate discarding of small terms, the equation of motion can be converted to the nonlinear Burgers' equation which, by an adroit substitution, can be put into linear form [81]. The resultant solutions, however, need the services of computers to interpret them [82].

The original work of Khokhlov was for an infinite plane wave. A later paper by Khokhlov and E.A. Zabolotskaya took beam spreading into account [83], and a subsequent refinement by Kuznetsov included acoustic absorption [84]. The resultant equation has come to be known as the KZK (Khokhlov–Zabolotskaya–Kuznetsov) equation [85].

The work of Lighthill, cited in the previous section, was used by Karl Uno Ingard (1924–   ) and David Pridmore-Brown (1928–   ) at MIT to interpret the results of an experiment in which they attempted to discern the existence of the scattering of one sound beam by another [86]. This work was quickly challenged by Westervelt [87]. In the years since, there have been volumes written on the subject as to whether there is any scattering outside the interaction region, other than that due to two collinear beams [88]. The problem is made more complicated because it has always been necessary to make assumptions as to the nature of the beam, the existence of dispersion, and other approximations [89]. Elements of the controversy still exist (See Chapter 10) [90].

Westervelt has also been responsible for a second major development in nonlinear acoustics. He first established the relations for the generation of sum and difference frequencies in the case of two collinear beams of different frequency, thus obtaining the theoretical justification for the results obtained in the previous century by Rücker and Edser (Chapter 6). At the same time he gave the name "parametric array" to the resultant combination frequency, because of its similarity to the effect of a so-called endfire array in underwater sound signaling [91]. This technique, which has the advantages of narrow directivity at low frequencies (the directivity of the frequency of the primary beam), low absorption (the absorption at the low-frequency rate of the difference frequency), and the virtual absence of sidelobes in the directivity pattern, has been widely exploited in underwater sound work in several countries [92].

9.6. Underwater Sound

Vigorous research on underwater sound continued throughout this period, fueled no doubt by the continuing Cold War between the United States and the Soviet Union. During World War II, there was a great deal of effort devoted to expressing the entire practice of sonar in engineering terms, with the development of the so-called sonar equations, which separated the different effects of source characteristics, behavior of the medium, and the effects of the target into a sum of terms, all in decibel notation [93]. The two primary modes of sonar reception have been *passive* (listening only), and *active* (transmitting a signal and listening for echoes). For the active mode, the echo-to-noise ratio, i.e., the amount in decibels by which the returned signal exceeded the noise level, would be given by

$$SL - 2TL + TS - (NL - DI),$$

where SL is the source level in db, referred to a reference level, TL is the one-way transmission loss, which consists of a beam spreading term and an absorption term, TS is the target strength, giving the ratio of the fraction of incident intensity reflected from the target, NL is the noise level at the receiver, and DI is the receiving directivity index, which measures the effective gain of the directionality of the receiver above that of a nondirectional one [94].

Both sending and receiving equipment became more and more sophisticated after World War II. By the use of arrays of transducers, appropriately delayed in time, it became possible to shape the beam so that a large fraction of its energy was projected in one direction (Fig. 9-13). In addition, the electronics could be designed as to "steer" the beam, i.e., a fixed set of transducers, attached to the hull of a ship or submarine, and could be made to aim its beam in various directions (the "steered array"). On the listening side, the German listening equipment remained considerably ahead of that used by the Allies. They "designed hull shapes and entire ship structures in accordance with the demands of their acoustic engineers [95]," whereas the American and British navies usually added the equipment as a kind of afterthought.

The equation above suggests that the higher the source level, the greater the advantage of the equipment, but it soon became clear that there were limitations on sonar power. Cavitation studies by Murray Strasberg (1917–   ) and others [96] indicated that sound intensities of the order of $1\,\mathrm{W\,cm^{-2}}$ would produce cavitation, and the air bubbles released by this phenomenon greatly reduced the intensity of the sound beam.

A second limitation, pointed out by Westervelt, was the effect of nonlinearity [97]. As the sound signal becomes more intense, the wave becomes more and more distorted, pushing energy into the higher harmon-

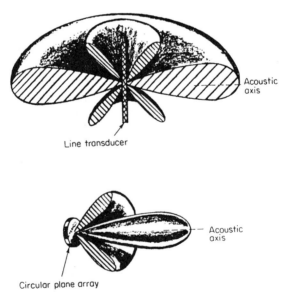

FIGURE 9-13. Three-dimensional views of the beam pattern of a line and a circular-plane array. (From Urick [94].)

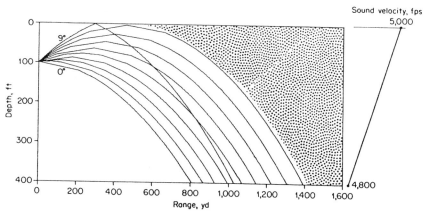

FIGURE 9-14. Ray diagram for a source at a depth of 100 ft in the linear gradient shown at the right. Stippled area is the surface shadow zone. (From Urick [94].)

ics, where the absorption is much greater. This effect was explored in some detail by Robert H. Mellen and Mark Beyer Moffett (1935–   ) at the Navy laboratory in New London, CT [98].

Systematic studies of the temperature and salinity of the oceans, together with ever-improving computer techniques for ray-tracing, led to detailed knowledge of how the sound beams progress through ocean waters. The existence of shadow zones (Fig. 9-14), and the effect of bottom and surface reflections (Fig. 9-15), all received extensive attention [99].

An alternative to the ray theory of propagation in shallow water is the method of normal modes. Instead of dealing with the propagation of rays, one examines the partial differential wave equation and solves it for its characteristic functions, called the normal modes [100]. These normal

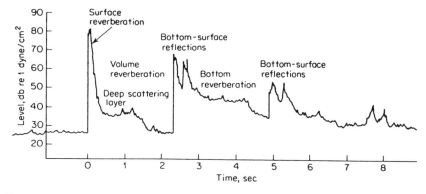

FIGURE 9-15. Reverberation following a 2-1b explosive charge detonating at 800 ft in water 6500 ft deep, as observed with a nearby hydrophone at a depth of 135 ft. Filter band 1 to 2 kHz. (From Urick [104].)

modes are then summed in such a way as to satisfy the boundary conditions. While the number of modes necessary to produce effective solutions of the entire problem in deep water is horrendous, the method, always making use of computer programs, is quite manageable for shallow water [101]. The application of normal-mode theory to shallow water was carried out by Pekeris as early as 1948 [102].

As soon as a sound signal of short duration—a "ping"—is projected into the ocean, the sound detector at the source location begins to receive signals that are replicas of the original ping. These signals are given the general name of reverberation [103]. It soon became clear that these reverberations came first from the volume of sea, where any inhomogeneities, whether of particulate matter, temperature fluctuations, or other small-scale changes in the medium, could act as sources of scattered sound. A second source is surface reverberation, from the roughnesses on the surface, and a third is bottom reverberation from irregularities on the bottom (Fig. 9-15 [104]).

The work on reverberation led to some interesting insights into marine life. Some volume scattering apparently occurred from a layer (soon to be called the deep-scattering layer or DSL). This was first reported by Eyring, Christensen, and Raitt [105]. This level apparently rose during the night and receded in the daytime, suggesting a biological, light-oriented connection (Fig. 9-16 [106]). Later studies have demonstrated that the source is a large number of various sea creatures (Fig. 9-17 [107], [108]).

The fluctuation of the properties of the ocean also received a great deal of study. The theory of Kolmogorov on the effects of turbulence in air, described in Chapter 8, was now applied to underwater sound. In a lengthy paper, D.C. Whitmarsh, Eugen Skudrzyk, and R.J. Urick examined forward scattering in the sea and its correlation with temperature microstructure [109]. If we assume that the temperature microstructure of the ocean is due to turbulent motion of the sea, then the Kolmogorov theory can be applied to the temperature distribution. Figure 9-18 compares the Kolmogorov 2–3 law with their experimental results. Other significant theory of scattering from the microstructure of the ocean was developed by David J. Mintzer [110].

The work of many scientists and engineers during World War II has been summed up in review pieces by Robert J. Bobber [111] and T. Bell [112]. More recently, the placing of bottom-mounted hydrophones—under the name Caesar (or, later, Sound Surveillance Systems (SOSUS)—for listening in coastal waters became an important operation in the Cold War [113].

While we shall consider the general problem of noise later, in Section 9.9, it is appropriate to mention here the fact of its significance in studies of underwater sound. One of the most prolific writers in this field has been Gordon M. Wenz. In 1962, he published a graph delineating the various noises that occur in the ocean (Fig. 9-19) [114]. This graph has been fre-

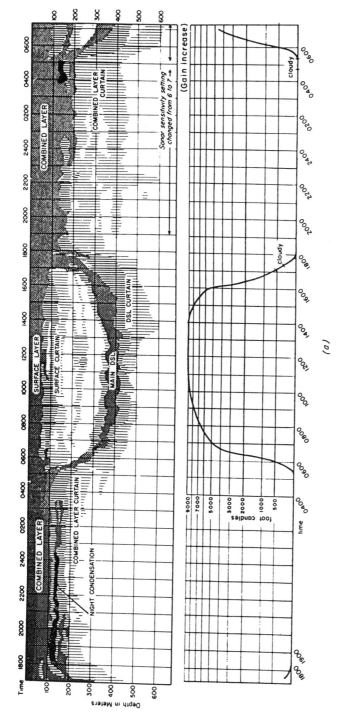

FIGURE 9-16. Above, diagram of scattering layers prepared from a 37 hour long echogram recorded in the Bay of Bengal at 06°10′N, 93°07′E, beginning 1700 hours November 24 and ending 0715 hours November 26. A Simrad sonar at 30 kHz was used. Stippled patterns indicate heavy scattering; medium and light vertical lines indicate medium and light scattering, respectively. Light scattering between the surface curtain and main deep scattering layer (DSL), centered at 250 m during the day, corresponds in position to the intermediate layer seen as heavy scattering on numerous other echograms. Below, curve indicating intensities of incident light for the period the echogram was being recorded. (From Clay and Medwin [107]; originally from Bradbury et al. [106].)

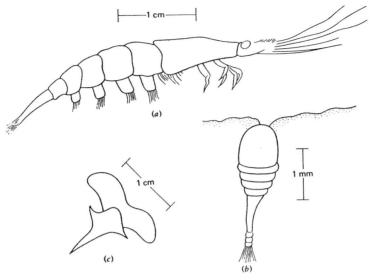

FIGURE 9-17. Zooplankton. (a) Euphasiid; (b) Copepod; and (c) Pteropod. (From Clay and Medwin [107].)

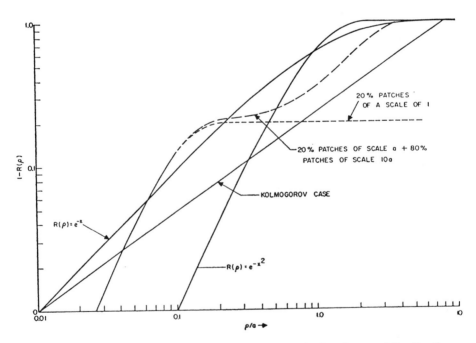

FIGURE 9-18. The function $1 - R(\rho)$ for an experimental and a Gaussian correlation function and the function $\langle (\Delta c/c)^2 \rangle (1 - R(\rho))$ for the Kolmogorov case. (From Whitmarsh et al. [107].) (R is a measure of the temperature fluctuation in the oacen at a distance ρ from the reference point.)

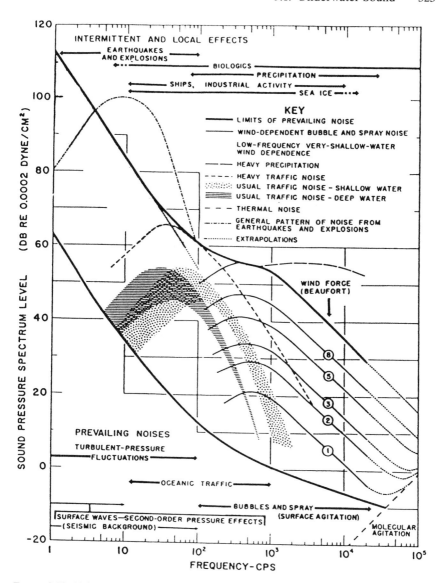

FIGURE 9-19. Noise sources in the ocean. A composite of ambient-noise spectra, summarizing results and conclusions concerning spectrum shape and level and probable sources and mechanisms of the ambient noise in various parts of the spectrum between 1 cps and 100 kc. The key identifies component spectra. Horizontal arrows show the approximate frequency band of influence of the various sources. An estimate of the ambient noise to be expected in a particular situation can be made by selecting and combining the pertinent component spectra. (From Wenz [114].)

quently reproduced both in American and Russian literature. In addition to the noises covered in this graph, there are noises of biological origin, snapping shrimp, for example, which can in certain locations drown out the signals involved in sonar.

9.7. Animal Echolocation

The rapid growth of the field of echolocation by animals is reflected in the book *Animal Sonar Systems*, published in 1980, with more than one thousand pages devoted to the subject [115]. The fact that the operations of the bat were indeed a form of sonar (and thus a classified subject) perhaps caused a delay of some publications until after World War II, but the real expansion of the field began in the late 1960s. According to a graph by Alan D. Grinnell, the number of papers on echolocation by bats published over a 3 year period increased from 6 in 1955–1957 to 106 in 1975–1977 [116].

The fact that dolphins (or porpoises [117]) and whales make sound has been known since ancient times, but the serious study of their character and their use in communication did not develop until World War II, when the sounds produced by underwater creatures became a serious interference problem in sonar operations [118]. After the war, serious studies of the sounds produced by these animals and of their hearing structures were carried out. A publication by W.N. Kellogg, Robert Kohler, and H.N. Morris in 1953 demonstrated that the bottlenose dolphin produced sound pulses with frequency components in the range 80–120 kHz [119]. A little later, Kellogg noted that these dolphins could avoid underwater obstacles by using their sound system [120]. And, in 1960, F.W. Reysenbach de Hahn wrote of

the so very interesting acoustic system in toothed whales...the brain is mainly built around the central acoustic system, which is a very remarkable feature in the mammalian kingdom. The nuclei reflex and coordinate centre, which possess a special significance for the interpretation and elaboration of information supplied by supersonic frequencies is particularly well developed. This suggests a striking parallelism with the...acoustic system of bats. [121]

It was soon well established that dolphins sent out echolocation signals in the range 50–150 kHz, that were highly directive in the forward direction [122]. K.S. Norris, W.E. Evans, and R.N. Turner [123] found that the bottlenose dolphin *Tursiops truncatus* could detect a difference in diameter of metallic spheres of as little as 10%. Since that time there have been numerous papers studying this dolphin ability in more detail [122].

The recording of sounds made by sperm whales goes back as far as 1952 [124]. The character of whale sounds, largely clicks with verying repetition rates and with center frequencies of the pulses in the neighborhood of

30 kHz have been studied in detail by William A. Watkins and William E. Schevill at the Woods Hole Oceanographic Institution [125].

It would appear that the whale sounds are a form of communication between individual members of the community. In an article in 1980, Watkins remarked

The acoustic behavior of sperm whales, though not apparently convenient for echolocation, does seem to fit a context of communication, perhaps to keep in contact with other sperm wales underwater. [126]

And with that, we shall leave the whales to their private conversations.

9.8. Signal Processing

The 1950s also saw the appearance of a new branch of acoustics—signal processing. In a review paper on the subject in 1971, Victor C. Anderson wrote "What a revelation it was to turn back to the report of the National Research Council of 1948 [127] of the basic problems in underwater acoustic research to find that there were no signal processing problems at that time; it fact, there wasn't even any signal processing [128]."

Up until 1950, the immediate problem of sonar research was thought to be the physico–psychological problems of the observer, rather than further processing of the signal. In the article just cited, Anderson defined signal processing as "the theory and practice of applying spatial and temporal transformations to samples of an acoustic wave field to enhance the measurement of a desired signal in the presence of an interfering background [129]." The growth of this field in the 1950s was generated by the co-development of experimental techniques in electronics (such as high-speed digital circuitry) and the advances in information and communication theory (which in turn were stimulated by research in radar). These developments included multiple channel spectral analysis [130], Fast Fourier Transforms (FFT) [131], and multiple preformed beam technology.

9.9. Structural Sound and Vibration

The introduction of the shape factor in vibrating beam theory by Timoshenko was justified in a paper by R.D. Mindlin and H. Deresiewicz in 1955 [132]. They used a technique of matching the approximate solutions from the Timoshenko theory with the exact limiting case of three-dimensional theory, developed by the same authors in an earlier paper [133]. Subsequent papers in beam theory in this period treated the necessity of using higher-order solutions when the elementary theory proves inadequate [134].

The problem of the behavior of the middle (central interior) surface of thin shells was finally resolved by E.W. Ross, Jr. in 1968, with the verdict mainly in Lord Rayleigh's favor. For a free edge of the shell, Rayleigh was correct. For other boundary conditions, the issue is still uncertain [135].

The work of Arnold and Warburton on the theory of the free vibration of a finite free shell, mentioned in Chapter 8, was followed up by a study of the effect of edge conditions on the natural frequencies of such vibrations by K. Forsberg [136]. His work, and that of Reissner, on the transverse vibrations of a shallow spherical shell [137] were some of a large number of such specialized studies. It soon became clear that any practical applications based of these analyses would require extensive computer use. Arturs Kalnins made such applications in a paper in 1964 [138].

9.10. Seismic Waves

A new element entered into seismology just before World War II and was exploited in the 1950s. This was the T wave, or tertiary wave. In 1940, D. Linehan [139] reported these waves in recordings from earthquakes in the West Indies and also those occurring in the equatorial region of the mid-Atlantic Ridge. These waves arrived long after the P and S waves. Their origins were not immediately identified.

Linehan and his colleagues returned to the study of this phenomenon after World War II [140], and an increasing number of other investigators joined in [141]. It became clear only gradually that the T wave was an acoustic wave, launched in the ocean by the arrival of the P and S waves at the boundary between the ocean and the ocean floor [142]. Since the T wave is an acoustic wave in water, its velocity is much smaller than the velocity of either the P or the T wave, thus accounting for the lateness of its arrival. A typical example of its reception is shown in Fig. 9-20 [143].

9.11. Noise

Hallowell Davis (1896–1992), whom we mentioned in Chapter 8, and who spent a long and distinguished career at Harvard and the Central Institute for the Deaf in St. Louis, used to tell an old riddle: "What comes

FIGURE 9-20. The T-wave. Recording by the Martinique station of the signal of the earthquake of 26 April, 1961, in the Caribbean region. (From Kadykov [143].)

with a carriage and goes with a carriage, is of no use to the carriage and yet the carriage cannot move without it [144]?" The answer is, of course, "noise." Noise is an asset to man and animal when it warns of approaching danger. But we are mostly concerned with the negative aspects of the phenomenon.

Noise, by its very definition, is a special part of acoustics that is unwanted. Most of the time we simply want to get rid of it. But in this riddance process, the field of noise and noise control contributes to every branch of acoustics. In the early history of the field, most attention was paid to the recording of noise levels, and to the materials that might absorb the noise or keep it from areas of interest, such as concert halls, offices, private residences, and the like. By 1950, however, the study of noise had become more sophisticated. One could divide the field into sources of noise, paths followed by noise, effect of noise on man and animal, noise reduction at the source, noise reduction along the path of the sound, and noise reduction at the ear. And each of these subjects became a minor field of its own.

Sources

Among man-made sources, the most widespread annoyances are provided by machinery, and motor vehicle and aircraft noise. Figure 9-21 shows the range of noise from different components of: (a) heavy trucks; and (b) light trucks and passenger cars. These figures were taken from a review paper by Ralph K. Hillquist and William N. Scott in 1974. It is clear that there are many sources of noise in each case and all of these need to be addressed before significant reduction of noise can be obtained [145].

For aircraft noise the traditional source has been the propeller. Blokhintzev, in his book in 1946 [146], gave some attention to propeller-generated noise, and numerous other theoretical expressions have been obtained since that time, the most significant one being that given in a paper by J.E. Ffowcs-Williams and D.L. Hawkings [147, [148]. A general review of the field was given in a tutorial paper by Alan Powell in 1995 [149].

We have already mentioned the contribution of Lighthill to the theory of sound generated by turbulence. In particular, he found that a jet aircraft traveling at relatively low Mach number M generated sound whose far-field intensity varies as M^8. But much experimental data suggested a proportionality to M^6, an exponent obtained earlier by Yudin [150] and by Blokhintzev [151]. A full discussion of this difficulty is given in a later essay by G.M. Lilley [152].

Noise Directionality

The earliest studies of noise did not concern themselves with the directions from which the noises came, but later work showed that this directionality

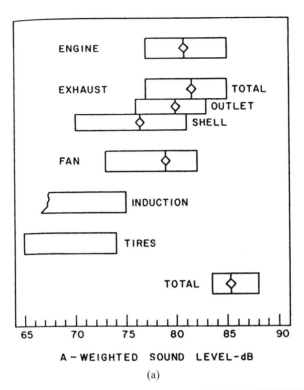

(a)

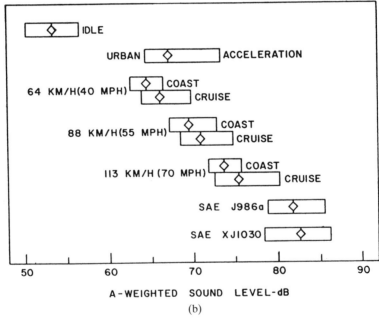

(b)

FIGURE 9-21. (a) Range and mean of component sound levels for a group of heavy-duty trucks accelerating at full throttle to rated engine speed and approximately 56 km hr⁻. (b) Range and mean exterior sound levels for a group of passenger cars and light trucks. (From Hillquist and Scott [145].)

could be significant. Of course, it was well known by this time that, as sound or noise travels through the atmosphere, it becomes attenuated because of various absorption mechanisms and because of beam spreading. The temperature variation of the atmosphere can cause sounds to bend away from the receiver, sometimes reducing the sound (both signal and noise) level. Still later, however, studies were made on the effect of reflection on such propagation. F.M. Wiener, C.I. Malms, and C.M. Gosos studied the effect of buildings on a street in reflecting sounds and maintaining the level of noise (Fig. 9-22) [153]. A general study of the effects of multipath transmission on urban sound was presented in 1974 by Richard Lyon [154].

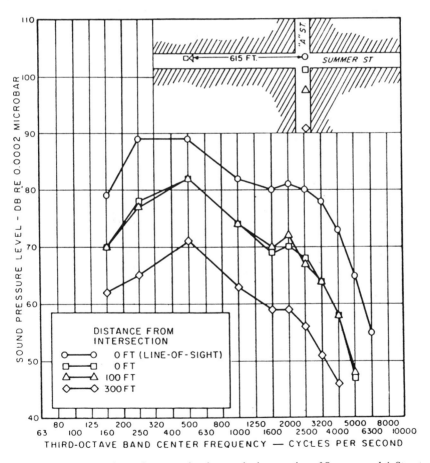

FIGURE 9-22. Measured sound pressure levels near the intersection of Summer and A Streets. A Street is 20 ft under Summer Street at right angles. (From Wiener et al. [153].)

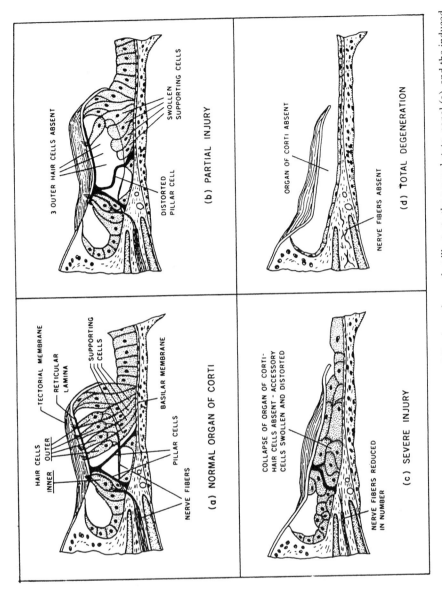

FIGURE 9-23. Drawings of the human organ of Corti are shown that illustrate the normal state, panel (a), and the induced permanent injury, panels (b), (c), and (d). (From J.D. Miller [144].)

Effects of Noise on Man and Animals

It been known scientifically that prolonged exposure to loud noise can cause temporary or even permanent ear damage. Figure 9-23 shows micro-photographs of the organ of Corti for normal ears and for ears exposed to loud noise for long periods [155]. Less serious effects are those of the temporary threshold shift (TTS), which is the level of the tone that can just be heard shortly after exposure to noise. TTS formed the basis for some of the first experiments on predicting the precise quantitative effects of noise. W. Dixon Ward (1924–1996) published extensively in this field in the 1960s, and showed that noises in the range 2000–6000 Hz were most significant [156]. If the loud noises are too loud or too-long endured, a permanent threshold shift will occur, i.e., the person is partially deafened. The develop-ment of electronically amplified music, rock bands, and boom boxes all contributed to such hearing losses.

Figure 9-24 shows the practical effect of loud noises on our ability to hear speech [157]. These measurements are the result of numerous

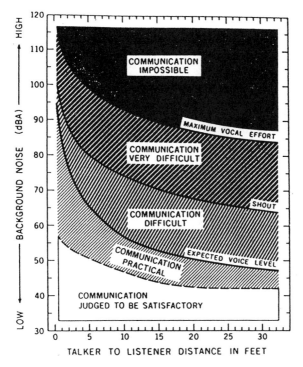

FIGURE 9-24. Simplified chart that shows the quality of speech communication in relation to the A-weighted sound level of noise (dBA) and the distance between the talker and the listener. (From J.D. Miller [155].)

studies, which include the effect of noise in disturbing sleep or other human occupations.

It was also at this time that some noises were recognized as more acceptable to humans than others, and that these noises could have a masking effect on the more obnoxious ones. Thus, a low background of air conditioning can mask such noises as airplanes or trains, or other noises that are short in duration. The term "acoustic perfume" has come into vogue to describe these disguising noises [158].

The effect of noise on animals began to be studied in detail in the 1970s. A report by the Acoustical Society of America, with support from the Alaska Eskimo Whaling Commission covered a study of the relation between man-made noise and vibration and arctic marine wildlife [159].

Detailed summaries of procedures involved in the war against noise may be found in publications of organizations such as the American Speech–Language–Hearing Association and the Council for Accreditation in Occupational Hearing Conservation [160].

Noise Reduction

The fight against noise has been pursued on many fronts. There are endless laws and legal ordinances against loud public noise (Fig. 9-25 [161]), efforts to turn off the sources, or move them elsewhere (the "not in my backyard" mentality). In the United States, the most important of these laws was the Noise Reduction Act of 1972. From the acoustic point of view, however, the battle has been carried out on many fronts. Machinery noise and vibration can be reduced by better mounting, by enclosures, and the use of absorbing materials. Flow noise can be passed through structured chambers, such as mufflers. These modifications of von Helmholtz's resonators were explored extensively by K.U. Ingard [162] and Miguel Junger [163].

In aircraft, most attention has been paid to the external sound produced by jets [164], but the improvement of quieting in the cabins of such aircraft has also been vigorously pursued [165]. Individual gains are small, but over time, the results are impressive. Unfortunately, more powerful jet engines are being built, and more and more flights are made, so that any gain is often literally "lost in the noise." On the ground, widespread use of ear protection equipment, and the development of barriers around airport runways, to intercept the sound and deflect it upward (Fig. 9-26), have led to significant quieting [166].

The efforts at noise and vibration control just described concentrated on: (a) removing the source; or (b) damping it in the progress of its activity. But in 1933, a new direction was taken when Paul Leug took out a patent on suppressing a low-frequency noise by producing an identical noise, but 180° out of phase with the source [167]. This waveform theoretically will cancel the original noise.

City of Bern (Switzerland)*

YEAR PASSED	BY-LAW
1628	Against singing and shouting in streets or houses on festival days
1661	Against shouting, crying or creating nuisances on Sunday
1695	Against the same
1743	For respect of the Sabbath
1763	Against disturbing noises at night
1763	Against noisy conduct at night and establishing regulations for night watchmen
1784	Against barking dogs
1788	Against noises in the vicinity of churches
1810	Against general noise nuisances
1878	Against noises near hospitals and the sick
1879	Against the playing of music after 10:30 p.m.
1886	Against the woodworking industry operating at night
1887	Against barking dogs
1906	For the preservation of quiet on Sundays
1911	Against noisy music, singing at Christmas and New Year's parties and against unnecessary cracking of whips at night
1913	Against unnecessary motor vehicle noise and blowing horns at night
1914	Against carpet-beating and noisy children
1915	Against beating carpets and mattresses
1918	Against carpet-beating and music-making
1923	For the preservation of quiet on Sundays
1927	Against noisy children
1933	Against commercial and domestic noises
1936	Against bells, horns and shouting of vendors
1939	Against excessive noises on holidays
1947	For the preservation of quiet on Sundays
1961	Against commercial and domestic noises
1967	For the preservation of quiet on Sundays

FIGURE 9-25. Antinoise ordinances. (From Schafer [161].)

Twenty years later, this work was further advanced by Harry F. Olson and Everett G. May. Calling their device the electronic sound absorber, they wrote that it "consists of a microphone, amplifier, and loudspeaker connected so that, for an incident sound, wave and sound pressure at the microphone is reduced. Thus it will be seen that the electronic sound absorber is a feedback system which operates to reduce the sound pressure in the vicinity of the microphone" (Fig. 9-27 [168]). This system was designed to operate near the head of the listener and led to many later applications, especially in aircraft. Noise has also been reduced by using ear protectors, ear plugs, etc. [169].

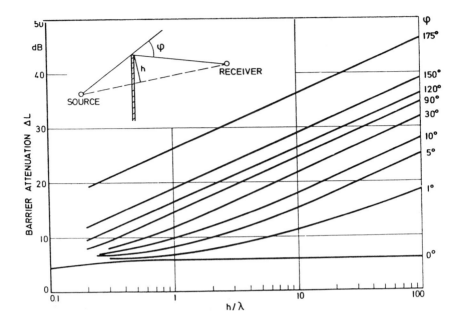

FIGURE 9-26. Effects of barriers. (From Kurze [166].)

9.12. Architectural Acoustics

The field of architectural acoustics in the early 1950s was characterized by optimism. Progress had been made in the period before World War II in the wave theory of rooms, while the increased use of electronics in studying rooms gave rise to a wealth of data on these rooms as well as to insights into their behavior. In 1949, Helmut Haas, reported extensive measurements on the influence of a single echo on the perceptibility of speech [170]. In particular, he measured the effect on intelligiblity when the echo occurred with a time delay from 0 to 160 milliseconds. This effect, variously known as the Haas effect, the precedence effect or the critical time difference, stimulated a number of researches, including one by R.H. Bolt and P.E. Doak [171] on the response of an auditorium to short tone bursts. These works attempted to correlate data with the subjective views of listeners in the rooms being studied, and encouraged the idea that the science of room acoustics must be considered equally with the art of music appreciation.

However, two new concert halls, the Royal Festival Hall, London, in 1951, and the Philharmonic Hall (later named Avery Fisher Hall), New York, in 1962, soon gave the architectural acoustics community an opportu-

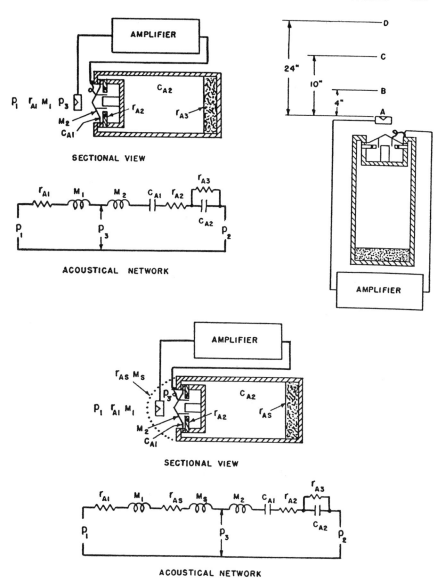

FIGURE 9-27. Active noise suppression. (From Olson and May [168].)

nity for practicing humility. In the case of the Royal Festival its reverberation time was clearly too short [172]. This weakness was corrected by means
of a technique newly developed by P.H. Parkin and K. Morgan [173]. This
technique relies on the ringing of loudspeakers after the sound has stopped,
brought about by having amplifiers operating just short of feedback instability. This introduction of a large number of microphones, amplifiers, and

loudspeakers into the hall, first applied over a narrow frequency band, and then over the range from 58 to 700 Hz, was first regarded solely as a stopgap measure, but it has now been used in the installation of a number of new auditoria [174]. As a result of the many studies that took place before and after this correction, it was said that the Royal Festival Hall was the most studied hall in the history of acoustics.

The second controversy, over Philharmonic Hall in New York, has been more enduring. The acoustical consultant on this project was Leo Leroy Beranek (1914–   ), who had built a distinguished reputation by his research in basic acoustics, textbook writing, and general consulting, first as a professor at MIT and later through the formation of the consulting firm of Bolt, Beranek, and Newman. In preparation for work on this Hall, Beranek collected information on some 54 concert halls, which he subsequently published as a book [175]. The book was indicative of the increasing attention paid to the attitude of the listeners in the various halls, which attention has only increased since that time.

Complaints arose against the quality of sound in Philharmonic Hall from its very beginning, for its "imbalance between reflected sound, which gives immediacy and presence, and reverberant 'symphonic' sound [176]." The controversy over this hall has lasted for more than 30 years. After many attempts at correction, the entire hall was redone, with Cyril M. Harris, professor at Columbia University, as the acoustical consultant, and reopened in 1976 as Avery Fisher Hall.

The author of this book is too far removed from the architectural acoustic scene in his technical expertise for him to venture far into this controversy. The original plan for Philharmonic Hall had been for a room with parallel side walls (the so-called shoebox design), but was modified (against Beranek's advice) to the form shown in Fig. 9-28. The final shape of Avery Fisher Hall reverted to one similar to the original plan, resembling that of the famed Symphony Hall in Boston (Fig. 9-29). Different points of view

FIGURE 9-28. (a) Philharmonic Hall, Lincoln Center, New York, by Max Abramovitz and acoustic consultants Bolt, Beranek, and Newman, 1962; plan as originally built. (Courtesy *Progressive Architecture*.) (b) Avery Fisher Hall, Lincoln Center, New York, by Philip Johnson and John Burgee, with acoustician Cyrill Harris, opened 1976. (Courtesy *Progressive Architecture*.) [176]. (c) Sectional view, schematic electrical diagram and acoustical network of an electronic sound absorber p_1 = sound pressure in free space. M_1 = inertance of the air load. r_{A1} = acoustical resistance of the air load. M_2 = inertance of the cone and voice coil of the loudspeaker. r_{AS} = acoustical resistance of the screen covering the microphone and cone. M_s = inertance of the screen. C_{A1} = acoustical capacitance of the suspension system of the cone. r_{A2} = acoustical resistance of the cloth over the apertures in the back plate. C_{A2} = acoustical capacitance of the volume of the cabinet. r_{A3} = acoustical resistance of the sound absorbing material in the cabinet. p_2 = driving sound pressure in the loudspeaker. p_2 = sound pressure at the microphone.

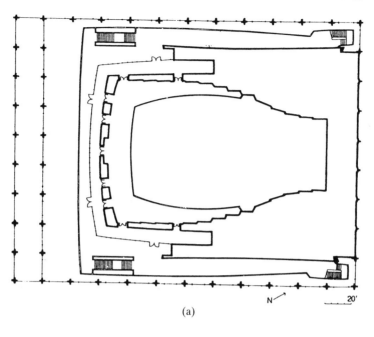

(a)

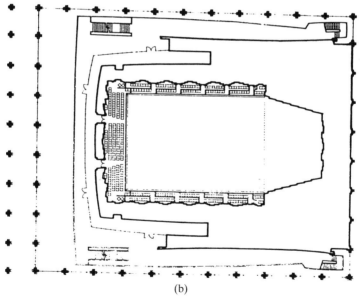

(b)

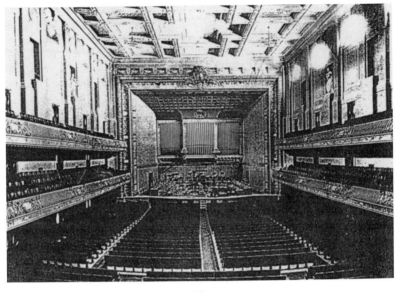

(a)

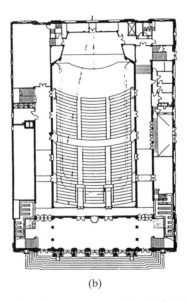

(b)

FIGURE 9-29. (a) Symphony Hall, Boston: interior. (Boston Symphony Hall); and (b) Symphony Hall, Boston. (*Monograph of the Work of McKim, Mead and White*, 1879–1915, Architectural Book Publishing Company, 1981 [1925].) [176].

can be found in Forsyth's book [177]. It is unfortunate that this controversy has obscured the fact that many fine halls have been designed in recent years, based on the acoustical ideas of Beranek, Harris, Parkin, and many others [178].

One outcome of this problem was the intensive work done by acousticians at the Bell Laboratories in measuring the properties of this hall. In 1966, B.S. Atal, Manfred R. Schroeder, Gerhard M. Sessler, and James E. West published a paper on the evaluation of the acoustical properties of rooms by means of digital computers [179]. They made tape recordings of the sound pressure at different points in the room, converted these analog signals to digital form, and performed filtering, envelope detection, and other forms of signal processing of the day. Using these techniques, they studied the forms of the reverberation, distiguishing direct signal, early and late reverberation, as well as the directional properties of the sound energy flux, which is now known as sound diffusion. This application of the computer to room acoustics, in the words of T.M. Northwood, "pioneered by M.R. Schroeder and colleagues at Bell Telephone Laboratories, has facilitated great leaps forward in all aspects of the subject [180]." In the very next paper in *JASA*, this extraordinary group applied their techniques to a study of Philharmonic Hall, reporting data from the hall in 1962, and after the intermediate corrections in 1963–1965 [181].

There is, of course, no such thing in human life as perfection. Even Symphony Hall in Boston has had its occasional critics, and we should not forget the storm of criticism by the audience in the first performances of many operas, such as *Carmen* and *La Bohème*. No matter how good a hall may be from the technical point of view, the performance still has to satisfy the tastes of the individual listeners and performers, tastes that do change with time.

A parameter affecting listener appreciation was dubbed by Beranek the initial-time-delay gap—the difference in arrival times (at the ears of the listener) between the direct sound and the first reflections (Fig. 9-30 [182]). The importance of the distribution of directions of the reflected sound, especially lateral reflections, was emphasized by A. Marshall [183] in 1968, and it was subsequently noted by his colleague, M. Barron [184] that, for the best halls, this first reflection came from lateral reflections. We shall see in the next chapter that a great effort was soon to be made in correlating the physical measurements with subjective tests.

Another important development of this period was a better appreciation of diffuse reverberation. It was generally thought that the mellow quality of many old halls was due to reflections from the many irregularities in the surfaces of their walls, but at one time there appeared to be no good way to quantify this attribute. Such measurements, however, were made by Erwin Meyer and his student R. Thiele in the 1950s [185].

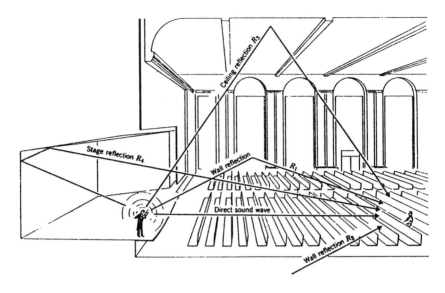

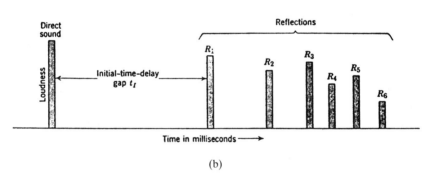

(b)

FIGURE 9-30. Various echoes in a concert hall. (a) Sketch showing direct and four reflected sound waves in a concert hall. (b) Time diagram showing that at a listener's ears, the sound that travels directly from the performer arrives first, and after a gap, reflections from the walls, ceiling, stage enclosure and hanging panels arrive in rapid succession. The height of a bar is related to the intensity of the sound. The initial-time-delay gap t_1 is indicated. (From Beranek [182].)

9.13. Physiological Acoustics

The major developments of physiological acoustics in this period were in studies of the characteristics of the ear passage from outer to inner, the dynamics of the cochlea, and the electrophysiology of the ear.

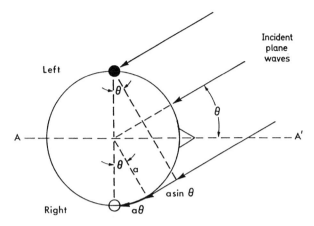

FIGURE 9-31. Coordinate system for horizontal plane through center of head showing azimuth θ of incident plane waves. (From Shaw [187].)

The Ear Passage

The diffraction around a sphere (serving as a model of the human head) has been shown to cause an increase of up to 6 dB above the free-field sound pressure [186]. We mentioned in Chapter 8 the work of Wiener and Ross in measuring the sound pressure distribution in the auditory canal. They had found that the sound pressure level in the auditory canal could be as much as 20 db higher than the free-field value, i.e., the outer ear serves as an amplifier. Detailed studies of the human ear and of models of the ear by Edgar A.G. Shaw and his colleagues at the National Research Council Canada laboratories in Ottawa have made clear the nature of this amplification (Fig. 9-31) [187].

The middle ear has also received its share of attention during this quarter-century [188]. A.H. Moller studied the acoustic impedance of the eardrum both in human subjects and in cats, and has also developed equivalent circuits for the middle ear cavities of different animals [189].

We recall von Helmholtz's idea that the hearing of combination tones was due to nonlinearities in the bones of the middle ear (Chapter 3). Measurements on the ears of cadavers have revealed some asymmetry in the inward and outward motions of the stapes for stimulation by high-intensity, low-frequency sounds [190] but later studies with anesthetized cats show linearity of the stapes for intensities up to 130 dB SPL (sound pressure level) for frequencies below 1500 Hz, and up to 140 dB for frequencies above that level [191]. The nonlinear effects of the ear must therefore be assigned to the cochlea.

Cochlear Mechanics

Since the role of the cochlea is to convert acoustic signals into neural signals, and then to perform some processing on the latter, the study of the cochlea can be divided into the mechano–acoustic behavior and the electro-physiological behavior. As was pointed out in Chapter 8, von Békésy iden-tified the problem as a hydrodynamic one. The vibrations at the oval window are transmitted through the cochlea, both to the basilar membrane and to the fluid. As we saw in Chapter 8, von Békésy's model experiments [192] indicated that the vibrations of the basilar membrane produce eddies within the fluid.

A later model of the operation of the cochlea, developed by Juergen Tonndorf [193], is shown in Fig. 9-32. By using aluminum dust particles in the fluid, Tonndorf was able to demonstrate particle motion in the fluid in response to a 50-Hz signal (Fig. 9-33) [194]. The particles of the fluid are under the effect to two forces resulting from the original motions of the oval window and the displacements of the basilar membrane. They thus execute simple Lissajous figures (the so-called trochoidal motion). From this Tonndorf inferred that surface waves propagated in the fluid. These waves were capillary waves (so that they are not affected by gravity), and corre-spond to shallow-water waves [195]. The character of these capillary waves is clearly nonlinear and provides an explanation for the origins of Tartini tones in the inner ear [196].

A wide variety of other problems of the cochlea were also studied, including the Q of the tuning curve of the basilar membrane, the nature of the stimulation of the hair cells, and the development of electrical network models [197].

We noted the birth of cochlear electrophysiology in Chapter 8. In 1950, H. Davis, C. Fernandez, and D.R. McAuliffe [198] noted that "the most widely accepted theory [of the initiation of nerve impulse in nerve fibers] has been that an electric potential known as the cochlear microphonic is generated by the hair cells in the organ of Corti and serves as a direct electrical stimulus to the peripheral terminations of the fibers of the audi-tory nerve." These, together with the summating potentials [defined by Peter Dallos (1934–   ) as "any stimulus-related dc electrical event that can be recorded from the cochlea" [199]] identified by these authors, form the so-called receptor potentials. The complexity of these signals may be inferred from the work of Dallos (Fig. 9-34) on a single hair cell. The figure shows waveforms of various signals and a block diagram of such a system as a whole [200].

Treatment of Hearing Impairment and Deafness

While the work in the 1930s emphasized fenestration of the horizontal semicircular canal as a relief of deafness, the 1950s marked a return to the

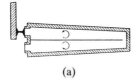

(a)

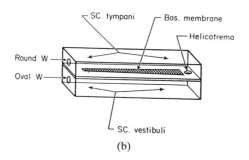

(b)

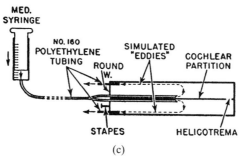

(c)

FIGURE 9-32. (a) Eddies in the cochlea. (From von Békésy [192].). (b) Tonndorf models of the cochlea [193]. (b) Cochlear model of the von Békésy type. This schematic drawing represents the model that I have used in my own experiments. It is very similar to von Békésy's original one. It has a transparent lucite shell enclosing two "perilymphatic" scalas. The perilymphatic fluid is a glycerin–water mixture controlled for viscosity. In a model five times larger than a human cochlea, a glycerin–water mixture of 30 centipoises provides approximately the same damping as obtained in a model the size of an actual cochlea with a fluid viscosity of 2 centipoises, which is that of human perilymph. Sandwiched between the two scalas is the cochlear partition, a thin metal frame giving support to the basilar membrane, which is made of latex, and, in typical cases, has an exponential stiffness gradient. Note also the helicotrema and the two cochlear windows. (c) Arrangement for production of simulated "eddies" in the cochlear model schematic). (From Tonndorf [194].)

Miot procedure (Chapter 8) for elimination of otosclerosis. In 1952, Rosen [201] brought back the Miot technique. However, a new world was opened up, first by J. Shea in surgical removal of the stapes, with appropriate modification of the connections with the other parts of the middle ear (stapedectomy) [202], and then by F. Zollner [203] and H. Wullstein [204] in the insertion of plastic replacements for the removed tissues (tympanoplasty).

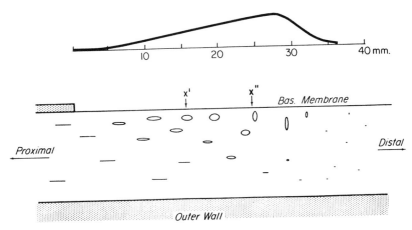

FIGURE 9-33. Particle motion in the fluids of scala vestibuli of a cochlear model in response to a 50-Hz signal. This drawing gives the type and relative magnitude of displacements in various locations: one-dimensional (longitudinal) in the vicinity of the window and along the outer wall; and two-dimensional (elliptical or circular) along the partition. Magnitudes are overstated for the sake of illustration. Each particle completes its pathway once each period; in the present example the rate was $50s^{-1}$. Such two-dimensional motion is known as trochoidal. Implied in the present drawing is the notion that the trochoidal fluid motion forms an orthogonal vectorial field with respect to the partition at rest. This notion (although correct for freely progressing surface waves) represents an oversimplification. The magnitude of displacement in the vertical direction of the two-dimensional, trochoidal pathways in close vicinity of the partition is shown in the curve at the top of the figure. It is identical to the envelope over the traveling waves along the partition itself. (From Tonndorf [194].)

The use of electronic, wearable hearing aids, which had begun in the 1930s with cumbersome battery packs and amplifier systems, underwent considerable modification in the 1960s and 1970s through the replacement of vacuum tubes by transistors and then by integrated circuits [205]. It was recognized that amplification alone was not sufficient, and the insertion of the hearing aid into the ear modifies the acoustical character of the auditory canal and produces its own distortions.

9.14. Psychological Acoustics

Sensitivity

As scientists produced quieter and quieter listening spaces, it became evident that the limiting factor on the minimum audible sound pressure (MAP) at the eardrum must be physiological noise—the idea advanced by Sivian and White [206] in 1933. This was supported by the research of Shaw and Piercy [207], who made extensive measurements of the noise level between the ear and the earphone. At frequencies below 500 Hz, they found that the sound pressure level in one-third octave bands was almost the same as that obtained by Sivian and White for the MAP level.

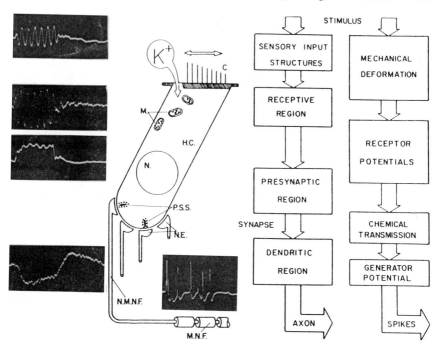

FIGURE 9-34. Single hair cell. Sketch of a hair cell and associated nerve endings along with structural (center) and functional (right) block diagrams of the system. The inserts demonstrate the waveforms of the various quantities indicated in the directional block diagram, namely, stimulus, CM, DIF SP, generator potential, and neural discharges. C, cilia; M., mitochondria; H.C., hair cell; N, nucleus of the cell; P.S.S., presynaptic structures; N.E., nerve endings; N.M.N.F., nonmyelinated nerve fibers (dendrites); M.N.F., myelinated nerve fibers (axons). (From Dallos [200].)

Another look at the minimum threshold was taken by K.J. Diercks and Lloyd A. Jeffress [208]. Their work, and later that of J.P. Egan [209], indicated that binaural detection possesses an advantage (in the sense of minimum threshold) only when there is some incoherence in the signal (or noise) at the two ears.

Finally (at least for this period), we should mention a paper by John Swets [210], which questioned whether any minimum threshold actually exists at all.

Efforts at defining the upper- and low-frequency limits of hearing were also made. N.S. Yoewart, M. Bryan, and W. Tempest [211] measured responses down to 1.5 Hz, while Northern et al. made extensive quantitative measurements up to 18 kHz [212]. A plot of these data, together with the earlier work on the threshold of feeling by von Békésy and others [213], is shown in Fig. 9-35. The continuing difference of 5–10 dB between the MAP and the minimum audible field (MAF), first noted by Sivian and White, and attributed in part by them to physiological noise, was still not completely accounted for.

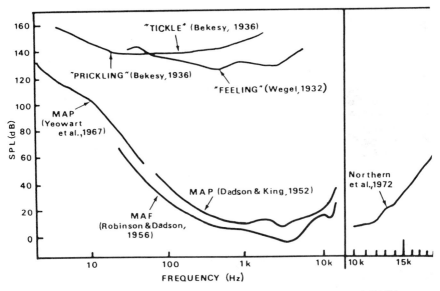

FIGURE 9-35. Minimal audible pressures (MAP). (From Yoerwart et al. [213].)

Pitch

In our treatment of pitch in the previous chapter, we mentioned the two theories of pitch—place and periodicity. In an attempt to embrace both theories, Joseph Carl Robert Licklider (1915–1990) presented arguments for a duplex theory of pitch perception [214]. A distinguishing feature of Licklider's work is that he introduced autocorrelation analysis. Noting that we can analyze sound waves both with frequency and periodicity, he pointed out that, while these quantities are reciprocal, our methods of measurement for each are quite different. To quote Licklider,

one—frequency analysis performed by an array of bandpass filters—has been incorporated into auditory theory. The cochlea is almost universally regarded as being an extended wave filter that distributes oscillations of different frequencies to different places. . . . Autocorrelational analysis is an analysis, carried out entirely within the time domain, that yields the same information as the power spectrum which is obtained analysis in the frequency domain. [215]

Masking and Time Resolution

The 1950s saw a considerable interest in time resolution, i.e., the shortest time interval between two acoustic events that the ear can detect. This can also be expressed in terms of the smallest interval of time, beyond which one notices a significant drop in the acoustic sensation. In his paper on the precedence effect (Chapter 8), Helmut Haas cited an estimate of the time

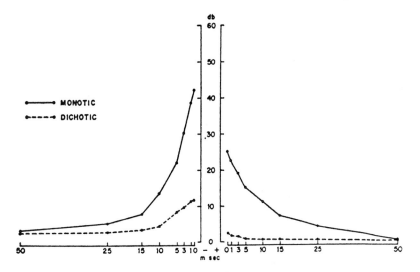

FIGURE 9-36. Results of backward and forward masking under conditions of 70-dB masking and 5-ms probe duration. (From Elliott [218].)

interval at 50 ms given by Petzold in 1927, and dubbed by Petzold the "threshold of masking [216]." Later research indicated this interval to be much shorter. But the problem is complicated by the fact that the ear accepts various cues that will distinguish one sound from another [217].

Masking research originally involved two signals thought to be heard simultaneously, but the field was broadened with the introduction of the concept of temporal masking—backward masking when the masking tone follows the signal, and forward masking if it precedes it (Fig. 9-36). These ideas led to vigorous work both in the United States and in the Soviet Union [218]. It is worth pointing out that the three distinguished acousticians who pioneered (independently) in this research were all women— Irina Samoilova, Ludmilla Chistovich, and Lois Elliott. It's been a long time since the days of Sophie Germain!

9.15. Speech

When Harvey Fletcher published his book on speech communication in 1953 [219], his speech interest was confined to the description of the human speaking mechanism, the measurement of the waveforms and the power of speech, and the frequency of occurrence of the different speech sounds— largely reflecting the viewpoint of the physicist and electrical engineer. But many more techniques were needed, and the various fields within the subject of speech would expand mightily in the following 20 years. In the preface to a collection of reprints on acoustical phonetics in 1976, E.D. Fry noted that, while

The use of memory stores for words, of grammatical and phonological forms, the organizing of messages at the semantic level and, in fact, most of what we regard as linguistic activity in speech is a matter of psychological functioning. The working of the nerves and muscles, broadly the articulatory aspect of speech, is material for physiological study, while the generation, transmission, and reception of sound waves are accessible only to the methods of physics. [220]

Speech Production

The duality between the physical and the nonphysical that we discussed in the section on architectural acoustics appears again in studies of speech and of hearing. We can pursue speech back through the mouth, throat, larynx, and lungs, but the control must be from the brain. As early as 1861, the French neurosurgeon Paul Broca had discovered that speech production was controlled in the frontal lobe of the left cerebral hemisphere [221], while in 1874, Wernicke [222] found that the understanding of speech was localized in the left temporal lobe (Fig. 9-37) [223].

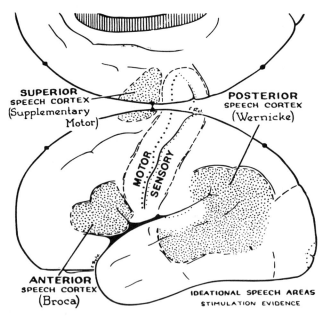

FIGURE 9-37. Broca and Wernicke regions of the brain. Summary map of the areas of the surface of the left cerebral hemisphere found by Penfield to be important for speech. The lower drawing shows the lateral surface, while the upper drawing shows the continuation of the areas on the medial cortical surface. (From Wilder Penfield and Lamar Roberts: Speech and Brait Mechanisms, © 1959 by Princeton University Press. Reprinted by permission of Princeton University Press.) (From Penfield and Roberts [223].)

The earlier work of Broca and Wernicke was performed mainly on cadavers. In the mid-twentieth century, Wilder Penfield and Lamar Roberts, neurosurgeons in Montreal, used electrical simulation on the brains of living subjects and were able to make extensive localization of many bodily activities, including speech (Fig. 9-38) [224].

One of the conclusions of the work of Penfield and Roberts was that speech "involves simultaneous action in many parts of the brain and is apparently too complex to be elicited by a single stimulus" [225] (i.e., man-made, from the outside).

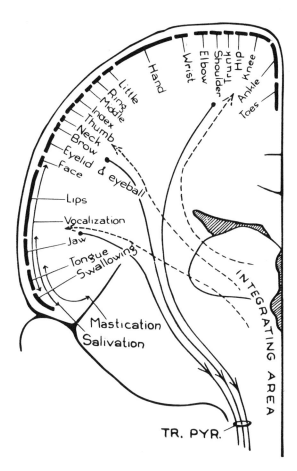

FIGURE 9-38. Location of response to electrical stimulation in the brain. Frontal cross section through the motor strip of the right hemisphere, showing the location of response to electrical stimulation. TR. PYR. refers to the pyramidal tract. (From Wilder Penfield and Lamar Roberts: Speech and Brain-Mechanisms, © 1959 Princeton University Press. Reprinted by permission of Princeton University Press.) (From Penfield and Roberts [224].)

Detailed studies by Brenda Milner et al. [226] have indicated that, for most people, both left- and right-handed, the naming of objects is produced primarily in the left side of the brain, while ordering of words, as in sentences, occurred in the opposite side. The studies of spoonerisms [227] indicate that speech production involves preplanning by the brain, As Borden et al. put it, "It is clear that speakers do not normally call forth and speak a sentence one word at a time." [228]

Work on the physics side of speech production has been equally intensive. Following on the earlier papers of J.Q. Stewart and H.K. Dunn (Chapter 8), various authors pointed out [229] the usefulness of electrical analogs in following the physics of speech production (which studies also lead to experiments on speech synthesis, see below). In connection with research on speech synthesis, James Loton Flanagan (1925–   ) of the Bell Laboratories also demonstrated the significance of the use of our understanding of fluid dynamics in analyzing the behavior of the glottis. In studying the bursts of air emitted by the glottis, he found the individual bursts to have a roughly triangular shape [230]. Flanagan has been one of the most distinguished workers in speech synthesis. After a full career at Bell Laboratories he has joined the faculty at Rutgers University. He was recently awarded the National Medal of Science.

The work of Flanagan has been paralleled by that of many other scientists, the most significant of which has been that of Jan Willem van den Berg at the Laboratory of Medical Physics, University of Groningen [231], who worked mainly with excised larynges.

Others working on this study of the structure of the air bursts include Kenneth N. Stevens and Arthur House [232], who noted that the triangular shape led to a signal that was rich in the harmonics of the fundamental (Fig. 9-39). Such signals then pass through the filter mechanism of the vocal tract between the lips and the glottis. If the tract was assumed to have formants centered at two frequencies, then the transfer function, viewed as an electrical analog, might take the form shown in Fig. 9-40. These same authors pointed out that one could now have a more accurate definition of the term formant:

it is proposed that a formant be interpreted as a normal mode of vibration of the vocal tract, and formant frequency be defined as the frequency of such a normal mode of vibration. The term formant . . . implies . . . a complex number, consisting of a real part (proportional to the formant bandwidth) and an imaginary part (the formant frequency). [233]

The goal of speech research has of course been the acoustic explanation of speech. One part of this has been the study of vowels. Stevens and House remark in their paper that current acoustical theory regards a vowel sound "to be the result of excitation of a linear acoustic system by a quasiperiodic volume velocity source." [234] One the other hand, as Katherine Harris has pointed out,

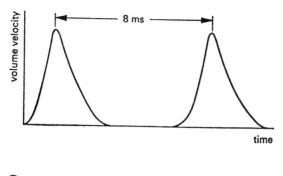

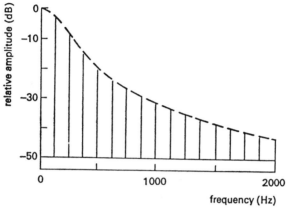

FIGURE 9-39. Frequency analysis of speech air bursts. (From Stevens and House [232].)

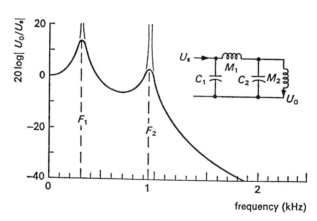

FIGURE 9-40. Formants in transfer functions. (From Stevens and House [233].)

as long as the glottal source operates and the mouth is open, vocal output has the form of a constantly changing formant stream. This stream is difficult to reconcile with the steady-state description of static entities such as a set of vowels. [235]

As one can judge, this pairing off of the static and the dynamic has made for vigorous disagreement in the speech community.

The development of the concept of the formant led to efforts to reproduce formants as a prelude to synthetic speech. As Flanagan and Lawrence R. Rabiner (1943–   ) wrote in 1973,

The engineering approach to speech synthesis has traditionally been one of modeling the frequency-transmission characteristics of the vocal tract between the glottis and the mouth. Since any device that synthesizes speech in this manner is an analog of the human speech-producing mechanism from a *terminal* point of view, such devices have been called terminal-analog or formant, synthesizers. [236]

The analog side of this research was carried out at Bell Laboratories by Flanagan, at Edinburgh University by Walter Lawrence, and at the Royal Institute of Technology in Sweden by Gunnar Fant [237].

Speech Intelligibility

The use of articulation testing and the articulation index (AI) was both supported and attacked in the 1950s. Licklider in 1959 summed up both the pros and the cons of the AI [238], concluding that "the difficulties (with the AI method) may be quite fundamental and that what constitutes a fair engineering approximation may not constitute a fair model for those whose interest is to understand the process of hearing." [239] On the other hand, Kryter, taking the engineering approach, developed the method further to became an ANSI standard in 1969 [240].

Speech Processing

The earlier model of an artificial larynx developed in the 1930s (Chapter 8) was replaced in the 1950s by a battery operated, electronic multivibrator with a telephonic diaphragm (Fig. 9-41) and an adjustable frequency [241]. As the figure indicates, the instrument provides the carrier frequency to replace the function of the larynx, allowing the use of the tongue and lips to shape the vocal tract and thus control its resonances so as to form speech. This has become a widely used instrument for people whose larynx has been removed.

Speech Perception

Work on speech perception was given a great push forward by Franklin Cooper (1908–   ). Cooper was the cofounder in 1935 (with Caryl

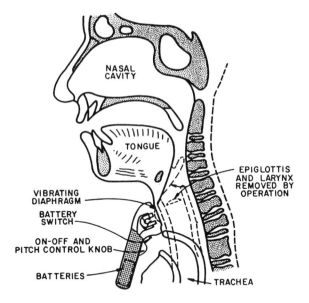

FIGURE 9-41. Use of the electronic artificial larynx. (From Flanagan [241].)

Haskins) of Haskins Laboratories [242]. He invented the so-called Pattern Playback speech synthesizer (Fig. 9-42) [243] to study speech perception in the days before computer-based synthesizers. This machine reversed the procedure of spectral analysis of speech, in which spectrograms of speech sound gave information about formants. The playback patterns (synthetic spectrograms) of formants for *different speech sounds* were painted on the Pattern Playback (Fig. 9-43 [244]–[246]) and the corresponding sounds were emitted. To quote a later review,

[with the Pattern Playback], the experimenter could see at a glance the whole acoustic pattern, could repeatedly hear how it sounded, and could easily modify it. Systematically varying an acoustic dimension thought to be important in perception, the investigators had listeners compare and label the synthesized stimuli. By such means, the Haskins group ... demonstrated the role of linguistic experience upon speech perception and the role of context in the perception of individual phonemes. [247]

Speech Recognition

Modern societal needs, including the military and the courts, led also to the study of speech recognition. Arthur J. Compton [248] studied the effects of filtering and vocal duration on speaker identification, from which he concluded that 25 ms was a sufficient time for voice identification, and that frequencies of the voice below 1020 Hz does not affect identification ability,

FIGURE 9-42. An acoustical artist at work. Franklin Cooper painting a syllable on his Pattern Playback synthesizer. (From Borden, Harris, and Raphael [243].)

but filtering of frequencies above that value reduces the ability of the listener to identify the speaker.

Attempts to identify voices on the telephone by speech spectra techniques became intensive in the 1950s and even earlier. Anyone who has read Solzhenitsyn's novel *The First Circle* will recall the laboratory described therein in which Stalin was attempting to identify his enemies by means of such techniques [249]. The practical breakthrough was provided in 1962 by Lawrence Kersta from Bell Laboratories [250]. Kersta distinguished between the sound spectrograph prints, which he called "bar voiceprints" (Fig. 9-44), and lines of constant intensity, which he called "contour voiceprints." In both figures, the horizontal axis is the time and the vertical axis is the frequency. The density in the figure of (a) represents the intensity of the sound. Since the patterns in the bar voiceprint depend on the dimensions and structure of the vocal tract, Kersta was confident that no two humans would have identical voiceprints, so that the technique could be used for identification in the same way as fingerprints. The contour voiceprint, Kersta believed, was more suitable for speaker identification at security checkpoints, while the bar

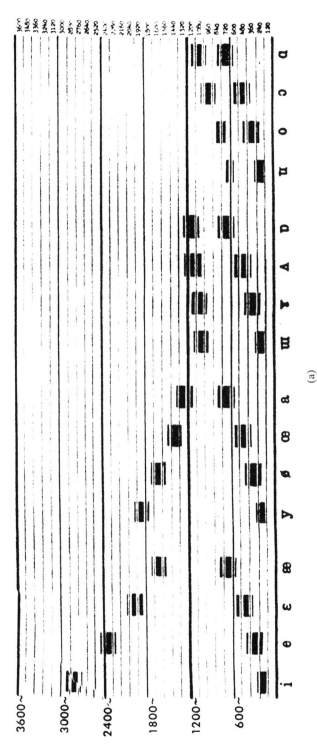

(a)

FIGURE 9-43. (a) Two-format synthetic vowels as patterns painted for the Haskins Pattern Playback. (From DeLattre et al. [218].) (b) Spectrographic patterns with varying transition duration. (From Liberman et al. [246].)

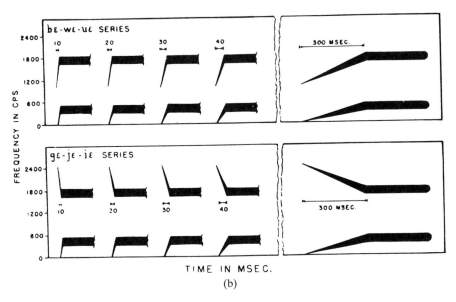

FIGURE 9-43. *Continued*

voiceprint could be used for identifying unknown voices in criminal actions [251].

These assertions by Kersta led to controversy, whereupon the Acoustical Society of America appointed a number of its members who were experts in the field to examine the process and make a report [252]. Their conclusions were that the technique might be useful in voice identification, but could not be relied on to allow an unequivocal conclusion as to the identity of the speaker. The debate continues, both in scientific journals [253] and in law reviews, and indeed, in the law courts [254].

9.16. Music

This period saw a number of significant developments in violin research. The first of these was the launching of the "Catgut Acoustical Society" in 1963 [255], an organization that began informally, with a few enthusiasts for violin research, but which now lists about a thousand members worldwide.

The second development was the creation of a new family of violins (Fig. 9-45) [256]. These instruments, produced by Hutchins and Schelleng, cover a wider frequency range than the four traditional instruments, and are approximately a half-octave apart. Not only are these instruments in use today, but there has also been a number of compositions written for their

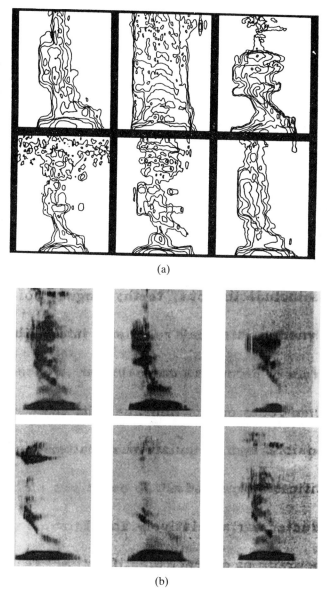

(a)

(b)

FIGURE 9-44. (a) Contour and (b) bar voiceprints. (From Kersta [250].)

use. A history of violin development is given by Carleen Hutchins in a recent publication [257].

A third development has been the use of the laser in producing holographic visualizations of the vibrations of the violin body. This holographic interferometry was discovered independently by a number of researchers in 1965 [258]. Its application to the violin was developed by Karl A. Stetson

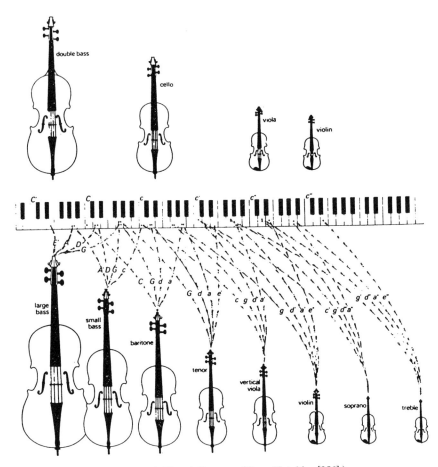

FIGURE 9-45. The violin octet. (From Hutchins [256].)

(one of the original discoverers) (1937–   ) [259] and by Lothar Cremer (1905–   ) and their colleagues [260]. A sample of such holograms is shown in Fig. 9-46 [261].

Overview

The third-quarter of the twentieth century was one of stupendous growth in the vigor, variety, and sophistication of the acoustical research. The field had come far from the days of a few individuals dominating the field, using simple but ingenious, home-made equipment. Any acoustician will note the failure of this author to include significant work, showing again that the history since mid-century had become too vast an undertaking for one individual to treat adequately. These remarks are even more appropriate for the final chapter. The author has therefore focused his "distorting lens"

TOP PLATE

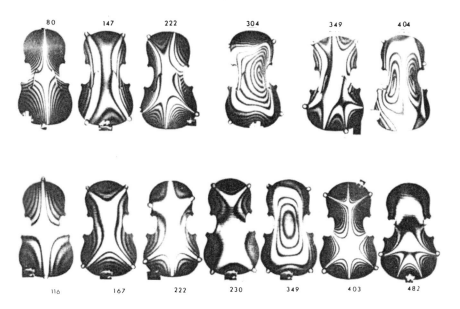

BACK PLATE

FIGURE 9-46. Holographs of violin plates. (From Hutchins et al. [261].)

(recall the remark in the Preface) on a selection of topics that are of great interest to himself, leaving the reader to look elsewhere for following up other parts of the field.

Notes and References

[1] R. Murray Schafer, *The Tuning of the World*, A.A. Knopf, New York, NY, 1977, p. 71.

[2] R. Bruce Lindsay, *J. Acoust. Soc. Am.* **37**, 357–381 (1965).

[3] And it continues to evolve; the Editor-in-Chief of *JASA* periodically issues a revision of the topics.

[4] The term "solid state physics" was gradually replaced toward the end of this interval by "condensed matter physics," largely because the general area of interest also included work in liquid helium, as well as other low-temperature phenomena.

[5] K.F. Herzfeld, *Phil. Mag.* **9**, 741–746 (1930).

[6] Philip M. Morse, *Vibration and Sound*, McGraw-Hill, New York, NY, 1936; 2nd ed., 1948. An even more extended analysis appears in P.M. Morse and H.

Feshbach, *Methods of Theoretical Physics*, vol. 2, McGraw-Hill, New York, NY, 1953.

[7] For example, J.J. Faran, *J. Acoust. Soc. Am.* **23**, 404 (1951); L.D. Hampton and C.M. McKinney, *J. Acoust. Soc. Am.* **33**, 664 (1961); Robert Hickling, *J. Acoust. Soc. Am.* **34**, 1582–1592 (1962). The work of Hickling served as a beacon for later attempts to identify scattering particles (Chapter 10).

[8] P.C. Waterman, *J. Acoust. Soc. Am.* **45**, 1417–1429 (1969).

[9] For the first of these, see P.M. Morse and H. Feshbach, *Methods of Theoretical Physics*, McGraw-Hill, New York, NY, 1953, pp. 494–523. For the second method, see H. Levine and J. Schwinger, *Phys. Rev.* **74**, 958–974 (1948); **75**, 1423–1432 (1949). For the third method, see D.S. Jones, *The Theory of Electromagnetism*, Macmillan, New York, NY, 1964, pp. 269–271; R.P. Banaugh and W. Goldsmith, *J. Acoust. Soc. Am.* **35**, 1590–1601 (1963).

[10] P.C. Waterman, *J. Acoust. Soc. Am.* **45**, 1417–1429 (1969).

[11] P.C. Waterman, *Phys. Rev. D.* **3**, 825–839 (1971).

[12] Vasundara Varatharjulu and Yih-Hsing Pao, *J. Acoust Soc. Am.* **60**, 556–566 (1976). Vasundara Varatharajalu later married Vijay K. Varadan, who also worked on scattering problems. They have continued to publish a large number of joint papers on the subject.

[13] P.C. Waterman, *J. Acoust. Soc. Am.* **60**, 567–580 (1976).

[14] W. Franz, *Z. Naturforsch.* **9A**, 705–716 (1954).

[15] Both figures are taken from the review by Werner G. Neubauer, "Observation of Acoustic Radiation from plane and curved surfaces," in *Physical Acoustics*, vol. x, 1973, pp. 61–126, but were part of earlier researches by Neubauer.

[16] For a detailed review of the electronics of the period, see R. Truell, C. Elbaum, and B. Chick, *Ultrasonic Methods in Solid State Physics* Academic Press, New York, NY, 1969; and R.T. Beyer and S. Letcher, *Physical Ultrasonics*, Academic Press, New York, NY, 1969, especially pp. 79–87, 310–311.

[17] H.B. Huntington, A.G. Emslie, and F.W. Hughes, *J. Franklin Inst.* **245**, 1–23 (1948).

[18] L. Bergmann, *Der Ultraschall*, S. Hirzel, Leipzig, 1954, 6th ed., pp. 85–199. See also W.P. Mason, *Physical Acoustics and the Properties of Solids*, Van Nostrand, Princeton, NJ, 1958, pp. 51–86. Mason developed equivalent circuits for the different transducer materials in this book.

[19] For further details on these transducers, see H. Jaffe and D. Berlincourt, *Proc. IEE-E* **53**, 1372 (1965); D.A. Berlincourt, D.R. Curran, and H. Jaffe, in *Physical Acoustics* (W.P. Mason, ed.), Academic Press, New York, NY, 1964, pp. 257–267.

[20] W.P. Mason, ed., *Physical Acoustics*, Academic Press, New York, NY, 1964.

[21] In 1970 (vol. 6), Mason's colleague at Bell Laboratories, Robert Thurston became coeditor of this series, and has continued it since Mason's death in 1986.

[22] I.G. Tamm, *Z. Physik* **60**, 345–363 (1930); J. Frenkel, *Phys. Rev.* **37**, 1289–1294 (1931).

[23] An excellent treatment of this phenomenon has been given by J. de Klerk in *Physical Acoustics* (W.P. Mason, ed.), vol. IVA, Academic Press, New York, NY, 1966, pp. 195–225.

[24] K.N. Baranskii, *Dokl. Akad. Nauk SSSR* **114**, 517 (1957); *Soviet Phys. Dokl.* **2**, 237 (1957).

[25] H.E. Bömmel and K. Dransfeld, *Phys. Rev. Lett.* **1**, 234 (1958); **2**, 298 (1959); **3**, 81 (1959).

[26] J. de Klerk and E.F. Kelly, *Appl. Phys. Lett.* **5**, 2 (1963); *Rev. Sci. Instr.* **36**, 506 (1965).

[27] M. Greenspan, *J. Acoust. Soc. Am.* **22**, 568–571 (1950); **28**, 644–648 (1956).

[28] E. Meyer and G. Sessler, *Z. Physik* **149**, 15 (1957).

[29] J.J. Markham, R.T. Beyer, and R.B. Lindsay, *Rev. Mod. Phys.* **23**, 353–411 (1951).

[30] Hans-Jörg Bauer, in *Physical Acoustics*, vol. IIA (W.P. Mason, ed.), Academic Press, New York, NY, 1964, pp. 48–131.

[31] L.I. Mandelshtam and M.A. Leontovich, *J. Exptl. Theoret. Phys. USSR* **7**, 438 (1937).

[32] J.J. Markham, *Phys. Rev.* **86**, 497 (1952).

[33] W.L. Nyborg, *J. Acoust. Soc. Am.* **25**, 68 (1953).

[34] J.E. Piercy and J. Lamb, *Proc. Roy. Soc. (London)* **A226**, 41 (1956).

[35] See, for example, R.W. Leonard, *J. Acoust. Soc. Am.* **12**, 241 (1940); E.F. Fricke, *J. Acoust. Soc. Am.* **12**, 245 (1940).

[36] J. Lamb and J.M. Pinkerton, *Proc. Roy. Soc. (London)* **A199**, 114 (1949). An exception to this was the earlier studies by P.A. Bazhulin (Chapter 8).

[37] The most extensive coverage of ultrasonic absorption and dispersion is given in the book by A.B. Bhatia, *Ultrasonic Absorption*, Clarendon Press, Oxford, UK, 1967. Reprinted by Dover, New York, NY, 1985.

[38] J.H. Andreae, R.L. Heasell, and J. Lamb. *Proc. Phys. Soc. (London)* **B69**, 625 (1956).

[39] L. Brillouin, *Ann. Phys. (Paris)* **17**, 88 (1922). Early work on this problem was performed by L.I. Mandel'shtam, *Zh. Russ. Fiz. Khim. Obshch.* **58**, 381 (1926), and the effect is known in Russia as Mandel'shtam–Brillouin scattering.

[40] A detailed review of work on Brillouin scattering is given in the book *Molecular Scattering of Light*, by I.M. Fabelinskii, Nauka, Moscow, 1965. English translation by R.T. Beyer, Plenum Press, New York, NY, 1968, from which Fig. 9-3 is taken.

[41] The success of these measurements was made possible by the use of a well-known optical method of measuring the frequency of the scattered light, the Fabry–Perot etalon, named for the nineteenth-century French physicists, Charles Fabry and Alfred Perot.

[42] C.V. Raman and C.S. Venkateswaren, *Nature* **142**, 791 (1938); **143**, 798 (1939). See also C.S. Venkateswaren, *Proc. Indian Acad. Sci.* **A15**, 362–370, 371–375 (1942).

[43] P.A. Fleury and R. Chiao, *J. Acoust. Soc. Am.* **39**, 7531 (1966).

[44] V.S. Starunov, E.V. Tiganov, and I.M. Fabelinskii, *Pis'mo Zh. Èksper. Teoret. Fiz.* **4**, 262 (1966) [*JETP Lett.* **4**, 176 (1966)].

[45] G. Kurtze and K. Tamm, *Acustica* **3**, 33 (1953); K. Tamm, G. Kurtze, and R. Kaiser, *Acustica* **4**, 380 (1954). For a general discussion see J. Stuehr and E. Yeager in *Physical Acoustics* (W.P. Mason, ed.), Academic Press, New York, NY, 1965, vol. II-A, pp. 354–476.

[46] R. Barrett and R.T. Beyer, *Phys. Rev.* **84**, 1060–1061 (1951).

[47] K. Tamm, G. Kurtze, and R. Kaiser, Ref. [45].

[48] M. Eigen and K. Tamm, *Z. Elektrochem.* **66**, 93, 107 (1962).

[49] V.O. Peshkov, *J. Phys. USSR* **8**, 318 (1944).

[50] K.R. Atkins, *Phys. Rev.* **113**, 962 (1959).

[51] C.W.F. Everett, K.R. Atkins, and A. Denenstein, *Phys. Rev.* **A136**, 1494 (1964).

[52] John Pellam, *Phys. Rev.* **73**, 608 (1948).

[53] K.A. Shapiro and I. Rudnick, *Phys. Rev. Lett.* **9**, 191–193 (1962); *Phys. Rev.* **A137**, 1383–1391 (1965); I. Rudnick et al. *Phys. Rev. Lett.* **19**, 488 (1967).

[54] G.A. Williams, R. Rosenbaum, and I. Rudnick, *Phys. Rev. Lett.* **42**, 1282 (1979).

[55] G.J. Jelatis, J.A. Roth, and J.D. Maynard, *Phys. Rev. Lett.* **42**, 1285 (1979).

[56] I. Rudnick, *J. Acoust. Soc. Am.* **68**, 36–45 (1980).

[57] L.D. Landau, *J. Phys. USSR* **5**, 71 (1941); **11**, 91 (1946).

[58] B.D. Keen, P.W. Matthews, and J. Wilks, *Phys. Lett.* **5**, 5 (1963); *Proc. Roy. Soc. (London)* **A284**, 125 (1965).

[59] H.J. Maris, *Phil. Mag.* **16**, 1331 (1967).

[60] H.J. Maris and J.S. Blinick, *Phys. Rev.* **A2**, 2139–2146 (1970).

[61] C. Quate, *Physics Today*, May 1979, p. 20.

[62] C. Quate, loc. cit.

[63] For a discussion of a number of these, see *Physical Ultrasonics*, by R.T. Beyer and S.V. Letcher, Academic Press, New York, NY, 1969, Chapter 8–10.

[64] A. Akhiezer, *J. Phys. USSR* **1**, 277 (1939).

[65] H. Bömmel and K. Dransfeld, *Phys. Rev.* **117**, 1245 (1960). See also T.O. Woodruff and H. Ehrenreich, *Phys. Rev.* **123**, 1535 (1961); and W.P. Mason and T.B. Bateman, *J. Acoust. Soc. Am.* **36**, 644 (1964).

[66] H. Bömmel, *Phys. Rev.* **96**, 220 (1954). See also L. Mackinnon, *Phys. Rev.* **98**, 1181 (1955).

[67] W.P. Mason. *Phys. Rev.* **97**, 557–558 (1955).

[68] A.V. Granato and K. Lücke, *J. Appl. Phys.* **27**, 583, 789 (1956). See also the review by the same authors in *Physical Acoustics* (W.P. Mason, ed.), Academic Press, New York, NY, 1966, vol. IV-A, pp. 225–276.

[69] J. Bardeen, L.N. Cooper, and J.R. Schrieffer, *Phys. Rev.* **108**, 1175 (1957).

[70] T. Tsuneto, *Phys. Rev.* **121**, 402 (1961).

[71] R.W. Morse and H.V. Bohm, *Phys. Rev.* **108**, 1094 (1957). Morse was a student of Bruce Lindsay, Professor at Brown University, Assistant Secretary of the Navy, and President of Case Western Reserve University. Henry Bohm has been a Professor at Wayne State University, and Vice-President at Wayne.

[72] A number of articles on these and related topics may be found in the series *Physical Acoustics* (W.P. Mason, ed.), Academic Press, New York, NY, 1966, especially vols. III and IV.

[73] R.W. Morse, in *The Fermi Surface* (Harrison and Webb, eds.), Wiley, New York, NY, 1966.

[74] M.J. Lighthill, *Proc. Roy. Soc. (London)* **A211**, 564–587; **A222**, 1 (1952); *Surveys in Mechanics* (G.K. Batcheler and R.M. Davies, eds.), Cambridge University Press, London, UK, 1956, pp. 250–351. Lighthill worked first at the University of Manchester and then became a professor at Imperial College, London. In 1969, he became the Lucasian Professor of Mathematics at the University of Cambridge. In 1979, he became Provost of University College, London. He was knighted for his work on aerodynamics.

[75] Rayleigh, *The Theory of Sound*, reprinted by Dover, New York, NY, 1945, vol. 2, p. 150. A brief presentation is given in R.T. Beyer, *Nonlinear Acoustics*, Naval Sea Systems Command, Washington, DC, 1974, pp. 7–8.

[76] This flowering of British science can be followed in the excellent reviews by J.E. Ffowcs-Williams, *Ann. Rev. Fluid Mech.* **1**, 197–222 (1969) and D.G. Crichton, *Prog. Aerospace Sci.* **16**, 31–96 (1975). Another paper of importance is that by M.S. Howe, *J. Fluid Mech.* **71**, 625–673 (1975), Ffowcs-Williams had been a professor at Imperial College, London, and is currently Rank Professor of Engineering (Acoustics) at the University of Cambridge, England. David Chrichton, who worked first at the University of Leeds, is also at Cambridge, Michael Howe is at Imperial College, London.

[77] F.E. Fox and W.A. Wallace, *J. Acoust. Soc. Am.* **26**, 994–1006 (1954); D.M. Towle and R.B. Lindsay, *J. Acoust. Soc. Am.* **27**, 530–533 (1955); V. Narasimhan and R.T. Beyer, *J. Acoust. Soc. Am.* **28**, 1233–1236 (1956); **29**, 532–533 (1957); V.A. Krasilnikov, V.V. Shklovskaya-Kordi, and L.K. Zarembo, *J. Acoust. Soc. Am.* **29**, 642–647 (1957).

[78] W.S. Keck and R.T. Beyer, *Phys. Fluids* **3**, 346 (1960); D.T. Blackstock, *J. Acoust. Soc. Am.* **34**, 9 (1962).

[79] S.I. Soluyan and R.V. Khokhlov, *Vestnik. Moskov. Univ.* **3**, 52–61 (1961) (English translation, *Benchmark*/19, pp. 193–206). See also R.V. Khokhlov and S.I. Soluyan, *Acustica* **14**, 241 (1964); *Soviet Phys. Acoust.* **8**, 170 (1962); A.L. Polyakova, S.I. Soluyan, and R.V. Khokhlov, *Soviet Phys. Acoust.* **8**, 78 (1962). Khokhlov was a prominent figure in Russian science, being, at the time of his death in 1977 Professor and Rector of the Moscow State University, Vice-President of the Soviet Academy of Sciences, and member of the Supreme Soviet, USSR. In 1977, when on a mountain-climbing expedition in the Soviet Pamirs, Rem Khoklov lost his own life in an attempt to save the life of an injured colleague. "Greater love no man hath than this, that the man lay down his life for his friends." (John, 15:13.)

[80] J.S. Mendousse, *J. Acoust. Soc. Am.* **25**, 51–54 (1953).

[81] The details of this process, and a number of applications of the result can be found in R.T. Beyer, *Nonlinear Acoustics*, Naval Ship Systems Command, 1974, Chapter 3. A reprinted version of this book, with some corrections and addenda, was published by the Acoustical Society of America, Woodbury, NY, 1997.

[82] See B.D. Cook, *J. Acoust. Soc. Am.* **34**, 941 (1962); D.T. Blackstock, *J. Acoust. Soc. Am.* **39**, 1019 (1966). The book, N.S. Bakhvalov, Ya.M. Zhileikin, and E.A. Zabolotskaya, *Nonlinear Theory of Sound Beams*, Russian edition, Nauka, Moscow, 1982. English translation by R.T. Beyer and Mark Hamilton, AIP Press, New York, NY, 1987, consists almost entirely of computer results of this material and of the equations of Refs. [69] and [70].

[83] E.A. Zabolotskaya and R.V. Khokhlov, *Akust. Zh.* **15**, 40–47 (1969) [*Soviet Phys. Acoust.* **15**, 35–40 (1969)].

[84] V.P. Kuznetsov, *Akust. Zh.* **16**, 548 (1970) [*Soviet Phys. Acoust.* **16**, 467–470 (1970)].

[85] The KZK equation is discussed in B.K. Novikov, O.V. Rudenko, and V.I. Timoshenko (Russia) *Nonlinear Underwater Acoustics*, Nauka, Moscow, 1981. English translation by R.T. Beyer and M.F. Hamilton, Acoustical Society of America, Woodbury, NY, 1987.

[86] K.U. Ingard and D.C. Pridmore-Brown, *J. Acoust. Soc. Am.* **28**, 367–370 (1956).

[87] P.J. Westervelt, *J. Acoust. Soc. Am.* **29**, 199–203, 934–935 (1957).

[88] A number of papers on this topic have been reproduced, and others referenced, in *Benchmark*/19.

[89] In their book, *Theoretical Foundations of Nonlinear Acoustics*, published in Russian, Nauka, Moscow, 1975. English translation by R.T. Beyer, Consultants Bureau, New York, NY, 1977. O.V. Rudenko and S.I. Soluyan make the point that there are two kinds of scattering—synchronous and diffractive. In the synchronous case, which corresponds to scattering in the ordinary sense of the term, there can be no scattering without dispersion. Perhaps the controversy can be resolved by a more rigorous examination of these conditions.

[90] For recent comments, see P.J. Westervelt, *J. Acoust. Soc. Am.* **97**, 3375 (A) (1995).

[91] P.J. Westervelt, *J. Acoust. Soc. Am.* **35**, 535–537 (1963). The effect was confirmed experimentally by J.L.S. Bellin and R.T. Beyer, *J. Acoust. Soc. Am.* **34**, 1051–1054 (1962). Westervelt's theory preceded the Bellin and Beyer work, but his paper came out later.

[92] For example H.O. Berktay (Britain), *J. Sound Vibration* **2**, 435–454, 462–470 (1965); G.A. Barnard, J.G. Willette, J.J. Truchard, and J.A. Shooter (USA), *J. Acoust. Soc. Am.* **52**, 1437–1441 (1972); B.K. Novikov, O.V. Rudenko, and V.I. Timoshenko (Russia) *Nonlinear Underwater Acoustics*, Nauka, Moscow, 1981 (Ref. [85]). Virtually this entire book is devoted to the subject of the parametric array.

[93] *Principles of Underwater Sound*, Natl. Def. Res. Comm., Div. 6, Sum. Tech. Rept. 7, pp. 175–199, 1946.

[94] This equation and a number of similar ones are discussed in great detail in R.E. Urick, *Principles of Underwater Sound for Engineers*, McGraw-Hill, New York, NY, 1983, 3rd ed., Chapter 2. See also R.E. Urick, *J. Acoust. Soc. Am.* **34**, 547 (1962). Sonar engineers have learned to think in decibels, but it was always a difficulty, not unlike a foreign-language one, for the author of this book to switch back and forth from decibels to what he has always thought to be the real world. Urick's book is an effective statement of our knowledge of sonar in the mid-1960s, and contains much valuable history of the subject. Since much of the research was classified, both in the Soviet Union and in Western countries, it is now quite difficult to assign credit for various improvements in equipment and method among US scientists and engineers, let alone those of other nations.

[95] Lee E. Holt, *J. Acoust. Soc. Am.* **19**, 678–681 (1947). The author gives a detailed review of German listening equipment as it existed at the end of World War II.

[96] M. Strasberg, *J. Acoust. Soc. Am.* **31**, 16 (1959); R. Esche, *Acustica* **4AB**, 208 (1952); L.D. Rosenberg, *Acustica* **12**, 40 (1962); G.E. Liddiard, US Navy Electron. Lab. Rept. 376, 1953.

[97] P.J. Westervelt, in *Proc. Third Intern. Congr. Acoust.* (L. Cremer, ed.), Elsevier, Amsterdam, 1960, pp. 316–321. Reference is made in that article to an earlier piece in an unpublished report of Bolt, Beranek, and Newman, Inc., by Westervelt in April 1959.

[98] R.H. Mellen and M.B. Moffett, *J. Acoust. Soc. Am.* **52**, 122, (A) (1972).

[99] R.J. Urick, *Principles of Underwater Sound for Engineers*, Chapters 5 and 6.

[100] The mathematical details of this method of analysis may be found in C.B. Officer, *Introduction to the Theory of Sound Transmission*, McGraw-Hill, New York, NY, 1958, especially Chapters 2–4.

[101] Some of the general references here are *Physics of Sound in the Sea*, Natl. Def. Res. Comm. 6, Sum. Tech. Rept. 8, Chapters 2 and 3, 1946; L.M. Brekhovskikh, *Waves in Layered Media* (translated from the Russian by David Liebermann), Academic Press, New York, NY, 1960. For special emphasis on propagation, see A.O. Williams, *J. Acoust. Soc. Am.* **51**, 1041–1048 (1972); for reverberation see Claude W. Horton, Sr., *J. Acoust. Soc. Am.* **51**, 1050–1061 (1972).

[102] C.L. Pekeris, "Theory of Propagation of Explosive Sound in Shallow Water," in *Propagation of Sound in the Ocean*, Geol. Soc. Amer. Mem. 27, 1948.

[103] R.E. Urick, *Principles of Underwater Sound for Engineers*, Chapter 8.

[104] R.E. Urick, *Principles of Underwater Sound for Engineers*, p. 188.

[105] C.F. Eyring, R.J. Christensen, and R.W. Raitt, *J. Acoust. Soc. Am.* **20**, 462–475 (1948).

[106] M.G. Bradbury et al., in *Proceedings of an International Symposium on Biological Sound Scattering in the Ocean* (G. Brooke Farquhar, ed.), Department of the Navy, Washington, DC, 1970, pp. 408–452. See also C. Clay and H. Medwin, *Oceanographic Acoustics*, Wiley, New York, NY, 1977, p. 238.

[107] C. Clay and H. Medwin, *Oceanographic Acoustics*, p. 240.

[108] H.R. Johnstone, R.H. Backus, J.B. Hersey, and D.M. Owens, *Deep Sea Res.* **3**, 256 (1956).

[109] D.C. Whitmarsh, E. Skudrzyk, and R.J. Urick, *J. Acoust. Soc. Am.* 29, 1124–43 (1957).

[110] D.J. Mintzer, *J. Acoust. Soc. Am.* **25**, 922–927, 1107–1111 (1953); **26**, 186–190 (1954). A general review of this type of scattering was given by Alan Berman and A.N. Guthrie (the first of a series of papers in *JASA* devoted to the status of underwater sound in 1970), *J. Acoust. Soc. Am.* **51**, 944–1009 (1971).

[111] R.J. Bobber, *Underwater Electroacoustic Measurements*, Naval Research Laboratory, Washington, DC, July, 1970, pp. 2–4.

[112] T. Bell, "Sonar and Submarine Detection," AD Report, USN Underwater Sound Laboratory, New London, CT, 1962.

[113] Jim Bessert, Defense Electronics, Oct., 1979, pp. 61–64. An earlier form of SOSUS, called project Artemis, the Greek goddess of the hunt, was so named to honor, with some subtlety, F.V. Hunt, whose brain child it was. This author can remember Hunt at a navy meeting in the 1950s, summing up the proposed operation by saying something like "I know it's expensive but, gentlemen, think of the scanning rate—one ocean per hour!"

[114] E.G.M. Wenz, *J. Acoust. Soc. Am.* **34**, 1936–1956 (1962). A later review was published by him in *J. Acoust. Soc. Am.* **51**, 1010–1024 (1972). Both articles contain extensive bibliographies.

[115] R.-G. Busnel and J.F. Fish, eds., *Animal Sonar Systems*, Plenum Press, New York, NY, 1980.

[116] Alan D. Grinnell, in Busnel and Fish, *Animal Sonar Systems*, p. xxi.

[117] The mixed use of the names dolphin–porpoise is discussed in Winthrop N. Kellogg, *Porpoises and Sonar*, University of Chicago Press, Chicago, IL, 1961, pp. 2–3. We shall not attempt to resolve the problem here.

[118] Kellogg, *Porpoises and Sonar*, pp. 32–34.

[119] W.N. Kellogg, Robert Kohler, and H.N. Morris, *Science* **117**, 239–243 (1953).

[120] W.N. Kellogg, *Science* **128**, 982–988 (1958).

[121] F.W. Reysenbach de Hahn, *Proc. Roy. Soc. (London)* **152B**, 54–52 (1960).

[122] A review of this subject, with a good bibliography, is given by Whitlow W.L. Au in *Animal Sonar Systems* (Busnel and Fish, eds.), pp. 251–282. The hearing structure of the porpoise is discussed in some detail in another article in *Animal Sonar Systems* (Busnel and Fish, eds.), (by J.G. McCormick, E.G. Wever, S.H. Ridgway, and J. Palin), pp. 449–467.

[123] K.S. Norris, W.E. Evans, and R.N. Turner, *Proc. Symp. Animal Sonar Systems* (R.-G. Busnel, ed.), Jouy-en-Josas, France, 1957.

[124] L.V. Worthington and W.E. Schevill, *Nature,* **180**, 291 (1957).

[125] W.A. Watkins and W.E. Schevill, *Deep Sea Res.* **19**, 691–706 (1972); **23**, 175–180 (1976); *J. Acoust. Soc. Am.* **62**, 1485–1490 (1977). This latter article also included a recording of whale sounds.

[126] W.A. Watkins, in *Animal Sonar Systems* (Busnel and Fish, eds.), pp. 283–289.

[127] Lyman Spitzer, Jr., Nat. Res. Council Comm. on Undersea Warfare Rep. of Panel on Underwater Acoustics (1 Sept. 1948), NRC CRW 0027.

[128] V.C. Anderson, *J. Acoust. Soc. Am.* **51**, 1062–1065 (1972).

[129] V.C. Anderson, loc. cit.

[130] J.L. Flanagan, *IEEE Trans. Audio Electroacoust.* **AU-15**, No. 2, 66 (1967); R.C. Singleton and T.C. Poulter, *IEEE Trans. Audio Electroacoust.* **AU-15**, No. 2, 104 (1967).

[131] W.T. Cochran, J.W. Colley et al., *IEEE Trans. Audio Electroacoust.* **AU-15**, No. 2, 45 (1967); *IEEE Trans. Audio Electroacoust.* **AU-17**, No. 2 (1969). The entire issue was devoted to the FFT.

[132] The reader must make the mental jump from "sound beams" as used in underwater sound signaling, and the "vibrating beams" of structure analysis. R.D. Mindlin and H. Deresiewicz, *Proc. 2nd US Nat'l Cong. Appl. Mech.* pp. 175–178, 1955; reprinted in *Benchmark/8*, pp. 10–13. This paper contains the following interesting footnote (p. 178)

   It is of some historical interest that the rotatory inertia correction, usually attributed to Rayleigh (*Theory of Sound*, Cambridge, England, first edition, 1877, current edition, Art. 162, and the transverse shear correction, usually attributed to Timoshenko (*Phil. Mag.* **41**, 744–746 (1921) and **43**, 125–134 (1922), are given by M. Bresse in his *Cours de Méchanique Appliquée*, Mallet-Bachelier, Paris, 1859, p. 126.

[133] H. Deresiewicz and R.D. Mindlin, *J. Appl. Mech.* **18**, 31–38 (1951).

[134] Work on high-order beam theory appears in R.W. Leonard and B. Budiansky, Natl. Adv. Comm. Aeron. Report 1173, pp. 1–27 (1954); J. Miklowitz, *J. Appl. Mech.* **20**, 511–514 (1953), and B.A. Boley, *J. Appl. Mech.* **22**, 69–76 (1955). All three of these papers, as well as the one cited in Ref. [133] are included in the *Benchmark* volume referenced in Ref. [132].

[135] A good discussion of the effect of Ross's work may be found in *Benchmark/8*, pp. 143–145.

[136] K. Forsberg, *AAIA J.* **2**, 2150–2157 (1964).

[137] E. Reissner, *Quart. Appl. Math.* **13**, 279–290 (1955).

[138] A. Kalnins, *J. Acoust. Soc. Am.* **36**, 1355–1365 (1964). Kalnins and Dym reviewed these and other developments in beam theory in Ref. [135, pp. 142–147].

[139] D. Linehan, *Trans. Geophys. Union*, 1940, part 2, pp. 229–232.

[140] L.D. Leet, D. Linehan, and P.R. Berger, *Bull. Seismol. Soc. Amer.* **41**, 123–141 (1951); L.D. Leet, *Bull. Seismol. Soc. Amer.* **41**, 165–167 (1951).

[141] I. Tolstoy and M. Ewing, *Bull. Seismol. Soc. Amer.* **40**, 25–51 (1950); M. Ewing, I. Tolstoy, and F. Press, *Bull. Seismol. Soc. Amer.* **40**, 53–58 (1950); M. Ewing, F. Press, and J.L. Worzel, *Bull. Seismol. Soc. Amer.* **42**, 37–51 (1952); M. Ewing and F. Press, *Ann. Geophys.* **9**, 248–259 (1953).

[142] An excellent coverage of the history of research on the T wave can be found in I.F. Kadykov, *Acoustics of Underwater Earthquakes (Seaquakes)* (in Russian), Nauka, Moscow, 1986.

[143] Ref. [142, pp. 28–29].

[144] Hallowell Davis, Introduction to *Sound and Hearing*, Time-Life Books, 1965. The riddle is cited at the start of a review article by James D. Miller, *J. Acoust. Soc. Am.* **56**, 729–763 (1974), which is a detailed coverage of our knowledge of the effect of noise on people at the end of the period covered by this chapter.

[145] R.K. Hillquist and W.N. Scott, *J. Acoust. Soc. Am.* **58**, 2–10 (1974).

[146] D.I. Blokhintzev, *Acoustics of a Moving, Inhomogeneous Medium*, Russian edition, Nauka, Moscow, 1946. English translation by R.T. Beyer and D.J. Mintzer, Brown University, 1952, pp. 119–121.

[147] J.E. Ffowcs-Williams and D.L. Hawkings, *Phil. Trans. Roy. Soc. (London)* Ser. A, **264**, 321–342 (1969).

[148] The two-volume work edited by Harvey H. Hubbard, *Aero-acoustics of Flight Vehicles: Theory and Practice*, NASA, 1991, vol. 1, 519 pp., vol. 2, 431 pp. (reprinted by the Acoustical Society of America, Woodbury, NY, 1994) presents an excellent review of all topics in aircraft-generated noise and its control, both for the period covered in this chapter and for that of Chapter 10.

[149] Alan Powell, *J. Acoust. Soc. Am.* **97**, 3261 (1995).

[150] E.Y. Yudin, *Zh. Tekhn. Fiz.* **14**, 561–567 (1944).

[151] D.I. Blokhintzev, loc. cit.

[152] G.M. Lilley, in Ref. [148, vol. 1, pp. 211–289].

[153] F.M. Wiener, C.I. Malms, and C.M. Gosos, *J. Acoust. Soc. Am.* **37**, 738–747 (1965).

[154] R.H. Lyon, *J. Acoust. Soc. Am.* **55**, 493–5503 (1974).

[155] J.D. Miller, Ref. [144, p. 732].

[156] Numerous references to Ward's work in this area are included in the paper by J.D. Miller cited in Ref. [144].

[157] J.D. Miller, *J. Acoust. Soc. Am.* **56**, 742 (1974).

[158] The author first heard the term used by Leo Beranek in 1957 in so labeling the sound of room air conditioners in a motel at which we were staying. The motel was surrounded by a busy highway, overflights by planes taking off or landing at a nearby airport as well as railroad trains, and Beranek, as an acoustic consultant, had recommended that an elevated level of air conditioning noise could reduce the annoyance to the guests from the other noise sources.

[159] *San Diego Workshop on the Interaction Between Man-Made Noise and Vibration and Artic Marine Wildlife*, 25–29 Feb., 1980. Acoustical Society of America, Woodbury, NY, 1981.

[160] Combatting Noise in the 1990s. American Speech–Language–Hearing Association, Dec. 1991; *Hearing Conservation Manual*, 3rd ed., Alice Suter, Council for Accreditation in Occupational Hearing Conservation, Milwaukee, WI, 1993.

[161] Table is from R. Murray Schafer, *The Tuning of the World*, A.A. Knopf, New York, NY, 1977, p. 190.

[162] A sampling of papers by Ingard and his colleagues include K.U. Ingard and R.H. Bolt, *J. Acoust. Soc. Am.* **23**, 509–516, 533–540 (1951); K.U. Ingard, *J. Acoust. Soc. Am.* **25**, 1037–1061, 1062–1067 (1953); **31**, 1202–1212 (1959).

[163] M.C. Junger, *Noise Control Engng.* **4**, 17–15 (1975).

[164] Work of this period is reviewed in the article by P.R. Gliebe, J.F. Brausch, R.K. Majjigi, and R. Lee in Harvey H. Hubbard, *Aeroacoustics of Flight Vehicles*, NA SA, 1991, vol. 2, pp. 207–270.

[165] See John S. Mixson and J.F. Wilby in Harvey H. Hubbard, *Aeroacoustics of Flight Vehicles*, vol. 2, pp. 271–356.

[166] A review of such work up to 1974 was given by Ulrich J. Kurze, *J. Acoust. Soc. Am.* **55**, 504–518 (1974).

[167] German patent, 1933. The American patent, also by Leug, was No. 2,043,416 (1936).

[168] Harry F. Olsen and Everett G. May, *J. Acoust. Soc. Am.* **25**, 1130–1136 (1953).

[169] Donald C. Gasaway, *Hearing Conservation*, Prentice Hall, Englewood Cliffs, NJ, 1985, Chapter 9.

[170] Helmut Haas, *Acustica* **1**, 49–58 (1950). The original article is reprinted in *Benchmark*/10, pp. 75–85, along with a brief English summary.

[171] R.H. Bolt and P.E. Doak, *J. Acoust. Soc. Am.* **22**, 507–509 (1950). Doak has long served as the editor of the British *Journal of Sound and Vibration*.

[172] Michael Forsyth, *Buildings for Music*, MIT Press, Cambridge, MA, 1985, p. 286. Chapters 6–8 of this book describe numerous concert halls, built during the past quarter century. For a detailed comment on the Royal Festival Hall, see P.H. Parkin, W.A. Allen, H.K. Purvis, and E.E. Scholes, *Acustica* **3**, 1–21 (1953).

[173] P.H. Parkin and K. Morgan, *J. Sound Vibration* **2**, 74–85 (1965); *J. Acoust. Soc. Am.* **48**, 1025–1035 (1970).

[174] Michael Forsyth, *Buildings for Music*, p. 294.

[175] L.L. Beranek, *Music, Acoustics and Architecture*, Wiley, New York, NY, 1962. This beautiful volume has set a standard for subsequent books on concert halls, such as the one by Forsyth just cited and series produced by the Acoustic Society of America, for its text, fine photography, and information on technical design. See also L.L. Beranek, *4th Int. Cong. Acoustics*, Copenhagen, 1962. Reprinted in *Benchmark*/10, pp. 89–103. Recently, Beranek has updated this material in *Concert and Opera Halls: How They Sound*, Acoustical Society of America, Woodbury, NY, 1996.

[176] Michael Forsyth, *Buildings for Music*, p. 286.

[177] Michael Forsyth, *Buildings for Music*. For Beranek's point of view, see L.L. Beranek, F.R. Johnson, T.J. Schultz, and B.G. Watters, *J. Acoust. Soc. Am.* **36**, 1247–1262 (1964).

[178] A final comment by Michael Forsyth is perhaps appropriate:

> Avery Fisher is much less reverberant [than Symphony Hall, Boston]; the tone is very different from the warm, rich, reverberant sound of Sabine's hall, being clear, precise, and cool, so that unlike nineteenth-century symphony halls it tends rather to reveal than to flatter. In fact, Avery Fisher is typical of modern North America hi-fi concert halls. (Ref. [172, p. 289].)

[179] B.S. Atal, M.R. Schroeder, G.M. Sessler, and J.E. West, *J. Acoust. Soc. Am.* **40**, 428–433 (1966).

[180] *Benchmark*/10, p. 118.

[181] M.R. Schroeder, B.S. Atal, G.M. Sessler, and J.E. West, *J. Acoust. Soc. Am.* **40**, 434–440 (1966). Shall we call them the Bell digital quartet?

[182] L.L. Beranek, *4th Int. Congr. Acoust.* 1962, pp. 15–29. Reprinted in *Benchmark*/10, pp. 89–103.

[183] A. Marshall, *Arch. Sci. Rev. (Australia)* **11**, 81–87 (1968). This paper was reprinted in *Benchmark*/10, pp. 104–111.

[184] M. Barron, *J. Sound Vibration* **15**, 474–494 (1971).

[185] R. Thiele, *Acustica* **3**, 291–302 (1953); E. Meyer and R. Thiele, *Acustica* **6**, 425–444 (1956).

[186] S. Ballantine, *Phys. Rev.* **32**, 988–992 (1928). The graph is taken from E.A.G. Shaw in *Handbook of Sensory Physiology*, vol. V/1 (W.D. Keidel and W.D. Neff, eds.), Springer-Verlag, New York, NY, 1974, p. 484.

[187] E.A.G. Shaw and R. Teranishi, *J. Acoust. Soc. Am.* **44**, 240–249 (1968). Shaw's extensive work in this field, along with that of other researchers, is recounted in the review article by Shaw cited in Ref. [186, pp. 454–490].

[188] A detailed review of the acoustics of the middle ear is given by A. Moller in *Handbook of Sensory Physiology*, vol. V/1, pp. 491–548.

[189] A.H. Moller, *J. Acoust. Soc. Am.* **33**, 168–176 (1962); A.H. Moller in *Foundations of Modern Auditory Theory*, vol. 2 (J. Tobias, ed.), Academic Press, London, UK, 1972, pp. 133–194.

[190] A.H. Moller in *Handbook of Sensory Physiology*, vol. V/1, p. 499.

[191] J.J. Guinan, Jr. and W.T. Peake, *J. Acoust. Soc. Am.* **41**, 1237–1261 (1967).

[192] G. von Békésy, *Experiments in Hearing* (translated and edited by E.G. Wever), McGraw-Hill, New York, NY, 1960. Reprinted by the Acoustical Society of America, Woodbury, NY, 1989, p. 420.

[193] J. Tonndorf, *J. Acoust. Soc. Am.* **30**, 929–937 (1958). The number of models of the cochlea is great. In a preface to an article by von Békésy in *Foundations of Modern Auditory Theory*, p. 305, Jerry Tobias wrote "Models of any complex system are necessarily imperfect unless, like a map that is exactly the same size as the country it represents, they become perfect replicas instead." Von Békésy describes yet another such model in this article.

[194] J. Tonndorf, in *Foundations of Modern Auditory Theory*, vol. 1, p. 213.

[195] The research on these waves has been considerable. A detailed review of the subject is given by Tonndorf in Ref. [194, pp. 205–254], along with more than 100 references.

[196] J. Tonndorf, *J. Acoust. Soc. Am.* **47**, 579–591 (1970).

[197] A discussion of these researches, with numerous references, may be found in *Benchmark*/15, pp. 98–104. Another excellent review of our knowledge of the human ear is contained in an article by Manfred Schroeder in 1975 [*Proc IEEE* **63**, 1332–1350 (1975)].

[198] H. Davis, C. Fernandez, and D.R. McAuliffe, *Proc. Nat. Acad. Sci.* **36**, 580–587 (1950).

[199] P. Dallos, in *Physiology of the Auditory System* (M.B. Sachs, ed.), National Education Consultants, Inc., Baltimore, MD, 1971, pp. 57–67.

[200] P. Dallos in *The Nervous System* (D.B. Tower, ed.), Raven Press, New York, NY, 1975, vol. 3, pp. 69–80.

[201] S. Rosen, *New York J. Med.* **53**, 2650 (1953).

[202] J. Shea, *Ann. Otol. Rhinol. Laryngol.* **67**, 932 (1958).

[203] F. Zollner, *J. Laryngol.* **69**, 637 (1955).

[204] H. Wullstein, *Laryngoscope* **66**, 1076 (1956). A review of these and other surgical developments is given by Victor Goodhill in *The Nervous System* (D. Tower, ed.), pp. 273–290.

[205] An account of these modifications and improvements is recounted by Raymond Carhart in *The Nervous System* (D. Tower, ed.), pp. 291–297. Other problems of the hearing handicapped and their resolution are discussed in the same volume by J.M. Pickett, pp. 299–304.

[206] L.J. Sivian and S.D. White, *J. Acoust. Soc. Am.* **4**, 288–321 (1933).

[207] E.A.G. Shaw and J.E. Piercy, *Fourth Inter. Congr. Acoust.*, Copenhagen, paper H46.

[208] K.J. Diercks and L.A. Jeffress, *J. Acoust. Soc. Am.* **34**, 981–984 (1962).

[209] J.P. Egan, *J. Acoust. Soc. Am.* **38**, 1043–1049 (1965).

[210] J.W. Swets, *Science* **134**, 168–177 (1961).

[211] N.S. Yoewart, M. Bryan, and W. Tempest, *J. Sound Vibration* **6**, 335–342 (1967).

[212] J.L. Northern, M.P. Downs, W. Rudmose, A. Glorig, and J. Fletcher, *J. Acoust. Soc. Am.* **52**, 585–595 (1967).

[213] G. von Békésy, *Experiments in Hearing*, pp. 219–220; R.L. Wegel, *Ann. Otol.* **41**, 740–779 (1932); R.S. Dadson and J.H. King, *J. Laryng. Otol.* **46**, 366–378 (1978); D.W. Robinson and R.S. Dadson, *British J. Appl. Phys.* **7**, 166–181 (1956).

[214] J.C.R. Licklider, *Experientia* **7**, 128–133 (1951). Licklider notes that "The suggestion has been made by R.M. Fano of the Research Laboratory of Electronics and by I.A. de Rosa of the Federal Telecommunications Laboratories, Inc., that the cochlea may operate more nearly as an autocorrelator than as a filter." This viewpoint is at variance with the ideas of Licklider. See also Licklider's article in *Psychology: A Study of a Science* (S. Koch, ed.), McGraw-Hill, New York, NY, 1959, pp. 41–144.

[215] J.C.R. Licklider, article in *Experientia* (Ref. [214, p. 155]).

[216] Ernst Petzold, *Elementare Raumakustik*, Bauwelt Verlag, Berlin, 1927, p. 8.

[217] A discussion of these problems is given in *Benchmark*/13, pp. 257–261. The role of "cues," especially in the understanding of consonants, has been pursued by Pierre C. Delattre, Alvin M. Libermann, and Franklin S. Cooper, *J. Acoust. Soc. Am.* **27**, 769–773 (1952) and Katherine S. Harris, *Language and Speech* **1**, 1–7 (1958).

[218] I.K. Samoilova, *Biofizika* **1**, 79–87 (1956); **4**, 550–558 (1959); L.A. Chistovich, V.A. Ivanova, and J.M. Pickett, *J. Acoust. Soc. Am.* **31**, 1613–1615 (1959); L.O. Elliott, *J. Acoust. Soc. Am.* **34**, 1108–1115, 1116–1117 (1962); **36**, 393 (1964); *Audiol.* **10**, 65–76 (1971).

[219] Harvey Fletcher, *Speech and Hearing in Communication*, Van Nostrand, New York, NY, 1953.

[220] *Acoustic Phonetics* (E.B. Fry, ed.), Cambridge University Cambridge, Cambridge, UK, 1976, p. 11.

[221] P. Broca, *Bull. Soc. Anatom. Paris VI* **36**, 330–357 (1861). A good historical account of these developments is given in *Speech Science Primer*, by Gloria J. Borden, Katherine S. Harris, and Lawrence J. Raphael, Williams & Wilkins, Baltimore, MD, 1994, 3rd ed., Chapter 4. This book contains a broad review of speech topics and is a pleasure to read.

[222] C. Wernicke, *Der Aphasische Symptomenocomplex*, Cohen and Weigert, Breslau, 1874.

[223] W. Penfield and L. Roberts, *Speech and Brain Mechanisms*, Princeton University Press, Princeton, NJ, 1959, p. 54.

[224] W. Penfield and L. Roberts, *Speech and Brain Mechanisms*, p. 52.

[225] G.L. Borden, K.S. Harris, and L.J. Raphael, *Speech Science Primer* (Williams and Wilkins, Baltimore, MD, 1994), p. 54. Dominance in speech by the left side of the brain occurred in 90% of the right-handers studied and 70% of the left-handers.

[226] B. Milner, C. Branch, and R. Rasmussen, "Observations on Cerebral Dominance," in *Psychology Readings: Language* (R.C. Oldfield and C. Marshall, eds.), Penguin Books, Baltimore, MD, 1968, pp. 366–378.

[227] V. Fromkin, *Sci. Am.* **299**, 110–116 (1973).

[228] Borden, Harris, and Raphael, *Speech Science Primer*, p. 54.

[229] J.L. Flanagan, *IEEE Trans. Commun. Technol.* **COM-19**, 1006–1008 (1971).

[230] K.N. Stevens, S. Kasowski, and C.G.M. Fant, *J. Acoust. Soc. Am.* **25**, 734–742 (1953); J.L. Flanagan and Lorinda L. Landgraf, *IEEE Trans. Audio Electroacoust.* **AU-16**, 57–64 (1968); P. Mermelstein, *Proc. Seventh Intern. Congr. Acoust.*, Paper 23C, 173–176 (1971); K. Ishizaka and J.L. Flanagan, *Bell. Syst. Tech. J.* **51**, 1233–1268 (1972).

[231] J. van den Berg, J.T. Zantema, and P. Doornenbal, Jr., *J. Acoust. Soc. Am.* **29**, 626–635 (1957); J. van den Berg, *J. Speech Hear. Res.* **1**, 227–244 (1958). Both these papers are reprinted in *Papers in Speech Communication*, vol. 1, *Speech Production* (Joanne Miller, ed.-in-ch.), Acoustic Society of America, Woodbury, NY, 1991.

[232] K.N. Stevens and A.S. House, *J. Acoust. Soc. Am.* **27**, 484–493 (1955).

[233] Loc. cit.

[234] Loc. cit.

[235] Katherine Harris, private communication to the author.

[236] *Benchmark/3*, p. 186. See also J.L. Flanagan, *IEEE: Trans. Commun. Technol.* **COM-10**, No. 6, 1006–1008 (1971).

[237] G.M. Fant in *For Roman Jakobson*, Mouton, The Hague, 1956, pp. 109–120; G.M. Fant, *Acoustic Theory of Speech Production*, Mouton, The Hague, 1960; J.L. Flanagan, *J. Acoust. Soc. Am.* **29**, 306–310 (1957); W. Lawrence, *Communication Theory*, Butterworth, London, UK, 1953, pp. 460–469. L.R. Rabiner, *J. Acoust. Soc. Am.* **43**, 822–828 (1968); L.R. Rabiner, L.B.

Jackson, R.W. Schafer, and C.H. Coker, *IEEE Trans. Commun. Technol.* **COM-19**, 1016–1020 (1971). See also the earlier article by J.L. Flanagan, C.H. Coker, and M. Bird, 15th Ann. Meeting Audio Engr. Soc., Preprint 307 (1963).

[238] J.C.R. Licklider, *Psychology: A Study of a Science* (Sigmund Koch, ed.). McGraw Hill, New York, NY, 1959, vol. 1, pp. 90–94. The cons were represented by I. Pollack. *J. Acoust. Soc. Am.* **20**, 295–266 (1948); **24**, 538–541 (1952); and by I.J. Hirsh, E.G. Reynolds, and M. Joseph, *J. Acoust. Soc. Am.* **26**, 540–539 (1954).

[239] J.C.R. Licklider, *Psychology: A Study of Science*, vol. 1, p. 94.

[240] K.D. Kryter, *J. Acoust. Soc. Am.* **34**, 1698–1702 (1962). American Standard Method for Measurement of Monosyllabic Word Intelligibility, S3.2-196.

[241] H.L. Barney, F.E. Haworth, and H.K. Dunn, *Bell Syst. Tech. J.* **38**, 1337–1356 (1959). See also J.L. Flanagan, *J. Acoust. Soc. Am.* **51**, 1375–1387 (1972).

[242] See citation for Franklin S. Cooper upon his being awarded the Silver Medal in Speech Communication by the Acoustical Society of America. *J. Acoust. Soc. Am.* **58**, insert following p. S78 (1975).

[243] G.J. Borden, K.S. Harris, and L.J. Raphael, *Speech Science Primer*, p. 21.

[244] P.C. DeLattre, A.M. Liberman, and F.S. Cooper, *J. Acoust. Soc. Am.* **27**, 769–773 (1955). Other papers on the topic include F.S. Cooper, *J. Acoust. Soc. Am.* **22**, 761–764 (1950); F.S. Cooper, P.C. DeLattre, A.M. Liberman, J.M. Borst, and L.J. Gerstman, *J. Acoust. Soc. Am.* **24**, 597–606 (1952).

[245] P. DeLattre, A.W. Liberman, F.S. Cooper, and L.J. Gerstman, *Word* **8**, 195–210 (1952). Figures 9–41 were reproduced in Ref. [221].

[246] A.M. Liberman et al., *J. Exp. Psych.* **52**, 122–137 (1956).

[247] G.J. Borden, K.S. Harris, and L.J. Raphael, *Speech Science Primer*, pp. 21–22.

[248] A.J. Compton, *J. Acoust. Soc. Am.* **35**, 1748–1752 (1963).

[249] Alexandr Solzhenitsyn, *The First Circle* (translated from the Russian), Harper & Row, New York, NY, 1968.

[250] L.G. Kersta, *Nature*, **196**, 1253–1257 (1962); *6th Int'l Congr. Acoust.* (Y. Kohasi, ed.), Maruzen, Tokyo, 1968, pp. 147–B150.

[251] The distinction made by Kersta between bar and contour spectrograms is not universally accepted.

[252] R.H. Bolt et al., *J. Acoust. Soc. Am.* **54**, 531–534 (1973). The author of this history was President-Elect of the Acoustical Society of America at the time of the appointment of this group of distinguished acousticians (by the Executive Council of the Society) to examine the voiceprint technique and report on its reliability. Bolt was later involved in the acoustical examination of the famous gap in the Nixon Watergate tapes.

[253] M.H.L. Hecker, American Speech and Hearing Association, Monograph No. 16, Washington, DC, 1971.

[254] Anonymous, "Voiceprint Identification," *Georgetown Law. J.* **61**, 703–745 (1973); E.J. Welch, Jr., *Trial Magazine* **9**, 45–47 (Jan.–Feb., 1973).

[255] Some details of the founding of this society by Robert Edward Fryxell (1923–   ), Carleen M. Hutchins, John C. Schelleng (1892–1979), and Frederick Saunders are given in a review and tutorial article by Hutchins [*J. Acoust. Soc. Am.* **73**, 1421–1440 (1983)]. The article also contains valuable information on the history of violin research. If a catgut can have a backbone, that bone has assuredly been Carleen Hutchins.

[256] C.M. Hutchins, loc. cit.

[257] *Research Papers in Violin Acoustics*, 1975–1993 (Carleen Maley Hutchins, ed., Virgiia Benade, assoc. ed.), Acoustical Society of America, Woodbury, NY, 1996, 3 vols. Vol. 1, pp. 3–12.

[258] The theory of holographic interferometry, along with the early history of the field, is given by G.M. Brown, R.M. Grant, and G.W. Stroke, *J. Acoust. Soc. Am.* **45**, 1166–1179.

[259] Carl-Hugo Agren (1931–   ) and K.A. Stetson. *J. Acoust. Soc. Am.* **51**, part 2, 1971–1983 (1972).

[260] Friedrich Ludwig Walter Reinicke (1939–   ) and L. Cremer, *J. Acoust. Soc. Am.* **48**, part 2, 988–992 (1970). For a detailed study of the violin from a physics viewpoint, see *The Physics of the Violin* by Lothar Cremer. English translation by John S. Allen, MIT Press, Cambridge, MA, 1984.

[261] C.M. Hutchins, K.A. Stetsona, and P.A. Taylor, *Catgut Acoust. Soc. Newsletter*, No. 16, 15–23 (1971). Reprinted in *Benchmark*/6, pp. 118–128.

10
Acoustics: 1975–1995 and Beyond. Fin de siècle—Again

> Acoustical science has always been a wrestling match between imagination and hard numbers.
>
> Bernard Holland, *New York Times*, March 5, 1996, p. C11

As the writing of this book has progressed, the chapter subheadings have come to resemble the names of the Technical Committees of the Acoustical Society of America, or, even more closely, the breakdown of acoustics in the detailed listing of the PACS (Physics and Astronomy Classification Scheme) [1]. In this final chapter we shall follow this index rather closely, but the reader should keep in mind the caveat expressed at the end of the previous chapter that this chapter would be a sampling of topics rather than a more or less complete coverage of the entire field.

10.1. Linear Acoustics. Scattering

The amount of work on scattering, from objects in air, water, and solids, continues to be strong. V.V. and V.K. Varadan at a number of locations, Larry Flax and his group at the Naval Research Laboratories, and Herbert Überall and Guillermo Gaunard at Catholic University, and the Naval Surface Weapons Center, all working both independently and in collaboration, have produced mountains of papers on the subject, much of which was reviewed in a paper by D. Brill and G. Gaunaurd in 1987 [2]. A new technique, the resonance scattering theory, was developed by L. Flax, L.R. Dragonette, and H. Überall in 1978 [3]. The work of Gaunaurd has more recently moved in the direction of attempting to identify the character of the scatterer from the information contained in the scattered signal [4]. One might expect that this form of identification will be pursued strongly in the future.

10.2. Nonlinear Acoustics. Radiation Pressure and Levitation

Radiation pressure has been a minor theme in nonlinear acoustics since the days of Lord Rayleigh, but the interest never seems to fade. Reviews of the subject appeared in 1978 [5] and in 1982 [6]. On the experimental side, the most interesting development has been the use of this radiation pressure to levitate small objects, such as glass spheres, and later, steel balls (Fig. 10-1) [7]. The relation of acoustic levitation to a classic of physics—the Botzmann–Ehrenfest principle—was explored by Seth Puttermann, Joseph Rudnick, and Martin Barmatz [8].

Radiation pressure has also been used effectively in making fine tuning adjustments in zero-gravity environments in space. Figure 10-2 shows a three-dimensional, piezoelectrically controlled mechanism, on which an object can be maintained in the center of the cube through the use of three piezoelectric drivers; this arrangement was used by Taylor Wang in a space flight [9]. The technique may well have a promising future in sensitive spatial control devices, including containerless experiments [10], and acoustic tweezers [11].

FIGURE 10-1. Levitation of a steel ball in a ground-based siren. (From Barmatz [7].)

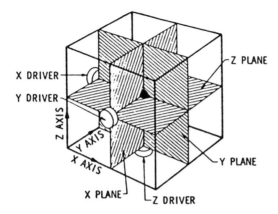

FIGURE 10-2. Triple-axis acoustic levitator. (From Wang et al. [9].)

10.3. Solitons

Two great watchwords of physical science at the end of the twentieth century have been solitons and chaos. Every special field or subfield has had some place in which to fit them and acoustics has been no exception. We shall review a few of these instances.

The principle of superposition—that the different solutions of a single differential equation can exist independently of one another—applies only to linear differential equations. A surprise has been the recent discovery that the solutions of certain nonlinear differential equations (known as evolution equations) have, in the words of G.L. Lamb, Jr., "a special type of elementary solution. These special solutions take the form of localized disturbances, or pulses, that retain their shape even after interaction among themselves, and thus act somewhat like particles [12]". In a paper on the nonlinear interaction of solitary wave pulses propagating in dispersive media, N.J. Zabusky and Martin D. Kruskal observed in 1965 that two distinct solitary waves of different amplitudes can interact nonlinearly and yet remain unchanged. Further, they noted that these "interacting localized pulses do not scatter irreversibly [13]," and gave to these pulses the name of "solitons" (Fig. 10-3).

As pointed out in Chapter 7, the study of solitary waves was related in the nineteenth century to the study of water disturbances in narrow shallow channels. Near the end of the nineteenth century, the defining relation for this phenomenon has been the Korteweg–deVries (K–dV) equation [14]. This equation differs from the Burgers equation, widely used in nonlinear acoustics, in that the term on the right-hand side of the Burgers equation is a second derivative, while the corresponding term in the K–dV equation is the third derivative [15]. The K-dV equation is very widely used in wave

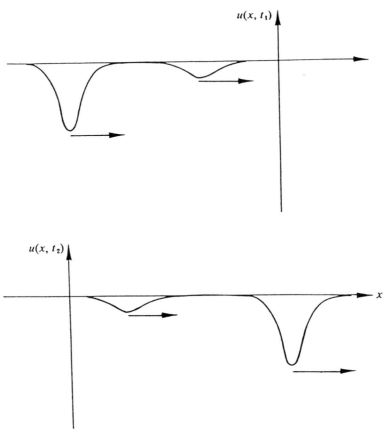

FIGURE 10-3. Interaction of solitons. (From Leibovich and Seebass [15].)

motion in plasma. An exact solution for it in an initial-value problem was provided by Gardner, Greene, Kruskal, and Miura in 1967 [16]. For that reason, it has proved to be very useful "as a model on which to test new techniques for solving nonlinear equations [17]."

Thus, a soliton can be described as a solitary wave that remains unchanged after collision with another solitary wave [18]. To date, it has been easier experimentally to find solitons in hydrodynamic waves than in acoustical ones and the research of acousticians has largely been in the former area. Isadore Rudnick and his colleagues at UCLA observed what they called the nonpropagating hydrodynamic soliton in a water tank at a wave frequency of 5 Hz [19]. Work similar to that of Rudnick and his group has been carried out in Nanjing, China, by Rong-jue Wei and his associates [20] (Fig. 10-4).

In condensed matter physics, acoustic solitons have been observed in paramagnetic crystals [21], in acousto-electromagnetic interactions in crys-

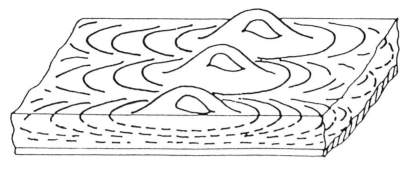

FIGURE 10-4. Array of solitons. (From Wei et al. [20].)

tals with a nonlinear electrostriction [22]; spinwave solitons have also been observed [23].

10.4. Chaos

The phenomenon in physics known as chaos represents the behavior of a system that is irregular and upredictable, but which is nevertheless governed by deterministic laws. Such systems are nonlinear in nature and are to be distinguished from noise, which obeys no such simple, deterministic law [24]. It was first noted in acoustics by Esche at Göttingen in 1952 [25]. Esche studied the appearance and collapse of cavitation bubbles in water. There is a noise emitted from such bubbles, and when the sound intensity is increased, there is a doubling of the period of the sound emitted (Fig. 10-5). As the intensity is further increased, there is more period doubling, so that regular repetition of the motion requires a longer and longer time interval and the emitted signal thus becomes irregular or chaotic throughout the period of observation. Examples of period doubling are shown in Fig. 10-6 from the work of Lauterborn [26], who describes them under the term of bifurcations.

10.5. Sonoluminescence

Most students of science are aware that both x-rays and radioactivity were discovered in a rather accidental manner, when the discoverers were looking for something else. A similar story can be told about sonoluminescence. H. Frenzel and H. Schultes, along with Hinsberg, observed the appearance of luminescence when air or oxygen had been dissolved in water and a sufficiently strong ultrasonic signal irradiated the liquid. In 1929, they exposed a photographic plate when acoustic waves were passing through the liquid, and observed a darkening of that plate [27], The study of this phe-

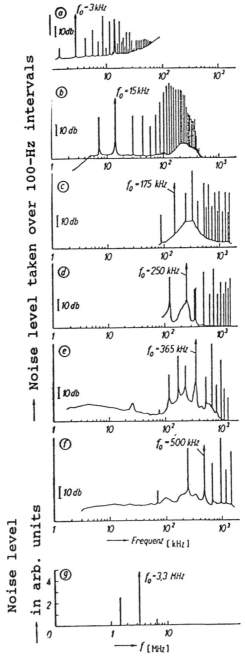

FIGURE 10-5. The historical spectra of acoustic cavitation noise. Note (e), where second period doubling has occurred. (From Esche [25].)

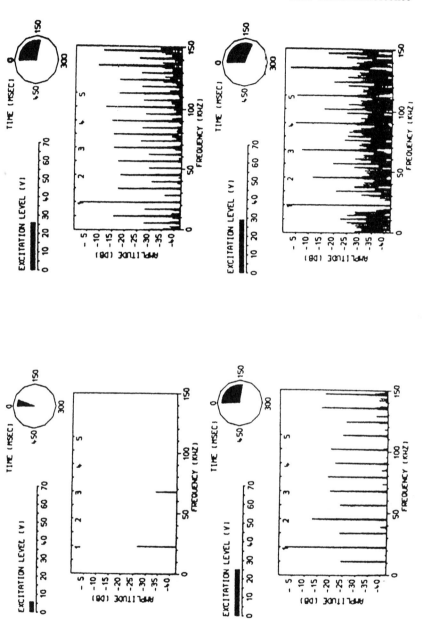

FIGURE 10-6. Power spectra of acoustic cavitation noise in water irradiated with a sound frequency of 23.56 kHz at different sound pressure amplitudes (indicated by the bar below "excitation level" in each plot). The sequence shows the successive "filling" of the spectra by the appearance of spectral lines exactly in the middle between the odd ones. (From Lauterborn [26].)

nomenon in recent years has separated into following the adventures of a single bubble (SBLS) and following those of multiple bubbles (MBSL). Examples of each of these are shown in Fig. 10-7 (taken from a recent paper by Larry Crum and Ronald Roy [28]). Out of such measurements has come a whole field of sonochemistry the status of which was reviewed by Kenneth S. Suslik in 1990 [29].

The emanation of light from a single bubble suggests the momentary creation of very high temperatures (see Puttermann [30]). The basis of much of the modern work is found in the paper by Felipe Gaitan et al. in 1988, in which the authors determined the conditions under which a single stable cavitation bubble would produce sonoluminescence during each acoustic cycle [31].

(a)   (b)

FIGURE 10-7. (a) Single bubble sonoluminescence; and (b) multiple-bubble sonoluminescence. Bright bubbles. An acoustic standing wave levitates a small gas bubble near the center of a glass cell (left) and drives that bubble to radial excursions of sufficient amplitude to generate sonoluminescence each and every acoustic cycle. Note the bright spot at the center of the cell, which can easily be seen without darkening the room; no chemical enhancement is required. This bubble was driven at a frequency of 22.3 kHz and at a pressure amplitude of about 1.3 bar. In an example of MBSL (right), the intense sound field near the tip of an acoustic source produces many transient cavitation bubbles that grow and collapse with such violence that they heat their respective interiors to incandescence. Because the individual bubbles persist for only a few acoustic cycles (at 22 kHz), they are not visible in this photograph. Luminol, a wavelength shifter, was added to enhance the light emission, which is normally too faint for unaided photography. (From Crum and Roy [28].)

The study of SBSL has been pursued on a number of fronts. Attempts at the measurement of the duration of the light flash have mainly shown the limits of equipment and technique rather than the true values of such times [32]. An upper bound for the duration of 50 ps (1 ps = 1 picosecond = 10^{-12} s) has been found by Barber and Puttermann [33].

At this date, the theory of the phenomenon remains obscure. The basic idea is the expansion and collapse of the cavitation bubble, and goes back to Rayleigh, but the mathematics are involved and checkable results have not appeared. Julian Schwinger (1918–1995), the famed quantum electrodynamicist and Nobel laureate, suggested something that is based on the dynamic Casimir effect [34] while others have advanced the idea of an electric discharge in which asymmetric bubble collapse brings about charge separation [35].

We mentioned above the creation of very high temperatures during the bubble collapse. Estimates of these have run as high as 10^8 K [36]. It has been noted that these temperatures are about that found at the surface of the Sun and near the level needed for nuclear fusion, so that cavitation bubbles and their collapse may still have a great research future before them [37].

The role of acoustics in cavitation, solitons, chaos, and sonoluminescence serves to demonstrate once again that acoustics remains a vigorous branch of physics.

10.6. Underwater Sound. Internal Waves

Benjamin Franklin made observations on the Gulf Stream during his many crossings of the Atlantic, and realized that there were boundaries within the ocean between layers of different salinity [38]. A somewhat similar phenomenon became known in the fjords of Norway, where layers of fresh water flowed over the surface of sea water. In 1925, Ekman [39] noted that ships whose keels and bottoms lay below this boundary level met with much greater resistance to their forward motion. It turned out that the ship was generating surface waves at the level of the sea water–fresh water interface. These were the first "internal waves" to be observed.

Such waves were of interest in oceanography, but acousticians did not pay much attention to them until the 1960s and 1970s. In this later time, it was observed that long-range propagation of sound in the ocean was subject to substantial amplitude fluctuations ("fading" was the term used), for which the period was of the order of minutes or longer. The problem was attacked by a large number of prominent physicists whose basic area of research was particle physics, and reasonable estimates of the nature of the internal waves were deduced [40].

10.7. Global Scale Acoustics

The exploitation of the SOFAR Channel in the 1940s led to the study of ocean behavior over long ranges. David Browning explored the SOFAR channel in a wide variety of oceans and lakes [41]. The most challenging experiment, however, is the one called the Heard Island experiment. Heard Island is a small islet at the southern extremity of the Indian Ocean. Because of its location, acoustic signals propagate from it through the SOFAR channel into the North Atlantic and the North Pacific Oceans. Thus, a sound path of about 12,000 miles can be studied (Fig. 10-8) [42].

Since the speed with which the sound wave travels in the ocean depends on the ocean temperature, the idea was proposed that the time of transmission of such signals be accurately recorded over a long period of time (years). If global warming exists, then there should be a small but measurable change in the time of propagation over the fixed distance, so that a change in the average temperature in the channel could be measured. Thus, a definitive test on possible global warming could be carried out.

This was the origin of the Heard Island Feasibility Tests of 1991 by Munk and others [43]. The operation has been put largely on hold, due to opposition from environmental and ecological groups, who feared harm to the animal species in the ocean. This has led to studies into the effect of continuous signal transmission on marine mammals [44].

The study of internal waves, mentioned above, demonstrated the large-scale nature of events taking place within the ocean. The result has been the birth of acoustic ocean tomography. An excellent review of the development of this field is found in the book by a major participant, Walter Munk, and his colleagues Peter Worcester and Carl Wunsch [45]. As they put it,

The problem of ocean acoustic tomography is to infer from precise measurements of travel time or of other properties of acoustic propagation the state of the ocean traversed by the sound field. [46]

The first use of the term apparently lay with Munk and Wunsch in a paper in 1979 [47], although they point out that a precursor was the work on the study of Rossby waves from acoustic signals by Elroy La Casce and John Beckerle in 1975 [48].

10.8. Ultrasonics. Optoacoustics

The field of optoacoustics (OA) or photoacoustics (PA) was begun by Alexander Graham Bell in 1880, as noted in Chapter 6, but remained dormant for many years. In 1971, it was revived in a paper by E.L. Kerr and J.G. Atwood [49], and a little later, L.B. Kreuzer employed laser-induced PA in the detection of minute quantities of gas constituents [50]. Shortly

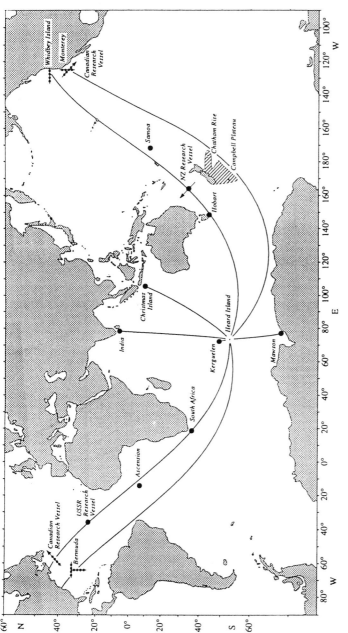

FIGURE 10-8. SOFAR paths from Heard Island. Heard Island feasibility test. The sources were suspended from the center well of the R/V Cory Chouest 50 km southeast of Heard Island. Black circles indicate receiver sites. Horizontal lines represent horizontal receiver arrays off the American West Coast and off Bermuda. Vertical lines designate vertical arrays off Monterey and Bermuda. Lines with arrows off California and Newfoundland indicate Canadian towed arrays. Ray paths from the source to receivers are along refracted geodesics, which would be great circles but for the Earth's nonspherical shape and the ocean's horizontal sound-speed gradients. Signals were received at all sites except the vertical array at Bermuda, which sank, and the Japanese station off Samoa. (Munk et al. [42].)

thereafter, William R. Harshbarger and Melvin B. Robin [51], A. Rosencwaig [52] and P.J. Westervelt and R.J. Larson [53]. The first two of these researches were extensions of the work of Kreuzer on gas-phase spectroscopy, while the Westervelt and Larson work concentrated on the physics aspects of the conversion of light energy into sound.

By the time the period covered by this chapter had begun, the field had exploded, with research being pursued in many directions, and soon review articles were appearing [54]. We can only summarize some of the leading directions of this exciting new field.

10.9. Weakly Absorbing Media

A typical experimental setup is shown in Fig. 10-9. A tunable laser signal is chopped at an acoustic frequency and passed through a liquid (or gas) cell (Patel and Tam [55]). The resultant acoustic signal is detected by the transducer and appropriately analyzed. The acoustic signal is shown in Fig. 10-10, along with the laser pulse that produced it. The intensity of the acoustic signal is proportional to the strength of the received laser beam. Thus the absorption properties of the medium can be deduced by varying the frequency of the laser and recording the sound signal.

These techniques can be used to analyze minute traces of materials in both gases (Sigrist [54]) and liquids (Tam [55]). Detailed studies of the phenomena of optoacoustics have appeared in a large number of books [56], [57]. In the optoacoustic process, the infrared (IR) radiation of the modulated laser is scattered or dissipated by the particles of a medium in

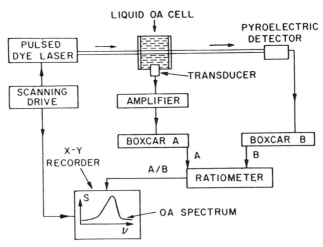

FIGURE 10-9. Block diagram of the simple experimental arrangement needed for linear spectroscopy of liquids with the pulsed OA detection method. (From Patel and Tam [55].)

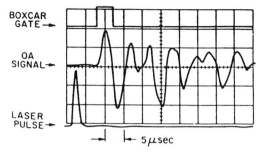

FIGURE 10-10. Oscilloscope picture shows the laser pulse (lower trace), OA signal (middle trace), and boxcar gate for the OA detection (upper trace) for a typical liquid. (Patel and Tam [55].)

various ways, including photochemical energy release through chain reaction in a gas mixture [58]. The profile of the radiation from single droplets of organic materials in water is shown in Fig. 10-11 [59]. Thus the technique of optoacoustics can be used for particle excitation and identification (signature).

A wide variety of techniques have been developed for producing square wave photoacoustic pulses [60]. The traditional volume change produced by thermal expansion of a heated medium is complemented by such optoacoustic processes as products of excited (vibrational or electronic) and photochemical processes. A review of this field has been given recently by S.E. Braslavsky (1942–   ) and G.E. Heibel (1963–   ) [61]. Still other techniques are being developed [62], all of which indicate a bright and continuing future for this area of research.

10.10. Structural Acoustics and Vibration

Fuzzy Structural Acoustics

In a paper at the Acoustical Society of America meeting in 1992, Christian Soize (1948–   ) defined the structural fuzzy as "the set of minor subsystems that are connected to the master structure but are not accessible by classical modeling [63]." By a study of the master system, he was able to do probabilistic modeling of this fuzzy to estimate the vibrations of the master system and its radiation. Figure 10-12 [64] is a stylized representation of the structural fuzzy.

The work of Murray Strasberg and David Feit concentrated on the existence of many oscillators in the structure, presenting a "simplified procedure for estimation of the effect of these substructures when they have many resonances in frequencies near the excitation frequency [65]." In a number of subsequent papers, Pierce, Sparrow, and Russell [66] studied the founda-

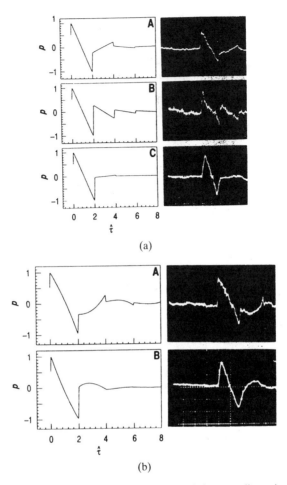

FIGURE 10-11. (a) Calculated pressure p (in arbitrary units) versus dimensionless time $\hat{t}$ from Eq. 2 for fluid droplets with $\hat{\rho} = 1$. (A) Hexane–carbon tetrachloride (CCl_4) droplet suspended in water ($\hat{\rho} = 1.01$, $\hat{c} = 0.645$, $a = 0.5$ mm). (B) Formamide droplet suspended in a hexane-CCl_4 mixture ($\hat{\rho} = 1.00$, $\hat{c} = 1.65$, $a = 1.5$ mm). (C) 1,2,3,4-Tetrahydronaphthalene (tetralin) droplet in water ($\hat{\rho} = 0.97$, $\hat{c} = 0.962$, $a = 1.4$ mm). The experimental time scales are 500 ns per division in the top oscillogram and 1 μs per division in the other two. The slight departure of the formamide wave form from the predicted shape is caused by the addition of extra dye, which was necessitated by the small expansion coefficient of formamide and the consequent low signal-to-noise ratio in the recorded wave. (b) Photoacoustic wave forms for droplets where $\hat{\rho} \neq 1$. (A) Hexane droplet in water ($\hat{\rho} = 0.66$, $\hat{c} = 0.708$, $a = 1$ mm). (B) CCl_4 droplet in water ($\hat{\rho} = 1.6$, $\hat{c} = 0.618$, $a = 1$ mm). The time scale in both oscillograms is 1 μs per division. (From Diebold et al. [59].)

tions of Soize's ideas and introduced the concept of internal mass per unit natural frequency bandwidth.

All of these studies are still in their infancy, and much more may be expected. In a philosophical way, they are reminiscent of the work on Thévenin's theorem in the analysis of electrical "blackboxes [67]."

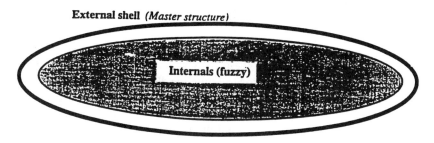

FIGURE 10-12. Stylized representation of a fuzzy structure. (From Norris [64].)

Active Sound and Vibration Control

The development of the electronic sound absorber by Olson and May in 1953 (Chapter 9) was followed by other applications of analog techniques in sound suppression [68]. When the electronic industry digitized, so did the patents on noise suppression [69]. The system of Olsen and May was actually designed for the removal of unwanted sounds, rather than broadband noise. Attention later shifted to the reduction of broadband noise [70]. In 1981, P.D. Wheeler developed an active noise reduction system for aircrew helmets [71]. An excellent review of this subject of "anti-noise" was given by Ffowcs-Williams in 1984 [72]. Figure 10-13 (from that review) is a fanciful illustration of the use of estimates of the strength of a background noise field for suppression purposes.

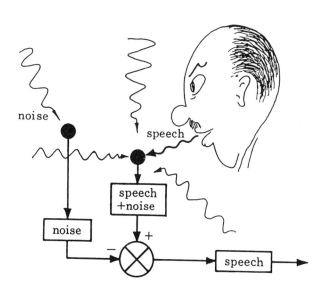

FIGURE 10-13. Schematic illustration of how multiple microphone estimates of a sound be combined to enhance signal levels. (From Ffowcs-Williams [72].)

10.11. Architectural Acoustics. Auditorium Design

In 1930, John von Neumann, writing in his *Mathematischen Grundlagen der Quantenmechanik* [73], gave a remarkable description of the duality of the measurement process, dividing the world into two parts— the observed and the observer. The first of these, the objective part, is amenable to physical measurement; the second, or subjective part, is not. In the observed portion, "we can follow up all physical processes (in principle at least) arbitrarily precisely ... [in the second, or observer portion] subjective perception leads us into the intellectual inner life of the individual, which is extra-observational by its very nature [74]." This duality was first pointed out by Niels Bohr in connection with his principle of complementarity [75]. von Neumann noted that we may shift the boundary toward the subjective zone as far as we like, but there is always the second, subjective, part to be dealt with. He added the remark that "it is a fundamental requirement of the scientific viewpoint that it must be possible so to describe the extra-physical process of the subjective perception as if it were in reality in the physical world [76]."

The same duality shows up in modern architectural acoustics, especially in the study of concert halls. In the early part of the century, architectural acousticians concerned themselves with measurement of the reverberation time. During the next quarter-century, they tried compare that time with subjective estimates of what was a good hall, acoustically speaking. One of the difficulties here turned out, as Beranek later put it, "incorrect data on audience absorption for listeners seated in modern upholstered chairs [77]." But it was in the latter half of the twentieth century that the problem began to be faced directly. We have noted (Chapter 9) Beranek's addition of the initial-time-delay gap" [78] to the parameters of concert halls. In the 1970s Manfred Schroeder and his colleagues and students [79] began systematic measurements of concert halls, and included the time delay gap along with reverberation time and the initial cross-correlation function between the response at the two ears to an impulsive sound as the three factors most important to the listening audience [80]. Schroeder wrote that, to gain the proper information, the way

is to record the program material in different concert halls by means of microphones at the "eardrums" of a *dummy head* [81], [82] and then to play these recordings back at a convenient time and place, allowing for rapid switching between different halls. To ensure identity of musical program material, a *recording* of an orchestra should be used and played for the stage—rather than repeated musical performances.

Naturally, the orchestra should be recorded [in an] acoustically neutral environment; in other words, in an anechoic chamber. Fortunately, such recordings are available. [83], [84]

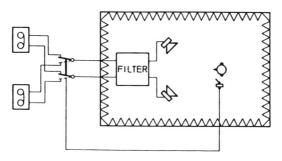

FIGURE 10-14. Setups for subjective comparison of concert halls. (From Schroeder et al. [80].)

These recordings, when suitably processed, allow, in Schroeder's words, "a realistic recreation of the acoustics of the hall at the position of the dummy" (Fig. 10-14) [80]. The recordings for different seats in different halls were played to a variety of listeners, both trained and untrained musically. By having these observers express a preference (yes or no) for individual seats and halls, Schroeder and his coworkers developed a correlation diagram in "preference space" (Fig. 10-15) [80]. Such an analysis clearly corresponds to von Neumann's description of "the extra-physical process of the subjective perception [85]."

Another example of this duality is described in the book by Yoichi Ando [86]. One can establish the acoustical characteristics of the outer ear, the middle ear, and the cochlea, and thus move from the acoustic field at the location of the outer ear, to the electrical presentation to the brain. Thus we steadily move the boundary between observed and observer into the brain itself (Fig. 10-16) [87]. These expressions can then be correlated with the subjective preferences of the listener.

A number of measurable quantities were developed between the 1960s and the 1980s that could be compared with listener preferences. These include the listening level (LL), and the interaural cross correlation (IACC) (defined as "the maximum value of normalized magnitude of the interaural cross correlation function within an interaural delay range of up to 1 ms") [88]. Both of these quantities may be associated with the right hemisphere of the brain.

To the parameters of auditorium design thus far mentioned, one should add the concept of the apparent source width (ASW). This term grew out of the work by M. Barron at Southampton and later at Cambridge, and by A. Herbert Marshall at the University of Auckland [89]. Their studies involved the spatial impression perceived from early lateral reflections in concert halls. In their joint paper [90], Barron and Marshall quote the manager of the Concertgebouw Orchestra of Amsterdam as stating that "the sensation of spatial impression corresponds to the difference between feeling inside the music and looking at it, as through a window."

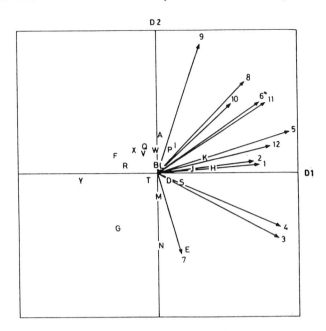

FIGURE 10-15. Preference space. The result of a factor analysis of individual preference judgments. Dimensions 1 and 2 (D1, D2), shown here, account for 45% and 16%, respectively, of the total variance. Capital letters signify 22 different halls included in the test. Numbered vectors represent 12 different listeners. Projections of hall positions on listeners vectors approximate different listener's individual preference scales. The projection of a hall on the abscissa (D1) represents "consensus preference" for that hall; projection on the ordinate (D2) reflects individual preference disparities. (From Schroeder et al. [80].)

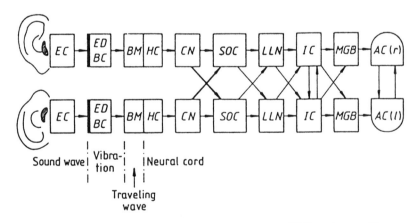

FIGURE 10-16. Auditory pathways. (EC: External canal; ED and BC: eardrum and bone c and HC: basilar membrane and hair cell; CN: cochlea nuclei; SOC: superior olivary (medial superior olive, lateral superior olive and nucleus of the trapezoid body); LL: lemniscus nucleus; IC: inferior colliculus; MGB: medial geniculate body; and AC: cortex of the right and left hemispheres). Feedback mechanisms are not considered. (From Ando [87].)

10.12. Acoustic Instrumentation

Ingenious acoustical devices seem to have been a characteristic of the nineteenth century with the world of von Helmholtz, König, and the great inventors, but the supply of such devices (or of the ideas behind them) is not yet exhausted. We shall look at a few of these ideas [91].

Acoustic Engines

In a brief historical review of this subject at the start of one of their papers, G. Swift et al. [92] noted that the first acoustic engine was probably the Sondhauss tube of 1850 [93], which formed part of the subject of the singing flame, and which was subsequently discussed by Rayleigh in the second edition of his book [94]. Swift et al. pointed out that it was Rayleigh who recognized (in his explanation of the effect) the "importance of the relative phasing between thermodynamic effect (temperature change) and motion, this phasing having been produced by the thermal time lag between the gas and the surrounding walls [95]."

The operation of an acoustic engine, using liquid sodium as the working fluid, is shown in Fig. 10-17 [95]. The tube, filled with liquid sodium, has a length equal to one-half the wavelength of the acoustic wave that will be generated. The sodium is called the primary medium. The secondary medium is a stack of plates made of a substance with a high acoustic impedance and a large value of the product of the heat capacity and the thermal conductivity. A heat exchange liquid passes through the ends of the plates, assuring a temperature difference $T_H - T_C$ between the sodium at the opposite ends of the stack. We thus have a setup for an engine, operating between the two temperatures as shown in (a) of the figure. This causes the sodium to oscillate spontaneously [92], [95]. The theory of operation of such engines has been worked out by Nicholas Rott and coworkers [96].

Ultrasonic Refrigerators

Like other thermodynamic engines, the acoustic engine can also be used as a refrigerator (Fig. 10-18) [95]. Examples of such refrigerators are shown in Fig. 10-19 [95], [97]. Thermoacoustic refrigerators have possible commercial applications, although they are currently of very low efficiency. A number of industrial laboratories are working on the improvement of their operation, and more may well be heard of them.

10.13. Physiological Acoustics

The Cochlea

The study of the dynamics of the cochlea, so vigorously pursued in the 1960s and 1970s, continued unabated as the century moved toward its conclusion. In a paper published in 1993, J.J. Zwislocki wrote

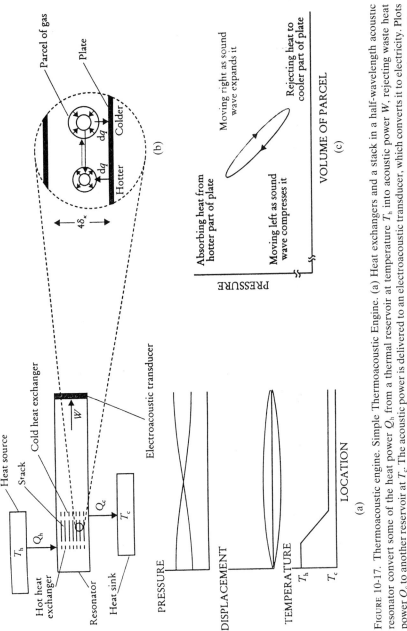

FIGURE 10-17. Thermoacoustic engine. Simple Thermoacoustic Engine. (a) Heat exchangers and a stack in a half-wavelength acoustic resonator convert some of the heat power Q_h from a thermal reservoir at temperature T_h into acoustic power W, rejecting waste heat power Q_c to another reservoir at T_c. The acoustic power is delivered to an electroacoustic transducer, which converts it to electricity. Plots below show gas pressure, gas displacement in the horizontal direction and average temperature as functions of location in the resonator. Pressure and displacement are each shown when the gas is at the leftmost extreme of its displacement (red), with density and pressure highest at the left end of the resonator and lowest at the right end, and 180° later in the cycle (blue). (b) Magnified view of part of the stack shows a typical parcel of gas (greatly exaggerated in size) as it oscillates in position, pressure, and temperature, exchanging heat dq with the nearby plates of the stack. Plates are separated by about four thermal penetration depths δ_K. (c) Pressure–volume (p–V) diagram for the parcel of gas shows how it does net work $\delta w = \oint p \, dV$ on its surroundings. (From Swift [95].)

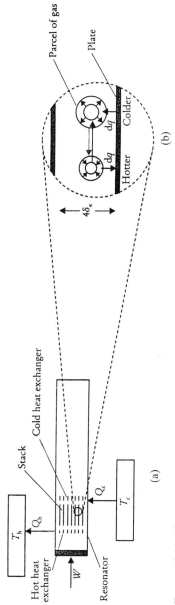

FIGURE 10-18. Thermoacoustic refrigerator. Simple Thermoacoustic Refrigerator. (a) Electroacoustic transducer at the left and delivers acoustic power W to the resonator, producing refrigeration Q_c at low temperature T_c and rejecting waste heat power Q_h to a heat sink at T_h. As in Fig. 10-17, this is a half-wavelength device with a pressure node at the midpoint of the resonator. The temperature gradient in the refrigerator's stack is much less steep than that in the stack for the engine shown in Fig. 10-17. (b) Magnified view of part of the stack shows a typical parcel of gas as it moves heat dq up the temperature gradient. Here $\oint p dV < 0$, so the pressure–volume cycle analogous to Fig. 10-17(c) goes counterclockwise, and the parcel absorbs work from its surroundings. (From Swift [95].)

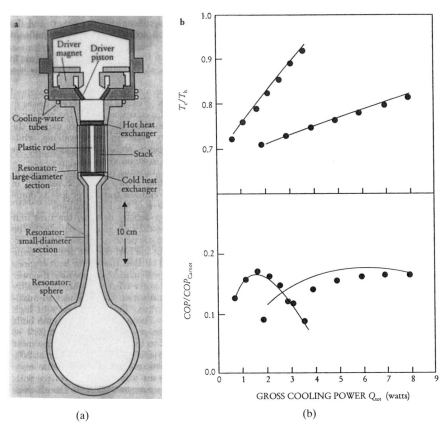

(a)   (b)

Figure 10-19. Thermoacoustic refrigerator and performance parameters. First Efficient Thermoacoustic Refrigerator (a) and some of its performance parameters; (b) as measured (data points) and calculated (curves) for operation with 500-Hz pressure oscillations in 10-bar helium gas, and with $T_h = 300$ K. Blue circles are data for 1.5% pressure oscillations; red circles, 3%. The gross cooling power Q_{tot} includes the deliberately applied load plus some small parasitic loads such as heat leak from room temperature. The coefficient of performance (COP) equals Q_{tot}/W, with W the acoustic power delivered to the resonator. (From Swift, after Wollan [95].)

Beginning in the late 1960s, a series of discoveries coming in rapid succession have radically changed our concepts of how the cochlea works. It is no longer regarded as a passive linear device but, rather as a nonlinear active one that can adapt to stimulus conditions. [98]

In 1967, Brian M. Johnstone and A.J.F. Boyle [99] made the discovery that the broad maximum in the wave along the basilar membrane was sharper in a living ear than in a dead one. New techniques with lasers, and use of the Mössbauer effect made such discoveries possible [100].

The revolution in our understanding of the cochlea is made evident in our knowledge of the hair cells. These cells, located in the organ of Corti in the cochlea, consist of three rows of outer hair cells and one row of inner hair cells [101]. Until recently, it had been thought that the outer hair cells were responsible for the ear's sensitivity, while the inner cells, which did not respond to sound levels below 50 dB, were responsible for fine frequency analysis [102]. More recent research has indicated, however, that the outer hair cells are more like muscles than sensory cells, and that their shape changes in response to various chemical, electrical, or sonic stimuli [103]. As Donald W. Nielsen writes,

[the inner ear] is an amplifier, driven by physiologic activities, mostly motile outer hair cells. These cells amplify the incoming acoustic signal, and feed the amplified signal to the inner hair cell, the real sensory cell or transducer of the cochlea. We don't know exactly how it works. [102]

This knowledge of the motility of the outer hair cells has found a practical research application in the subject of otoacoustic emissions.

Otoacoustic Emission

While the first intimations of otoacoustic emission (OAE) were sounded by T. Gold in 1948 [104], [105], who suggested that the sharp frequency selectivity of the cochlea resulted from a feedback system with production of sound within the cochlea itself, the idea was not verified experimentally until 1978, when D.T. Kemp published the first of a series of papers on otoacoustic emissions in the ear canal [106].

The otoacoustic emissions may be spontaneous or evoked. An example of spontaneous emission is shown in Fig. 10-20, where three spikes of OAE rise above the ambient noise level (noise produced by various body sounds) [107].

Among the evoked otoacoustic emissions, the most prominent are those produced by the use of brief acoustic stimuli [108]. Still other forms are stimulus-frequency otoacoustic emissions (Kemp and Chum, 1980 [109]), resulting from the introduction of a low-level constant tone, and distortion-product otoacoustic emissions DPOAE) (Kemp, 1979) [110]. These emissions are the rest of the intermodulation of the tones of two primary frequencies introduced into the ear. An example of the DPOAE is shown in Fig. 10-21 [111].

It is now generally accepted that these emissions occur in the cochlea, and a substantial branch of study has involved the correlation of damage to the cochlea from various causes, including specific drugs, with reduction in the OAE [112]. Thus, the OAE can be used to monitor the physical condition of the outer hair cells.

In a recent paper, Killion has demonstrated the connection of the DPOAE with Tartini tones, see Chapters 1 and 3, and applied this knowledge to the study of nonlinear distortions in hearing aids (Fig. 10-22 [113]). In the experiment, B and D were produced by oscillators and the

FIGURE 10-20. A commercially interesting thermoacoustic system. Half-wavelength refrigerator with two stacks driven by two loudspeakers was built at CSIR in South Africa. It operates at 120 Hz with 15-bar neon. The heat exchangers are located where the water lines connect to the green resonator body. (Courtesy of Peter Bland, Quadrant.) (From Martin [107].)

otoemission spectrum of the ear, recorded by means of a probe fixed in the ear canal. The Tartini tone at F is clearly visible.

Aids to the Deaf and Hard of Hearing

Mankind's first attempts at compensating for deafness was most probably the resort to talking (or shouting) louder. All of us must have had one or more deafened relatives, usually an older person, for whom this technique was frequently applied. Such a method must have been followed rapidly by the cupping of the hand of the listener around his or her ear. This was the forerunner of the "ear trumpet," which was essentially an inverted megaphone, which amplified the sounds approaching the ear, and which came into use in the latter part of the seventeenth century [114].

The development of electric amplifiers suggested the usefulness of a further increase in the intensity of the sound entering the outer ear, and the invention of the vacuum tube amplifier further accelerated the use of this type of hearing aid, but it was not until the invention of the transistor that it was possible to produce a hearing aid small enough not to be a nuisance in lugging it around, and the hearing aid industry was born.

Assistance to the deaf has developed along three lines in recent years. For the immediate needs of the hard of hearing, there has been the intensive competition among hearing aid manufacturers. In the case of profound deafness, a more radical solution has been that of cochlear implants, and for

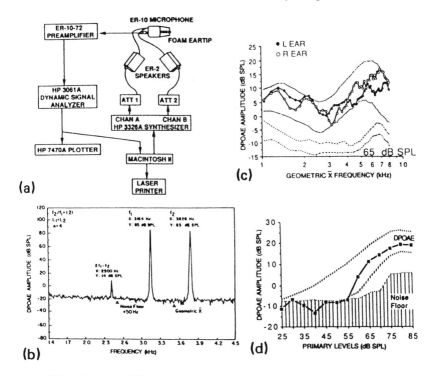

FIGURE 10-21. Example of distortion-product otoacoustic emission. Equipment used to measure distortion-product otoacoustic emissions. (a) DPOAE setup showing the miniature microphone which is inserted into the ear canal of the patient for detecting emissions. Two tones are presented through miniature ear speakers to provide the stimulating sounds that evoke the DPOAE. All levels of the stimuli and measurements are accomplished under computer control. (b) The emitted response at $2f_1 - f_2$ is shown in the spectrum of the ear-canal signal recorded from the sensitive microphone. The various types of measures that can be obtained with this method are depicted in portions (c) and (d) of the figure. (c) DPOAE "audiogram" in which the level of the emission is plotted as a function of the geometric mean of the stimulating tones. This plot is comparable to the traditional audiogram, but provides a measure of outer hair cell receptor function rather than hearing level. Thin lines above denote the ranges of normally hearing individuals, whereas the lower pair of dashed lines show the limits or noise floor of the recording system beyond which no responses can be measured. (d) A response/growth function obtainable with DPOAEs that measures the dynamic range of the emission and thus provides an estimate of "threshold" and DPOAE growth at suprathreshold levels. Dashed lines and stripes indicate the normal DPOAE levels and noise floors, respectively. (From Kemp [111].)

the long-range hopes, there are the possibilities of hair cell regeneration. We shall look at each of these in turn.

Hearing Aids

By and large, the acoustical community has concentrated more on the study of deafness than on means of alleviating such a disability. The second

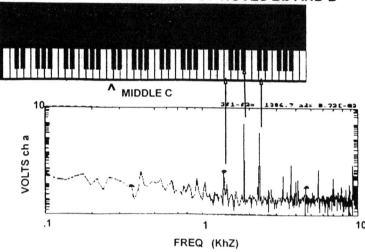

FIGURE 10-22. Connection of distortion-product otoacoustic emission with Tartini tones. Generation of music note F played—literally—by ear. *Note*: In these figures, the piano keyboard has been scaled and positioned to line up with the frequency axis on the graph. The lowest octave on the piano—extending down to 27.5 Hz—has been omitted. (From Killion [113].)

edition of Harvey Fletcher's *Speech and Hearing* [115] devotes most of two chapters to hearing loss and its measurement, but only a small portion of one of these is concerned with the measurement of the effect of hearing aids. In a small book published in 1947, Hallowell Davis reported on a detailed study of the usefulness of the various hearing aids of the time [116]. During the next quarter-century, there were many hearing aids produced, but little objective study of their qualities. Early hearing aids mainly increased the sound level over the entire frequency range. This led to irritations from making sounds at or above the threshold of hearing. To avoid this, clipping was employed, but this produced new harmonics. In addition, the presence of intense signals also led to interaction of sounds of different frequencies, the production of combination tones, and, thus, a rise in the noise level.

Efforts to reduce this noise level have often weakened the original signal. The kind of problem faced by the deaf is summed up by Edgar Villchur:

A deaf person with recruitment [see Chapter 8] perceives sound as though listening through a volume expander followed by an attenuator, the expander ratio and attenuation being typically frequency dependent. The subject is often prevented from using enough hearing-aid gain to bring weak consonants into the useful dynamic range of his hearing, because this amount of gain would make lower-frequency, high-amplitude vowels intolerably loud. [117]

Villchur's solution, following up on earlier ideas by Steinberg and Gardiner [118] was the use of wide-dynamic-range compression (WDRC) in the hearing aid. In the original work, the speech underwent signal processing, in which a two-channel amplitude compressor with adjustment of the frequency compression ratio to offset recruitment of the individual subject. As Villchur put it, the "compressed speech ratio is subjected to frequency-selective amplification similarly adapted to the subject [119]."

The researches that followed up on this technique have led to considerable improvement in the modern hearing aid, with the aid itself becoming smaller and smaller, so that it has become more and more acceptable. The present status of hearing aids is discussed in articles by Mead Killion [120] and by Hawkins and Naiboo [121].

Cochlear Implants

The history of electric stimulation of hearing goes far back into time [122]. The beginnings may be found as far back as the work of Volta in 1800, when he connected his newly invented "voltaic pile" (30–40 of his voltaic cells connected in series) to his two ears, and heard a continuing crackling sound in his ears—as well as experiencing a severe shock) [123]. Much later, F.J.J. Buytendijk in Holland in 1910 [124] discovered local electric currents (action-currents) in the auditory nerve of rabbits when a pistol was fired nearby. In the 1920s, radio engineers found that stimulating electrodes with a modulated ac current, held near the ear, produced auditory sensations, while it was discovered by Wever, Bray, and Willey [125] in 1937 that electrical potentials arise in the cochlea as a result of acoustic stimulation. This latter phenomenon was given the name of cochlear microphonics, because the cochlea acts like a microphone in converting sound waves to electric currents [126].

These early researches were not connected with the idea of improvement of hearing, but in 1957, A. Djourno and C. Eyriès [127] performed the direct stimulation of the auditory nerve (Fig. 10-23) [128]. While the amount of hearing obtained was minimal, it was a start.

Beginning in 1960, William F. House tried electric stimulation of a number of patients, with mixed results [129]. The same was true of work performed by Simmons [130], in which he implanted a hard-wire cluster of six electrodes (stainless steel) in direct electrical stimulation. He found that the patients could hear a number of sounds, but were unable to produce speech discrimination. An evaluation of the work of House and Simmons was performed in 1977 by R.C. Bilger et al. While some hearing could be obtained, speech discrimination was still not possible [131].

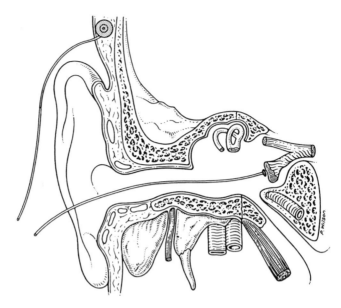

FIGURE 10-23. Djourno and Eyriès stimulation of the auditory nerve in 1957. The induction coil is not shown. (From Luxford and Brackmann, after Djourno and Eyriès [127], [128].)

The conclusion of all this research, according to Luxford and Brackmann, was that

the implant provides, among another things, an awareness of environmental sounds, improved speechreading, and possibly some minimal speech discrimination. [122]

Major differences in cochlear implants have been the number of channels being used. In the House group, emphasis has been on single channel systems, Fig. 10-24 [132], while others have developed multichanneled systems (Fig. 10-25) [133]. The sequence of operations involved in a cochlear implant is shown in Fig. 10-26 [134]. At the present time, this technique is a very imperfect one, but for the totally deaf it can be a great boon. The situation is summarized by Derek Sanders:

For selected children and adults for whom conventional amplification does not allow useful benefits, the cochlear implant provides considerable potential for developing and improving speech communication. . . . The full potential of the cochlear implant as technology advances has yet to be determined. [135]

The comments by Luxford and Brackmann, and by Sanders, cited above, would appear to be somewhat pessimistic. More recent findings, by Mary J. Osberger for children [136] and by Richard C. Dowell for adults [137], indicate considerable improvement of hearing perception, especially with the use of multiple channeled systems. A more optimistic report, given by

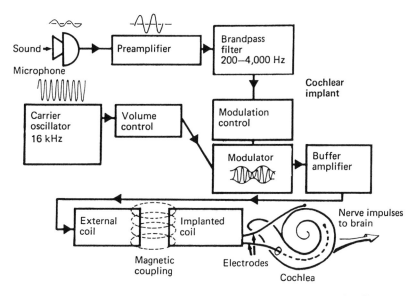

FIGURE 10-24. Single-channel excitation. (From W. House et al. [132].)

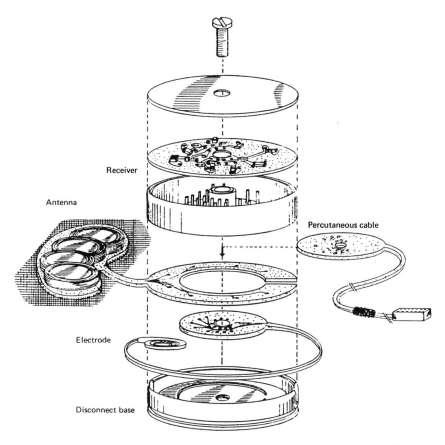

FIGURE 10-25. Multichanneled systems of stimulation. (From Merzenich et al. [133].)

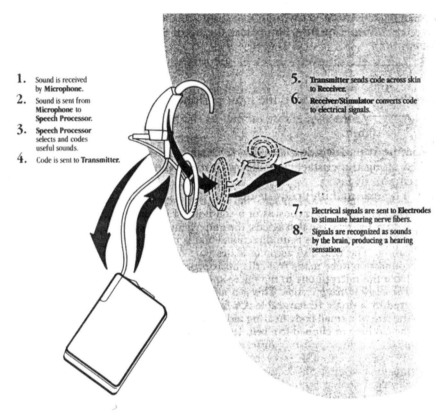

1. Sound is received by Microphone.
2. Sound is sent from Microphone to Speech Processor.
3. Speech Processor selects and codes useful sounds.
4. Code is sent to Transmitter.

5. Transmitter sends code across skin to Receiver.
6. Receiver/Stimulator converts code to electrical signals.

7. Electrical signals are sent to Electrodes to stimulate hearing nerve fibers.
8. Signals are recognized as sounds by the brain, producing a hearing sensation.

FIGURE 10-26. A depiction of the sequence by which a cochlear implant provides direct stimulation of the auditory nerve fiber. (Cochlear Corporation. From Sanders [134].)

Anita Wallin, who became totally deaf as an adult, sums up her gains as giving her the chance to

Hear environmental sounds.
Hear what one person says in a quiet environment.
Listen to music again, at least to some extent.
Feel silence, for the implant interrupts the sounds of tinnitus. [138]
Get feedback of my own voice. [139]

Hair Cell Regeneration

Perhaps the most exciting results in recent cochlea research, from the long-range viewpoint, has been the observation of outer hair cell regeneration in certain animals. Until 1988, it was regarded as sure that the hair cells in warm-blooded vertebrates were established by the time of birth of the animal, and that any damage or destruction could not be reversed. However, in 1988, two teams, Brenda M. Ryals and her colleagues at James

A. CONTROL   **B.   0 DAYS**

C. 5 DAYS   **D. 14 DAYS**

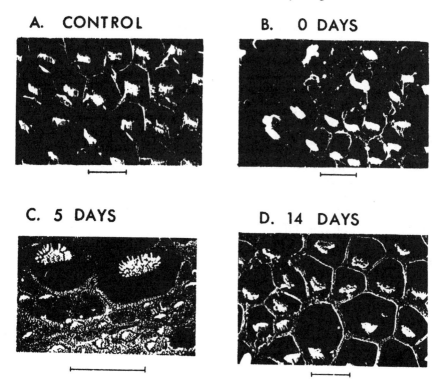

FIGURE 10-27. Hair-cell regeneration. Four photomicrographs are presented and the calibration bar accompanying each indicates 10 μm. The control picture (A) is from the equivalent location of the lesion on a nonexposed papilla 17-days-old. The important aspect to note is the orderly organization of the hair cells on the unexposed sensory surface. Immediately after removal from the exposure (B, 0 days' recovery), the papilla of this 3-day-old chick shows a greatly reduced number of hair cells. C illustrates the papilla after 5 days of recovery. New and old hair cells can be seen. After 14 days of recovery (D), the lesion is filled with hair cells whose surface area is quite variable. A comparison of panels A and C indicates much less organization of the receptor cell field after 2 weeks of recovery. (From Henry et al. [141].)

Madison University [140], and James C. Saunders and others at the University of Pennsylvania found that the hair cells of birds could be regenerated (Fig. 10-27) [141]. In a hopeful note for the future, Ryals and Rubel wrote

Our study and their study [142] demonstrate hair cell regeneration after terminal mitosis in the inner ear of birds. . . . The regenerative process is retained in adult animals, suggesting that a dormant stem cell population is retained throughout life. Although the location and mechanism of activation of these precursor cells are yet to be identified, the potential may exist to restore inner ear sensory elements after injury or disease. [140]

Or, as Nielsen wrote in a recent paper,

We now know that supporting cells can become hair cells postembryonically. . . . We are getting very close to understanding what triggers hair cell regeneration in birds. . . . This progress leads me to the exciting conclusion that hair cell regeneration shows promise as a cure for the devastating effects of sensorineural hearing loss. [102]

10.14. Psychological Acoustics

Minimal Audible Pressure

The determination of the minimal sound pressure that can be detected by the ear has been of interest to the acoustics community since the days of Rayleigh. A repeated theme has been the missing 6 dB, repeatedly studied since the original work by Sivian and White in 1933. One might hope that the final word on this difference between MAP (minimum audible pressure) and MAF (minimum audible field) was given in a paper by Mead Killion in 1978. He summed up his findings by writing

Eardrum pressures at hearing threshold have been calculated from both earphone data (ISO R389-1964 and ANSI S3.6-1969) and free-field data (ISO R226-1961). When ear diffraction, external-ear resonance, and an apparent flaw in ISO R226 are accounted for in the free-field data, and real-ear versus coupler differences and physiological noise are accounted for in the earphone data, the agreement between the two derivations is good. [143]

Speech Perception in the Young

An important development in the study of speech perception was the discovery by workers in the Haskins Laboratory (Alvin Liberman and his colleagues) that, if a sound was gradually changed from, say, /ba/ to /pa/, the listener identified the sound as /ba/ until a sort of acoustic boundary was reached, after which the sound was identified as /pa/. There was no intermediate range of ambiguity. This phenomenon has come to be known as categorical perception [144].

It was found in 1971 by Peter Eimas and his colleagues at Brown University that this same phenomenon occurs in infants [145]. The ways in which we can learn about what infants perceive form a fascinating picture of modern research. The ingenuity of researchers in this exciting area of modern acoustics rivals that of the great figures of the nineteenth century. The first problem in dealing with the very young is to develop a process in which we can be sure that the infant is recognizing the difference between two alternatives proposed to the child and the nature of the child's decision. The new-born child has an extraordinary ability to respond to the human face and the human voice. Infants as young as 45 minutes after birth will imitate gestures such as mouth opening or tongue protrusion [146]. Of the response to the human voice, Patricia Kuhl, a distinguished researcher at the University of Washington, writes, given a choice, young infants prefer to listen to *Motherese*, a highly melodic speech signal adults use when addressing infants (Fernald [147]). It is not the syntax or semantics of Motherese that holds infants' attention—it is the acoustic signal itself. [148]

Two techniques for learning about an infant's ability to perceive speech are high-amplitude sucking (HAS) [149] and headturning (HT) [150]. In the first of these (HAS), the pressure on a nipple is recorded. When a sound is presented to the infant there is an increase in the sucking rate. When the same stimulus is applied a number of times, the infant's interest lags and the sucking rate decreases. However, if there is a sound change—a different syllable or frequency—the sucking rate again increases. A sample of the results is shown in Fig. 10-28 [149]. This technique is useful in the age range 1–4 months. After that time, the second method (HT) is used (age range 6 months–1 year). The child can be trained to turn its head whenever a speech sound is changed [150].

A typical result of a change in the presented sound is shown in Fig. 10-29 [150]. As can be seen the sucking rates of the two infants were identical before the sound shift but quite different afterward. Kuhl sums up the results as follows:

These studies show that infants come into the world with the basic abilities to hear the distinctions important for language. The acoustic cues that differentiate phonetic units in the language are very subtle. It is considered very helpful that infants are born with the capacity to attend to and distinguish these acoustic cues. [151]

The number of discoveries in this young field is great, and we can only touch on a few. It has been shown by Janet Werker of the University of British Columbia [152] that infants lose the ability to distinguish contrasting sounds in a foreign language that they once could discriminate. Apparently, this is another case of "use it or lose it." A case in point is the difficulty for an adult Japanese to distinguish between /r/ and /l/, whereas the American

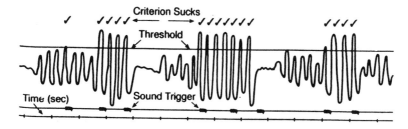

FIGURE 10-28. High-amplitude sucking (HAS). Young infant being tested using the high-amplitude sucking (HAS) technique. The infant sucks on a nonnutritive nipple and the pressure changes inside the nipple are recorded. Sucking responses that exceed a threshold result in the presentation of a speech sound. (From Jusczyk [149].)

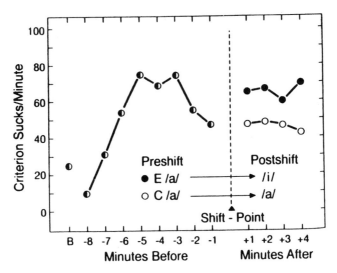

FIGURE 10-29. Infant detection of a sound change. The number of sucking responses produced in each minute of the experiment by an experimental and control infant. The two infants are treated identically prior to the shift point; after the shift point, the experimental infant hears a new sound while the control infant continues to hear the first sound. (From Kuhl [150].)

listener can easily do so. As an infant, the Japanese did not listen to a collection of words in which these two sounds were clearly marked, whereas the American infant was so exposed.

Kuhl has developed the Native Language Magnet (NLM) theory to account for some of this behavior. As she puts it,

infants form mental representations of the sounds they hear. These representations constitute the beginnings of language-specific speech perception, account for infants' perception of both native- and foreign-language sounds, and serve as a blueprint which guides infants' initial attempts to produce speech. [153]

According to the NLM theory, infants partition a "vowel space" into perceptual divisions. Sounds that are separated by a boundary line are distinguishable (Fig. 10-30) [154]. These perceptual maps will be different for infants being raised in Sweden, England, and Japan, for example. As infants experience the vowel sounds of one language, their ability to distinguish vowel sounds not common in their own language decreases. Figure 10-31 shows the individual vowel sounds in the three languages, while Fig. 10-32 shows the simplified structure of the auditory boundaries. "Thus, by six months of age, exposure to the ambient language alters infants' perception of the phonetic units of language [155]." What takes place earlier in the speech learning of the infant is still to be determined.

FIGURE 10-30. Infants' natural auditory boundaries. NLM Theory: At birth, infants perceptually partition the acoustic space underlying phonetic distinctions in a language-universal way. They are capable of discriminating all phonetically relevant differences in the world's languages. (From Kuhl [154].)

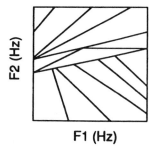

Infants' Natural Auditory Boundaries

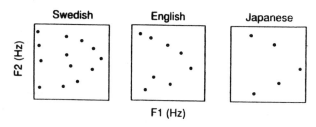

FIGURE 10-31. Vowel space in different languages. NLM Theory: By 6 months of age, infants reared in different linguistic environments show an effect of language experience. They exhibit language-specific magnet effects that result from listening to the ambient language. (From Kuhl [154].)

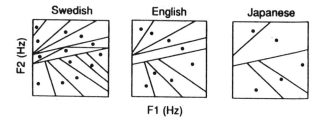

FIGURE 10-32. Disappearance of certain phonetic boundaries. After language-specific magnet effects appear, certain phonetic boundaries "disappear"; magnet effects alter the perceived distance between stimuli, making certain distinctions more difficult to discriminate. (From Kuhl [154].)

A Final View

Much of the excitement in modern acoustics lies in the pursuit of extrema— the very small [(the acoustic microscope), the very high-frequency (hypersound), the very detailed examination (scattering problems)], in pushing back the dividing line between the acoustically measurable and the mentally immeasurable (architectural acoustics, speech, and hearing), and in service to humanity (noise reduction, hearing and speech impairment aids, artificial cochlea, medical ultrasonics). As has always been the case, while old giants have passed from the scene (Georg von Békésy, Smitty Stevens, Hallowell Davis, Warren Mason, Vern Knudsen, Bruce Lindsay, Ted Hunt, Erwin Meyer) new ones have come to the fore, and one can perceive, in the work presented at technical meetings, the moving up of the young, the giants of the future.

Dayton Miller, in his *Anecdotal History of the Science of Sound*, began his final chapter by remarking that "More progress has been made in the realm of sound in the first third of the twentieth century than in all of the preceding centuries," whereas Bruce Lindsay remarked (Chapter 7) that the first two decades of this century formed "a rather stagnant period in acoustics," one can only conclude that it all depends on the point from which the historian makes his estimates. Lindsay had lived through the multitudinous developments of the mid-century Miller had not. In the time since Lindsay, however, acoustics has continued to grow at an even faster rate. In 1995, under the editorship of Daniel W. Martin, the *Journal of the Acoustical Society of America* (*JASA*) has grown to 7700 pages, while the annual "References to Contemporary Papers on Acoustics" of *JASA* (which publication does not include articles published in *JASA* itself) listed approximately 15,000 titles. The field has indeed become too large to be viewed in its entirely by any individual. The very recent publication, *Encyclopedia of Acoustics* [156], has been written by more than 200 authors and contains 3100 pages. And yet, there remains a fundamental unity in the field. We still talk to one another, listen to another (at least in part), and use the same core of mathematics and physical principles. Lindsay's wheel is still rolling along. Long and far may it roll!

Notes and References

[1] Physics and Astronomy Classification Scheme, 1995. AIP Publ. R-261.13. American Institute of Physics, Woodbury, NY. The acoustics portion of this scheme is the outgrowth of the classification system started in the *Journal of the Acoustical Society of America* (*JASA*), by Floyd Firestone. In a Foreword to the first Cumulative Index (vol. 1–10), Firestone, who had become editor, wrote "Through flattery and other seductive means of persuasion, Editor F.R. Watson induced the present editor to undertake the preparation of the analytical subject index." Succeeding editors have added to this index so that it covers in considerable detail all the current interest areas in acoustics, and provides a resource of inestimable value to researchers and historians.

[2] D. Brill and G. Gaunaurd, *J. Acoust. Soc. Am.* **81**, 1–21 (1987).

[3] L. Flax, G. Gaunaurd, and H. Überall, "The Theory of Resonance Scattering," in *Physical Acoustics*, Academic Press, New York, NY, vol. XVI, 1981, pp. 191–194; G. Gaunaurd and H. Überall, "Resonances in Acoustic and

Elastic-Wave Scattering," in *Recent Developments in Classical Wave Scattering* (V.K. Varadan and V.V. Varadan, eds.), Pergamon, New York, NY, 1979, pp. 413–429. A sentence in the paper by Brill and Gaunaurd (Ref. [2]) may be applied to all of this work—"Recastings, reformulations, summaries, and computations of these classical solutions have appeared in recent years." And still they come!

[4] G.C. Gaunaurd, *J. Acoust. Soc. Am.* **84**, S168 (1988); **85**, S134 (1989). The author of this book was not unaffected by this brand of scattering; see US Patent No. 4,339,944, "Particle Identification by Ultrasonic Means," issued July 1982 to L.R. Abts and R.T. Beyer.

[5] R.T. Beyer, *J. Acoust. Soc. Am.* **63**, 1025–1030 (1978).

[6] B.T. Chu and R.E. Apfel, *J. Acoust. Soc. Am.* **72**, 1673–1687 (1982). This article led to its own controversy with J.A. Rooney and W.L. Nyborg. The Nyborg–Rooney papers appeared in *J. Acoust. Soc. Am.* **40**, 1825–1830 (1972); **75**, 263–264 (1984), with a rejoinder from Chu and Apfel in *J. Acoust. Soc. Am.* **75**, 1003–1004 (1984). The ease with which one can fall into error or at least, dispute, on this subject makes even more real the statement made by the author of this book in his 1978 article, quoted by Chu and Apfel in 1982: "Radiation pressure is a phenomenon that the observer thinks he understands—for short intervals, and only now and then." It appears to be virtually impossible to set up the theoretical conditions for an experiment that can actually be carried out, without making some approximation, or neglecting some detail, that brings the result into conflict with someone else's theory and experiment.

[7] C.R. Allen and I. Rudnick, *J. Acoust. Soc. Am.* **19**, 857–865 (1947); M. Barmatz, in *Materials Processing in the Reduced Gravity Environment of Space* (G.E. Rindone, ed.), North-Holland, New York, NY, 1982, p. 25.

[8] S. Putterman, J. Rudnick, and M. Barmatz, *J. Acoust. Soc. Am.* **85**, 69–71 (1988). The passage of generations can be marked by noting that Joseph Rudnick is the son of the oft-mentioned (in this volume) Isadore Rudnick.

[9] T.G. Wang, M.M. Saffren, and D.D. Elleman, AIAA Paper No. 74–155 (1974); T.G. Wang in *Proceedings of the International School of Physics, "Enrico Fermi,"* (D. Sette, ed.), North-Holland, Amsterdam, 1986.

[10] Examples of this containerless research may be found in T. Baykar et al., *High Temp. Sci.* **32**, 92–93, 113–154 (1991); J.K.R. Weber et al. *Rev. Sci. Instr.* **65** (2), 456–465 (1994).

[11] Junru Wu, *J. Acoust. Soc. Am.* **89**, 2140–2143 (1991).

[12] G.L. Lamb, Jr., *Elements of Soliton Theory*, Wiley, New York, NY, 1980, p. 1.

[13] N.J. Zabusky and M.D. Kruskal, *Phys. Rev. Lett.* **15**, 240–245 (1965). The illustration is taken from *Nonlinear Waves* (S. Leibovich and A.R. Seebass, eds.), Cornell University Press, Ithaca, NY, 1974, p. 24.

[14] D.J. Korteweg and G. de Vries, *Phil. Mag.* **39**, 422 (1895).

[15] The Burgers equation may be written in reduced notation ($u_x = \partial u/\partial x$, $u_{xx} = \partial^2 v/\partial x^2$, etc.) as

$$u_z - uu_t = \delta u_{tt},$$

while the Korteweg–de Vries equation has the form

$$u_t + \alpha uu_z + \beta u_{xxx} = 0,$$

where α, β, and δ are appropriate constants. A brief discussion of these and related equations is given in A. Jeffrey and T. Kawahara, *Asymptotic Methods in Nonlinear Theory*, Pitman, Boston, MA, 1982, pp. 9–13, while a much more detailed treatment appears in *Nonlinear Waves* (S. Leibovich and A.R. Seebass, eds.), Chapters IV and VIII.

[16] C.S. Gardner, J.M. Greene, M.D. Kruskal, and R.M. Miura, *Phys. Rev. Lett.* **19**, 1095–1097 (1967).

[17] R.M. Miura in S. Leibovich and A.R. Seebass, Ref. [15, p. 213].

[18] D.J. Acheson, *Elementary Fluid Dynamics*, Oxford University Press, New York, NY, 1990 p. 110. See also P.G. Drazin and R.S. Johnson, *Solitons: An Introduction*, Cambridge University Press, Cambridge, UK, 1989.

[19] Junru Wu, R. Keolian, and I. Rudnick. *Phys. Rev. Lett.* **52**, 1421–1424 (1984); Junru Wu and I. Rudnick, *Phys. Rev. Lett.* **55**, 204–207 (1985). For theoretical studies of this phenomenon, see A. Larazza and S. Putterman, *J. Fluid Mech.* **148**, 443–449 (1984); J.W. Miles, *J. Fluid Mech.* **148**, 451–460 (1984).

[20] R.-J. Wei, B.-R. Wang, Y. Mao, X.-Y. Zheng, and G.-Q. Miao, *J. Acoust. Soc. Am.* **88**, 469–472 (1990). Wei had been a student of Rudnick's in the 1950s.

[21] G.T. Adamashvili, *Soviet Phys. Solid State* **33**, 900–901 (1991).

[22] G.N. Burlak, O.N. Bulanchuk, and V.V. Gimal'skii, *Soviet Phys. Tech. Phys.* **37**, 469–472 (1992).

[23] A.F. Popkov, *J. Exptl. Theoret. Phys.* **77**, 48–491 (1993).

[24] W. Lauterborn, *Physics Today*, Jan. 1986, pp. S-4, S-5.

[25] R. Esche, *Acustica*, **AB208**, 1952. See also V. Akulichev, in *High Intensity Ultrasonic Fields* (L.D. Rozenberg, ed.), Plenum Press, New York, NY, 1971, pp. 203–259.

[26] W. Lauterborn, *Acustica* **82**, S46–S55 (1996). This article describes many special features of acoustical chaos and contains an extensive bibliography. See also the earlier work of W. Lauterborn and J. Holzfuss, *Internat. J. Bifurcation and Chaos*, **1**, 13–26 (1991).

[27] H. Frenzel, Hinsberg, and H. Schultes, *Z. Ges. Exp. Med.* **89**, 246 (1929); H. Frenzel and H. Schultes, *Z. Phys. Chem.* **B27**, 421–424 (1934).

[28] L. Crum and R.A. Roy, *Science*, **266**, 233–234 (1994).

[29] K.S. Suslick, *Science* **247**, 1439–1445 (1990). See also Kenneth S. Suslick, Edward B. Flint, Mark W. Greenstaff, and Kathleen A. Kemper, *J. Phys. Chem.* **97**, 3098–3099 (1993).

[30] S. Putterman, et al., *Phys. Rew. Lett.* **74**, 5276–5279 (1995); **78**, 799–802 (1997). See also R. Hiller, S.J. Putterman, and B.P. Barber, *Phys. Rev. Lett.* **69**, 1182 (1992); R. Hiller and B.P. Barber, *J. Acoust. Soc. Am.* **94**, 1794 (1993).

[31] F. Gaitan, L.A. Crum, R.A. Roy, and C.C. Church. *J. Acoust. Soc. Am.* **91**, 3166–3183 (1992).

[32] The world never changes. In the 1600s, Galileo's attempts to measure the speed of light by exchanging lantern flashes between two hills primarily measured the reaction time of his assistant!

[33] B.P. Barber and S.J. Putterman, *Nature* **352**, 318–320 (1991); see also R. Hiller, K. Weninger, S.J. Putterman, and R.P. Barber, *Science* **266**, 248–250 (1994).

[34] J. Schwinger, *Lett. Math. Phys.* **A1**, 43–47 (1975); *Proc. Nat. Acad. Sci.* **89**, 4091–4093, 11118–11120 (1992); J. Schwinger, I. De Raad, and K. Milton,

Ann. Physik **115**, 1–23 (1978). See also H. Casimir, *Proc. K. Nederl. Akad. Wetensch.* **51**, 793–795 (1948); H. Casimir and D. Polder, *Phys. Rev,* **73**, 360 (1948). Casimir derived expressions for the interaction between a perfectly conducting plate and an atom with a given static polarizability, and also for the interaction between two perfectly conducting plates, all based on quantum mechanical considerations (see references to Schwinger). Schwinger also extended these calculations to dielectric media. The forces involved varied as the inverse of the fourth power of the separation distance. In his "dynamic Casimir effect," Schwinger analyzed the interaction of the contraction and expansion of a bubble as a hole in the dielectric medium.

[35] L. Lepoint and F. Mullie, *Ultrasonics and Sonochemistry* **1**, S13 (1994); M.A. Margolis, *Ultrasonics* **30**, 152–155 (1992).

[36] C.C. Wu and Paul H. Roberts, *Phys. Rev. Lett.* **70**, 3424–3427 (1993).

[37] C.C. Wu and P.H. Roberts, *Phys. Rev. Lett.* **72**, 1380–1383 (1994); see also Peter Jarman, *Proc. Phys. Soc. (London)* **72**, 68 (1959); *J. Acoust. Soc. Am.* **32**, 1459–1462 (1960).

[38] For a discussion of Franklin's work, see C.M. Daugherty, *Searchers of the Sea*, Viking Press, New York, NY, 1951, pp. 43–59.

[39] C.W. Ekman, *Sci. Results Norweg. North Polar Exped.* **5**, No. 15, 1–152 (1925).

[40] A historical view of the problem, with many references, can be found in James Lighthill, *Waves in Fluids*, Cambridge University Press, Cambridge, UK, 1978. See also S.M. Flatté, R. Dashen, W. Munk, K. Watson, and F. Zachariasen, *Transmission Through a Fluctuating Ocean*, Cambridge University Press, Cambridge, UK, 1979. For later work, see S.M. Flatté and R.B. Stoughton, *J. Acoust. Soc. Am.* **84**, 1414–1424 (1988).

[41] David J. Browning, Project CANUS, NUSC Rpt. 422, 1 (1971); D.G. Browning et al., NUSC Rept. 4501 (1973).

[42] Walter Munk, Peter Worcestor, and Carl Wunsch, *Ocean Acoustic Tomography*, Cambridge University Press, Cambridge, UK, 1995, p. 337.

[43] Walter Munk, Peter Worcestor, and Carl Wunsch, *Ocean Acoustic Tomography*, Chapter 8.

[44] W. John Richardson, Charles R. Greene, Jr., Charles I. Malme, and Denis A. Thomas, *Marine Animals and Noise*, Academic Press, San Diego, CA, 1995, especially pp. 309–311, 421, 422.

[45] Walter Munk et al., *Ocean Acoustic Tomography*.

[46] Walter Munk et al. *Ocean Acoustic Tomography*, p. 1.

[47] W. Munk and C. Wunsch, *Deep Sea Res.* **26**, 123–161 (1979).

[48] E.O. LaCasce, Jr. and J.C. Beckerle, *J. Acoust. Soc. Am.* **57**, 966–967 (1975). Rossby waves are low-frequency, very long-wavelength water waves that are a result of Coriolis forces. See I. Tolstoy, *Wave Propagation*, McGraw-Hill, New York, NY, 1973, pp. 160–165.

[49] E.L. Kerr and J.G. Atwood, *Appl. Opt.* **7**, 915–921 (1968).

[50] L.B. Kreuzer, *J. Appl. Phys.* **42**, 1934 (1971); L.B. Kreuzer and C.K.N. Patel, *Science* **173**, 45–47 (1971).

[51] W.R. Harshbarger and M.B. Robin, *Acc. Chem. Res.* **6**, 329 (1973).

[52] A. Rosencwaig, *Opt. Commun.* **7**, 305–308 (1973).

[53] P.J. Westervelt and R.S. Larson, *J. Acoust. Soc. Am.* **54**, 121–122 (1973).

[54] C.K.N. Patel and A.C. Tam, *Rev. Mod. Phys.* **53**, 517–550 (1981); Andrew Tam, *Rev. Mod. Phys.* **58**, 381–431 (1986); Markus W. Sigrist, *J. Appl. Phys.* **60**, R83–R121 (1986).

[55] C.K.N. Patel and A.C. Tam, loc. cit.

[56] These include *Ultrasonic Laser Spectroscopy* (David S. Kliger, ed.), Academic Press, New York, NY, 1983 (a study of photoacoustics in various media, one-photon excitation spectroscopy and the thermal lens); L.M. Lyamshev, *Lazernoe Termoopticheskoe Vozbuzhdenie Zvuka (Laser Thermo-Optic Excitation of Sound)*, Nauka, Moscow, 1986 (a detailed coverage of early Russian work in the field); V.P. Zharov and V.S. Letokhov, *Laser Optoacoustic Spectroscopy*, Springer-Verlag, New York, NY, 1986; *Photoacoustic, Photothermal and Photochemical Processes in Gases* (P. Hess, ed.), Springer-Verlag, New York, NY, 1989; *Air Monitoring by Spectroscopic Techniques* (Markus W. Sigrist, ed.), Wiley, New York, NY, 1994 (especially strong on gas-phase trace detection); *Photoacoustic and Thermal Wave Phenomena in Semiconductors* (Andreas Mandelis, ed.), North-Holland, New York, NY, c. 1987 (extensive coverage of semiconductor research, including solid-gas techniques); V.E. Gusev and A.A. Karabutov, *Laser Optoacoustics*, AIP Press, New York, NY, 1993 (special attention is paid to nonlinear properties).

[57] F.V. Bunkin, Al.A. Kolomensky, and V.G. Mikhalevich, *Lasers in Acoustics*, Harwood Academic, Reading, MA, 1991.

[58] M.T. O'Conner and G.J. Diebold, *Nature* **30**, 321–322 (1983).

[59] G.J. Diebold et al., *Science* **250**, 101–104 (1990). See also P.J. Westervelt, *J. Acoust. Soc. Am.* **84**, 2245–2251 (1988).

[60] T. Sun and G.J. Diebold, *Nature* **355**, 806–808 (1992). See also Refs. 1–8 in the paper of Chen and Diebold, *Science* **170**, 963–966 (1995).

[61] S.E. Braslavsky and G.E. Heibel. *Chem. Rev.* **12**, 1381–1410 (1992).

[62] H.X. Chen and G.J. Diebold, *J. Chem. Phys.* **104**, 6730–6741 (1996).

[63] C. Soize, *J. Acoust. Soc. Am.* **92**, 2365 (A) (1992). His earlier work on the subject appeared in *Rech. Aérospat.* **5**, 23–49 (1986).

[64] Andrew N. Norris, *J. Acoust. Soc. Am.* **97**, 3342 (A) (1995), and private communication to the author.

[65] M. Strasberg and D. Feit, *J. Acoust. Soc. Am.* **94**, 1814–1815 (A) (1993). A more detailed presentation of their work is to be found in M. Strasberg and D. Feit, *J. Acoust. Soc. Am.* **88**, 335–344 (1996).

[66] D.A. Russell and V.M. Sparrow, *J. Acoust. Soc. Am.* **91**, 2440 (A) (1992); D.A. Russell, J.I. Rochat, A.D. Pierce, and V.M. Sparrow, *J. Acoust. Soc. Am.* **93**, 2412–2413 (A) (1993).

[67] For a discussion of Thévenin's theorem as applied in acoustics, see L.L. Beranek, *Acoustics*, Acoustical Society of America, Woodbury, NY, 1986, pp. 78–80.

[68] B. Widrow et al., *Proc. IEE-E* **63**, 1692–1716 (1975); M.J.M. Jessel, Brevet Francais, No. 1,494,967 (1967); M.A. Swinbanks, US Patent No. 4,044,203 (1977).

[69] G.B.B. Chaplin, US Patent No. 4,152,815 (1977).

[70] C.F. Ross, *J. Sound Vibration* **61**, 473–480 (1978); O. Jones and R.A. Smith, *Proc. Internoise* **83**, 375 (1983).

[71] P.D. Wheeler, *An Active Noise Reduction System for Aircrew Helmets.* NATO AGARD cp311 (1981).

[72] J. Ffowcs-Williams, *Proc. Roy. Soc. (London)* **A395**, 63–88 (1984).

[73] J. von Neumann, *Mathematische Grundlagen der Quantenmechanik*, J. Springer, Berlin, 1930. English translation by R.T. Beyer, *Mathematical Foundations of Quantum Mechanics*, Princeton University Press, Princeton, NJ, 1955, pp. 418–421.

[74] Ibid.

[75] N. Bohr, *Naturwiss.* **16**, 245–257 (1928); **17**, 483–486 (1929).

[76] J. von Neumann, loc. cit.

[77] L.L. Beranek, *J. Acoust. Soc. Am.* **92**, 1–40 (1992). This review paper contains an extraordinary amount of information on the recent history of the concert halls acoustics.

[78] L.L. Beranek, *Music, Acoustics and Architecture*, Wiley, New York, NY, 1962, pp. 417–425, 556–561.

[79] M. Schroeder, D. Gottlob, and K.F. Siebrasse, *J. Acoust. Soc. Am.* **56**, 1195–1201 (1974).

[80] M. Schroeder et al. also noted the defintition of the parameter—"the energy in the first 50 ms of the impulse response, divided by the total energy (Ref. 7)" [see E. Meyer and R. Thiele, *Acustica* **6**, *Akust. Beih.*, Heft 2, 425 (1956)] but also remark in the same paper that it is not an independent parameter, since in all the halls tested, there was "a high negative correlation between the reverberation time T and definition D: the greater the T, the smaller the D, and vice versa."

[81] V. Mellert, *J. Acoust. Soc. Am.* **51**, 1359–1361 (1972).

[82] P. Damaske, *J. Acoust. Soc. Am.* **50**, 1109–1115 (1971).

[83] A.N. Burd, *Rundfunk. Mitteil.* **13**, 200 (1969).

[84] M. Schroeder, D. Gottlob, and K.F. Siebrasse, loc. cit.

[85] M. Schroeder, *J. Acoust. Soc. Am.* **68**, 22–28 (1980).

[86] Y. Ando, *Concert Hall Acoustics*, Springer-Verlag, Berlin 1985, Chapters 2–4.

[87] Y. Ando, *Concert Hall Acoustics*, p. 33.

[88] Y. Ando, *Concert Hall Acoustics*, especially Eq. (3.7) and the surrounding discussion.

[89] A.H. Marshall, *J. Sound Vibration* **5**, 100–112 (1967); **7**, 116–118 (1968); M. Barron, *J. Sound Vibration* **15**, 474–494 (1971); Ph.D. Thesis, University of Southampton, 1974.

[90] M. Barron and A.H. Marshall, *J. Sound Vibration* **77**, 211–232 (1981). The quotation is on p. 214.

[91] The author is indebted to Ilene Busch-Vishniac for supplying a copy of her paper on "New Acoustical Products and Processes," given at the 127th meeting of the Society in Cambridge, MA, in June 1994 [*J. Acoust. Soc. Am.* **85**, 2910–2911 (1994)].

[92] J.C. Wheatley, T.J. Hofler, G.W. Swift, and A. Migliori, *J. Acoust. Soc. Am.* **74**, 153–170 (1983); G.W. Swift, A. Migliori, T. Hofler, and J. Wheatley, *J. Acoust. Soc. Am.* **78**, 767–781 (1985). This latter article has a good bibliography for earlier work on acoustic engines.

[93] C. Sondhauss, *Ann. Physik* **79**, 1–34 (1850).

[94] Rayleigh, *The Theory of Sound*, vol. II, pp. 224–231.

[95] G. Swift, *Physics Today*, **48**, No. 7, 22–28 (July, 1995).

[96] N. Rott, *Z. Angew. Math. Phys.* **20**, 230–243 (1969); **24**, 54–72 (1973); **25**, 417–421 (1974); **25**, 43–49 (1975); N. Rott and Gerasismos Zouzoulas, *Z. Angew.*

Math. Phys. **27**, 325–344 (1976) (all these papers are in English). See also the tutorial on thermoacoustic engines by G.W. Swift, *J. Acoust. Soc. Am.* **84**, 1145–1180 (1988); T.J. Hofler, Ph.D. Dissertation, University of California, San Diego, CA, 1986; T.J. Hofler, *Proc. 5th Int. Cryocoolers Conf.* (P. Lindquist, ed.), Wright Patterson Air Force Base, OH, 1988, p. 93. S.L. Garrett, D.K. Perkins, and A. Gopinath, in *Heat Transfer* 1994, Institute of Chemical Engineers, Rugby, UK, 1994, p. 375.

[97] Peter Bland, *Quadrant* (figure taken from Ref. [95]).

[98] J.J. Zwislocki, *J. Acoust. Soc. Am.* **93**, 2345 (1993); and private communication.

[99] B.M. Johnstone and A.J.F. Boyle, *Science* **158**, 389–390 (1967); **160**, 1139 (1968).

[100] L.U.E. Kohllöffel, *Acustica* **27**, 49–65, 66–81, 82–89 (1972).

[101] S.T. Neely, *J. Acoust. Soc. Am.* **94**, 137–146 (1993). This article contains a review of work on outer hair motility and models for cochlear mechanics.

[102] D.W. Nielsen, from a paper read at the *98th Meeting of the Acoustical Society of America*, St. Louis, MO, 17 Nov.–1 Dec. 1995; *J. Acoust. Soc. Am.* **98**, 2935 (1995); and private communication to the author.

[103] W.E. Brownell et al., *Science* **22**, 194–196 (1985).

[104] T. Gold, *Proc. Roy. Soc. (London)* **B136**, 492–498 (1948).

[105] R. Probst, B.L. Lonsbury-Martin, and G.K. Martin, *J. Acoust. Soc. Am.* **89**, 2027–2067 (1991). This article gives a detailed review of the field, with hundreds of references.

[106] D.T. Kemp, *J. Acoust. Soc. Am.* **64**, 1386–1391 (1978).

[107] G.F. Martin et al. *Ear Hear.* **11**, 106–129 (1990).

[108] This process is known by many names in the literature. These names are discussed in Ref. [101], where its authors settled on the name "transiently evoked otoacoustic emissions (TEOAE)."

[109] D.T. Kemp and R.A. Chum, in *Psychophysical, Physiological, and Behavorial Studies in Hearing* (D. van den Brink and F.A. Bilsen, eds.), Delft University, Delft, Holland, 1980, pp. 34–42.

[110] D.T. Kemp, *Arch. Otorhinolaryngol.* **224**, 37–45 (1979).

[111] Ref. [105, p. 2033].

[112] Ref. [105, p. 2024].

[113] M.C. Killion, in *Hair Cells and Hearing Aids* (C.I. Berlin, ed.), Singular Press, San Diego, CA, 1994.

[114] A brief history of the development of aids to hearing may be found in *Encyclopedia Americana*, Grolier, Danbury, CT, vol. 13, p. 919. The author of this book remembers giving a lecture to a (largely older) women's club early in his career, in which at least three of the members used this device.

[115] Harvey Fletcher, *Speech and Hearing*, 2nd ed., Van Nostrand, New York, NY, 1953.

[116] Hallowell Davis, S.S. Stevens, and R.H. Nichols, Jr., *Hearing Aids*, Harvard University Press, Cambridge, MA, 1947.

[117] Edgar Villchur, *J. Acoust. Soc. Am.* **47**, 1646–1657 (1973). For a more recent commentary, see M. Killion, *J. Acoust. Soc. Am.* **98**, 2882 (A) (1995).

[118] J.C. Steinberg and M.B. Gardiner, *J. Acoust. Soc. Am.* **9**, 11–23 (1937).

[119] E. Villchur, *J. Acoust. Soc. Am.* **47**, 1646 (1973).

[120] M.C. Killion, Ref. [109]; *The Hearing Review*, **1**, 40, 42 and 43 (1994).

[121] D.B. Hawkins and S.V. Naiboo, *J. Amer. Acad. Audiol.* **4**, 221–228 (1993).

[122] W.M. Luxford and D.E. Brackmann, "The History of Cochlear Implants," in *Cochlear Implants* (Roger F. Gray, ed.), College Hill Press, San Diego, CA, 1985, pp. 1–26. See also William F. House, "Cochlear Implants," in *Ann. Otol. Rhinol. Laryngol.* **85**, Suppl. 27, 3–6 (1976).

[123] Volta's work is described in a historical article by Graeme M. Clark in *Cochlear Implants* (Ref. [122]), pp. 165–218.

[124] F.J.J. Buytendijk, *Proc. Roy. Soc. Amsterdam* **13**, 649–652 (1910).

[125] E.G. Wever, C.W. Bray, and C.F. Willey, *J. Exper. Psychol.* **20**, 336–349 (1937); see also W.M. Luxford and D.E. Brackmann, Ref. [122].

[126] S.S. Stevens and H. Davis, *Hearing*, Harvard University Press, Cambridge, MA, 1937. Reprinted by the Acoustical Society of America, Woodbury, NY, 1983, Chapter 13.

[127] A. Djourno and C. Eyriès, *Presee. Med.* **35**, 14–17 (1957). See also A. Djourno, C. Eyriès, and B. Vallancien, *C. R. Soc. Biol. (Paris)* **151**, 423–425 (1957); *Bull. Nat. Acad. Med. (Paris)* **141**, 481–483 (1957).

[128] Luxford and Brackmann, Ref. [122, p. 3].

[129] B.G. Edgerton, W.F. House, J.A. Brimacombe, and L.S. Eisenberg, in *Cochlear Implants in Clinical Use* (W.D. Keidel and P. Finkenzeller, eds.), S. Karger, New York, NY, 1984, pp. 68–89.

[130] F.B. Simmons, *Arch. Otolaryngol.* **84**, 2–54 (1966).

[131] A.C. Bilger et al. *Ann. Otol. Rhinol. Laryngol.* **86** (Suppl. 38), 1–176 (1977).

[132] W.F. House, Ref. [122].

[133] M.M. Merzenich, S.J. Rebscher, G.E. Loeb, C.I. Byers, and R.A. Schindler, Ref. [122, pp. 119–144].

[134] Derek A. Sanders, *Management of Hearing Handicap*, Prentice Hall, Englewood Cliffs, NJ, 1993, 3rd ed., p. 208.

[135] Derek A. Sanders, *Management of Hearing Handicap*, p. 212. Perhaps the next stage in hearing improvement will be cochlear transplants. Stay tuned.

[136] Mary Joe Osberger, in *Profound Deafness and Speech Communication* (Geoff Plant and Karl-Erik Spens, eds.), Whurr, London, UK, 1995, pp. 231–261.

[137] Richard C. Dowell, in *Profound Deafness and Speech Communication*, pp. 262–284.

[138] Tinnitus, or ringing in the ears, has many causes, including overstimulation by a loud sound or by hyper-irritability of the ear tissues. See S.S. Stevens and H. Davis, *Hearing*, reprinted by the Acoustical Society of America, Woodbury, NY, 1983, pp. 351–352.

[139] Anita Wallin, in *Profound Deafness and Speech Communication*, pp. 219–230.

[140] B.M. Ryals and E.W. Rubel, *Science*, **240**, 1774–1776 (1988).

[141] W.J. Henry, M.O. Makaretz, J.C. Saunders, M.E. Schneider, and P. Vrettakos, *Otolaryngology—Head and Neck Surgery* **98**, 607–611 (1988). See also H.J. Adler and J.C. Saunders, *J. Neurocytol. (UK)* **24**, 112–116 (1995).

[142] J.T. Corwin and S.A. Cotanche, *Soc. Neurosci. Abstr.* **13**, 539 (1987); J.T. Corwin and D.A. Cotanche, *Science* **240**, 1172 (1988).

[143] Mead C. Killion, *J. Acoust. Soc. Am.* **63**, 1501–1508 (1978). Killion's references in the quotation are to various standards: American National Standards Institute (ANSI) "Specifications for Audiometers," S3.6-1969; International Standards Organization (ISO) R226-1961 and R389-1964. These standards

are available through the ANSI office in New York or the office of the Acoustical Society of America.

[144] P.K. Kuhl in *Communication Sciences and Disorders*, Singular Publishing, San Diego, CA, 1994, p. 108ff.

[145] P.D. Eimas, E.R. Siqueland, P. Jusczyk, and H. Vigorito, *Science* **171**, 303–306 (1971).

[146] A.N. Meltzoff and M.K. Moore, *Child Development* **54**, 702–709 (1983).

[147] A. Fernald, *Infant Behavior and Development* **8**, 181–195 (1985).

[148] P. Kuhl, in *Communication Sciences and Disorders* (Fred D. Minifie, ed.), Singular Publishing, San Diego, CA, 1994, Chapter 3. See also A. Fernald and P. Kuhl, *Infant Behavior and Development* **10**, 279–291 (1987); D.L. Grieser and P.K. Kuhl, *Developmental Psychology* **24**, 14–20 (1988).

[149] P.W. Jusczyk in *Measurement of Audition and Vision in the First Year of Postnatal Life* (G. Gottlieb and N.A. Krasnegor, eds.), Ablex, Norwood, NJ, 1985, pp. 195–222.

[150] P.K. Kuhl, in *Measurement of Audition and Vision in the First Year of Postnatal Life*, pp. 223–251.

[151] P.K. Kuhl, *Communication Sciences and Disorders*, p. 108.

[152] J.F. Werker and R.C. Tees, *Infant Behavior and Development* **7**, 49–63 (1984).

[153] P.K. Kuhl, *Communication Sciences and Disorders*, p. 132.

[154] P.H. Kuhl, *Communication Sciences and Disorders*, p. 133. Figures 10-30 and 10-31 are taken from the same article, pp. 114 and 136, respectively.

[155] P.H. Kuhl, *Communication Sciences and Disorders*, p. 126.

[156] Malcolm J. Crocker, ed., *The Encyclopedia of Acoustics* (4 vols.), Wiley, New York, NY, 1997.

A Book Review by Herman von Helmholtz, *Nature* **17**, 237–239 (Jan. 24, 1878).

Rayleigh's *Theory of Sound*. By J.W. Strutt, Baron Rayleigh, FRS, vol. I, Macmillan, London, 1877.

The author [Lord Rayleigh], who already by a series of interesting treatises belonging to different branches of mathematical physics has acquired a respected name in the domain of science, undertakes to give a complete and coherent theory of the phenomena of sound in the work above-mentioned, the first volume of which has recently been published; and he does this with the application of all the resources furnished by mathematics, since without the latter a really complete insight into the causal connection of the phenomena of acoustics is altogether impossible. We must confess that, even in spite of the most intense exertion of the powers of mathematical analysis, in the present state of its development several problems remain unsolved, for which, indeed, the conditional equations are known, but for which it has not yet been found possible to carry out the calculations.

The author will merit in the highest degree the thanks of all who study physics and mathematics if he continues his work in the manner in which he has begun it in the first volume. The separate treatises, in which the acoustic problems that have been solved hitherto are discussed, are for the most part dispersed in the publications of academies or of scientific societies which can be found only in larger libraries, and which frequently are not at all easily traced. But even if one has found a treatise of this kind and reads it, it happens often enough that the author refers in his quotations to other works quite as difficult of access, the knowledge of which is necessary for understanding his treatise. Thus the zeal of the student is paralysed by a number of purely external difficulties, and the ordinary result at which an intelligent student arrives after a few attempts in this direction, is that for problems in which he takes great interest he prefers starting anew to find the solution, rather than trying to hunt for it in libraries. Even if we must admit that the insight into the essence of a problem for which one has found the solution oneself is much deeper and clearer than when one has obtained the solution from some other author, yet an enormous amount of time is thus lost, and the survey of the whole extent of solvable problems remains incomplete. A survey of this kind, however, is necessary for all who wish to work at the progress of science themselves. For in order to obtain decisive results by new scientific investigations it is necessary above all things to be quite clear with regard to the question for which forms of experiment or of observation the theoretical deduction from principles can be carried through as purely as the experiment itself. I know by experience that a number of young physicists lose their time and their zeal by trying to solve problems which, taken by themselves, are very interesting, but for which at present the deductions from the theoretical principles of the given case can only be drawn in coarse approximation, and where the experiments cannot be free from important sources of error.

While praising Lord Rayleigh's book as a means of overcoming the difficulties described, I do not at all wish to designate it as a mere compilation. On the contrary, it is a perfectly coherent deduction of the special facts from the most general principles, according to a uniform method and in a consequent manner. The mechanical principles of the doctrine of minute oscillations are contained in the present volume and are developed in greater generality than in any other book known to me. For this purpose, the author in the first chapter explains that the general physical principles of sound, of its propagation, of pitch and its dependence on the rapidity of vibration, of the musical scale, of the quality of sound and its dependence on the harmonic over-tones and in the second one the doctrine of the composition of harmonic motions of either equal, or nearly equal, or consonant number of vibrations, and further illustrates them, by the description of the physical phenomena and methods in which the principles developed are applied, and to which belong the doctrine of musical beats and of the physical methods to render the forms of vibration visible.

Then follows the development of the most general peculiarities of oscillating motions, first, in the third chapter, for mechanical systems to the motion of which only one degree of freedom is allowed, and then, in the fourth chapter, for systems with a finite number of degrees of freedom. There is a great multitude of peculiarities common to the most heterogeneous sounding bodies, which up to the present have mostly been found in certain instances only, but which can also be deduced from the most universal form of the motion-equations of systems of one or more degrees of freedom. The author in the form of the equations and in the manner of denotation, closely follows the "Natural Philosophy" of Thomson and Tait; in fact, the whole manner of treatment of the mathematical problems corresponds so closely to that adopted in the work just mentioned, that Lord Rayleigh's book may be looked upon as the acoustic part of the excellent handbook of the two celebrated physicists named.

With all systems of this kind if there are no exterior forces acting upon them, we find, on the whole, a number of proper tones equal to the number of degrees of freedom, and the pitch of which does not depend on the amplitude of the vibrations as along as this one remains small enough. Exceptionally, however, several of these proper tones may be of equal pitch. If there is no friction or dissipation of energy the amplitude of every kind of oscillation remains constant. To each separate proper tone a certain form of motion of the whole system belongs; so that the directions and magnitudes of the displacement of the separate points of the system are different in each case. Each arbitrary motion of the system produced in any arbitrary manner, may be regarded as a superposition of these forms of vibrations belonging to the various proper tones of the system. In order to find the amplitude and phase of these different vibrations for a given original displacement and of given velocities of its different parts, quite similar methods are adopted, as those which are employed to develop a given periodical function into one of Fourier's series; only the whole method here becomes far more intelligible and has a thoroughly certain foundation, because we have to do with a finite number of unknown factors instead of with the infinite number of continuously succeeding values of a function, with finite sums instead of with integrals or with infinite series. Of course for Fourier's series as well as for the developments of Laplace by means of spherical harmonic functions the proof for the correctness of their values can also be furnished in the case of continuous functions. For a large number of other functions which are given by differential equations of the second degree this proof results under certain suppositions regarding the continuity of the functions and the limit conditions, from the theorems of Sturm and Liouville, which Lord Rayleigh explains when speaking of the vibrations of strings of unequal thickness. Yet, in mathematical physics we are still compelled to employ a great number of series-developments of functions which do not belong to this class; and even the vibrations of rods and plates are cases in point. In this respect, the treatment of the problems mentioned with a finite but arbitrarily large number of degrees of freedom of motion is interesting also with regard to analysis.

For vibrating systems of one degree of freedom the oscillations of which are subjected to damping, the doctrine of the laws of resonance is developed in the third chapter. The author calls the vibrations which are continuously maintained by the influence of a periodical force acting externally, forced vibrations. In all cases, their intensity is greatest when their period of vibration, which equals the period in which the force changes, is also equal to the period of the system vibrating freely and without friction. For the relations between the intensity and the phase of the co-vibration, between the breadth of the co-vibration in case of small alterations in the pitch and the degree of damping, which I had myself proved for certain instances and used for certain observations, the general proof is given here. The author has further employed these chapters to set up certain general maxims respecting the direction and magnitude of the corrections which must be made in cases where one cannot completely solve an acoustic problem, but can only find the solution of a somewhat altered vibrating system. These are like the outlines of a "theory of perturbations" applied to acoustic problems. The author illustrates these maxims by many various examples. Thus, for instance, he replaces a string by an imponderable stretched thread which carries weights either in the middle only or at certain distances from each other; or a tuning-fork by two imponderable springs with weights at the ends.

For vibrations of very small amplitude, the forces which tend to lead the moving points back to their position of equilibrium may always be considered proportional to the magnitude of their distance form the position of equilibrium. As long as this law holds good, the motions belonging to different ones are superposed without disturbing one another. But when the vibrations become more extensive, so that the law of proportionality just named no longer applied, then perturbations occur which become manifest by the appearance of new tones, the combination tones. In my book on acoustic sensations (*Die Lehre von den Tonempfindungen*) I have myself explained this manner of origin of the combination tones, only for the motion of but a single material point. In Lord Rayleigh's book this explanation is given with reference to any compound vibrating system of one degree of freedom, and it is further amplified with regard to the manner in which the forces deviate with the displacements from the law of proportionality.

Certain laws of reciprocity, of which I had given instances in my investigations on the vibration of organ pipes, may be proved in a general way for all kinds of vibrating elastic systems. If on the one hand, at point A an impulse is given and the motion at point B is determined after the time t has elapsed, and if on the other hand an impulse is given at point B in the direction of the motion, which occurred there, and, after the time t, the motion-component falling into the direction of the first impulse is examined at point A, then the two motions in question are equal if the impulses were equal.

Chapters VI to X of Lord Rayleigh's book treat the vibrations of strings, rods, membranes, and plates. The vibrations of strings have played an important part in acoustics; their laws are simple, and the physical conditions which the theory demands are fulfilled with comparative facility, different modes of producing the tones may be employed and a number of various motions may be produced. It is just because the physical phenomena in connection with strings were well known that the observation of the way in which the ear is affected by their various modes of vibration has materially facilitated the solution of the problems of physiological acoustics. The material importance of strings rests on the circumstance that the series of their proper tones corresponds to that of the harmonics, the vibration-numbers of which are entire multiples of those of the fundamental tone. For this reason, if the motions of many proper tones are superposed on one string, a periodical motion results, and this is the cause why on strings we can produce notes of the most varied quality. We need to remember how differently the same string sounds according to whether it is plucked with the finger or with a metallic point, whether a violin bow is drawn across it or whether it is caused to vibrate by means of a tuning-fork.

In this chapter less new work remained to the author, however, this example shows how much easier it is to understand all these separate problems if they are treated separately but developed in coherent representation, which, after the most general principles, the validity of which is independent of the special peculiarity of the case have been first explained.

The short Chapter VII gives the laws for the longitudinal and torsional vibrations of rods; the laws are simple and resemble those of the open and stopped organ pipes. The lateral vibrations of rods during which they bend, give more complicated analytical expressions; the proper tones do not form a harmonic series, but are given by the roots of a transcendental equation. The tones are different according to whether one or both ends of the rod are free to rotate and to move, or free to rotate but hindered from moving (supported), or hindered from rotating and moving (damped). With this more complicated problem the advantage of first treating of the general principles becomes clearly apparent. The forms of the simple vibrations are calculated and represented graphically. The mode of vibration of a stretched rod, for which Seebeck and Donkin have already given the solution, is also treated here in order to determine the influence of rigidity upon the vibrations of strings.

Then the vibrations of a uniformly stretched membrane are investigated. This investigation is of more theoretical than physical importance, since it shows in a case which may be treated in an easier way, the peculiarities of vibrations which are capable of spreading in two dimensions. Unfortunately we have not yet succeeded up to the present in obtaining good membranes which would be fit for experiments of measuring in order to investigate, with some degree of exactness, how far theory corresponds with the experiment.

On the contrary, in the case of elastic plates, the vibrations of which the author treats in the last chapter of the present volume, the experiments can be made with more accuracy while the analytical difficulties are so great that, on the whole, only few cases permit of a solution of the problem. Indeed, even the formulae expressing the conditions which must be fulfilled at the edge of the plate have given rise to discussions. Poisson had thought that three conditional equations were necessary for the edge; Kirchhoff has shown that in reality only two are required. Lately M. Mathieu opposed this view. Lord Rayleigh has adopted Kirchhoff's views, and no doubt with perfect right. He gives the analysis of the latter of the vibrations of a circular plate, and has made an important condition of his own to the solvable cases, by teaching us how to deduce theoretically a series of vibration forms of square plates, at least for that case where they consist of an elastic substance the resistance of which to change of volume may be neglected; and these theoretical deductions sufficiently correspond with the forms observed. Also for elastic rings and for cylinders vibrating in the manner of bells, he has improved the theory in an essential point, by proving theoretically and experimentally, that the node lines of such plates execute vibrations in a tangential direction. These tangential vibrations are the ones which are the first produced if the edge of a drinking-glass is rubbed with the wet finger.

The above survey will give an idea of the numerous contents of the book. As in the treatment of the separate problems it touches everywhere the limits of our present knowledge, it cannot but demand sound mathematical knowledge on the part of the reader. Yet the author has rendered it possible, by the very convenient systematic arrangement of the whole, for the most difficult problems of acoustics to be now studied with far greater ease than hitherto. He thus proves himself to be a philosopher who does not lose the liberty of intellectual supervision, even when he is occupied with the most obstruse calculations.

H. Helmholtz

Lord Rayleigh's *Theory of Sound* [a review by Herman von Helmholtz, *Nature* **19**, 117–118 (Dec. 12, 1878)].

The Theory of Sound. By J.W. Strutt, Baron Rayleigh, FRS, vol. II, Macmillan, London, 1878.

The second part of Lord Rayleigh's highly instructive work on acoustics contains the mechanics of oscillatory motions in liquids and gases. Atmospheric air is that medium by which by far the greater number of sound-waves are conveyed to our ear, since it is only exceptional that this happens through solid bodies which come in contact with our teeth or with the bones of the skull. But it is just for this reason that all circumstances are of considerable importance, which influence the transmission of sound-waves in the air, i.e., change either their velocity, their direction, or their intensity. This part of the theory has been worked out very minutely and completely by the author. We find here the compilation and demonstration of a large number of acts which, in other works on acoustics, are hardly mentioned. The author, after having first developed (in Chapter XI) the general law of the motion of liquids as expressed in hydrodynamical equations, and then explained the difference between rotational and irrotational motion of liquids, passes on to the simplification of the equations, which is determined by the circumstance that with sound, as a rule, we have to do with oscillations of extremely small amplitude. First, the motion of plane waves is investigated, and it is shown that with waves which move only in one direction half their equivalent of work consists in the vis viva of motion, and the other half in the potential energy of the medium. Then follows the explanation of the influence which the change of temperature, taking place with compression or dilatation of gases, exercises upon the velocity of transmission of sound. It is shown in the manner first employed by Professor Stokes, that if a perceptible quantity of heat could be exchanged between the compressed and dilated layers of the waves during the lapse of one oscillation, the intensity of the sound-waves would very quickly decrease in their transmission and they would die away.

The subjects treated up to this point are generally known among physicists; less known are a series of other results of the theory. The author next gives a comparatively very elegant and easily intelligible demonstration of the results at which Poisson and Riemann arrived, when investigating the propagation of sound-waves for which the velocities of oscillation are no longer infinitesimal when compared to the velocity of transmission. It appears that the different layers of the wave transmit their phases with different velocities, viz., with a velocity which represents the sum of the ordinary velocity of transmission of the smallest waves and of the dilatation velocity of the particles of air oscillating in the same direction. The compressed layers of the wave, therefore, are propagated quicker than the dilated ones, and as they must gradually change the shape of the wave, and finally overtake the preceding dilated layers. What could happen in that case, whether perhaps a breaking of the waves of air would take place, is not yet clear since the hydrodynamical equations apply only to velocities changing continuously.

These circumstances have not always been considered in experimental researches concerning the velocity of sound. A precise answer to the question regarding the magnitude of this velocity can only be given if we confine ourselves to oscillations of extreme smallness.

The author has also investigated under what conditions a sound-wave of finite amplitude can move forward without changing its form. It appears that this could happen only under the supposition of a special law for the compressibility of the medium which does not correspond with the law applying to gases.

The propagation of sound in the atmosphere is subjected to yet other perturbations, which partially arise from the different temperatures and moistures of the superposed state, and partly from the different force of winds. At the surface of water or extensive masses of solid substance, the sound-waves of the air are totally reflected even under very small angles of incidence; under perpendicular incidence their reflection although not total in the strict sense of the word, is nearly as complete. For that part of the sound which enters the new medium, the same law of refraction holds good which applies to waves of light. But also from a surface of hydrogen one-third of the sound coming through air at a vertical incidence is reflected and the angle of incidence for total reflection is not larger than $15\frac{1}{4}$ degrees.

The problem to determine theoretically how the propagation of sound in the atmosphere is changed by the different temperatures of its strata cannot yet be solved completely. However, it can be ascertained in what direction the most powerful effect must travel. On account of the great dimensions of the strata of the atmosphere, compared to which the wave-length of the large majority of audible tones disappears entirely, the conditions of the propagation of sound are similar to those of light. We may imagine the sound-waves dissolved, as it were, into rays of sound, and then look upon each separate ray as being almost completely independent of the motion of its neighbouring rays. This is no longer admissible if obstacles are in the way of the travelling sound, the dimensions of which exceed the sound wave-lengths only in moderate proportions, as is the case in our houses and rooms, with the transmission of sound through windows and doors. Then, as in the case of light under similar circumstances, diffraction takes place. The great difference in the propagation of sound and light, as it becomes evident in ordinary experiences, has its cause in the very different magnitude of wave-lengths. The greater the wave-length the greater the diffraction on the passage through the same aperture. These circumstances, which are forgotten so frequently, the author considers in Chapter XIV. When sound is propagated in the unbounded space of the atmosphere, the conditions of the problem are such, that they allow its decomposition into rays of sound. If a source of sound is near the ground, then its sound rays are all bent into an upwards direction, as Professor Osborne Reynolds first pointed out, and those which travel in a direction parallel to the ground are mostly annihilated through friction or other obstacles. The sound proceeding from a source near the ground is, therefore, not heard far off by an observer standing on the ground. It is heard at a much greater distance if the observer of the source of sound be in an elevated position. The state of the atmosphere will have great influence upon these conditions. In dry air and sunshine the deflection of sound upwards will be greater than in moist air which forms clouds above, or during rain.

The decrease of temperature in the upper strata of the atmosphere causes sound to travel at a lesser speed than in the lower ones. Now if a wind is blowing, the velocity of which increases with the height then, as Professor Stokes remarks, this causes an increase in the velocity of the sound-waves in the direction of the wind, and a decrease in velocity in the opposite direction; thus, for sound which travels in the former direction, the retarding effect of decrease of temperature is neutralised, and for that travelling in the latter it is augmented. We therefore hear better if the wind blows towards us from the source of sound than if the contrary takes place, indeed, by an upper wind layer, sound produced in the lower tranquil air may be totally reflected. This influence of wind is remarkable also because it forms an exception to the law demonstrated by myself, viz., the law of reciprocity in the propagation of sound if the sound-source and the observer change places.

The problems of the reflection of sound by fixed walls, for instance, the phenomena of whispering galleries, speaking-trumpets, and the echo, are treated in the same manner. Although here the admissibility of the decomposition of sound into sound-rays does not as a rule appear quite so unquestionably justified, yet the phenomena observed agree with this hypothesis on the whole.

An essential progress in the application of the theory upon experiments has been made by the author in the calculation of the influence of the open apertures of organ pipes and resonators upon their pitch. In my own demonstration of this part of the theory, I had started out from those forms of motion which did not render the calculation too difficult, and had then derived the corresponding forms of pipes; finally I had so determined the optional constants of my hypotheses, that the form of the pipe approached the form wished for, the cylindrical one, for instance; yet I remained confined to a few forms if I did not wish to complicate the calculation too much. Lord Rayleigh, on the contrary, supposes a given form of pipe and has employed the maxims, developed in the first volume of this work regarding the variation of conditions under which sound-motion takes place, to determine the limits with which the true value of the desired magnitude must lie, and had indeed been able to draw these limits so narrowly for the most important problems like that of the cylindrical open pipes that practically the solution is perfectly sufficient. In this way he has been able to treat simultaneously a number of problems which hitherto had not even received an approximate solution, for instance, the determination of the proper tones of resonators of the shape of bottles with wide body and narrow neck.

Besides the problems mentioned, which are of direct importance to experimental physics, a series of others are worked out, where the mathematical solution can be completely given, such as the propagation of sound in balls, spherical layers and rectangular boxes filled with gas, the reflection of sound from the outer surface of a ball and the communication of sound to air by oscillating balls and strings. These problems are valuable not only as theoretical exercises, but also with regard to our understanding of physical phenomena. They are examples affording to the mental eye of the physicist a particularly perfect insight into the essence of sound-motion and the changes it undergoes, when the conditions under which it occurs are changed. Thus he obtains quite as good as conception of the typical behaviour of sound as if he had actually seen the phenomena, and this conception will also guide him safely in cases of observation where the exterior conditions are not as simple as they are in the theoretical example.

At the end of the volume, Lord Rayleigh has placed the words: "The End." We hope that this may be only the provisional, not the definite end. There is still an important chapter wanting, viz., that on the theory of reed-pipes, including the human voice. For the former, at least, the principles of their mechanics can already be given and the methods the author employs seem to me to be particularly well adapted for further progress in these domains.

After reed-pipes, we could mention the theory of singing flames, and the blowing of organ-pipes. In the latter case the leaf-shaped current of air, which comes from the wind-case, forms a sort of reed, which oscillates under the influence of the oscillating column of air in the interior of the pipe, and which throws its air now into the interior of the pipe, and now outside.

Altogether, the whole of this important class of motions, where oscillatory movements are kept up through a cause which acts constantly, deserves detailed theoretical consideration. The action of the violin bow, and the sounding of the Aeolian harp, also belong to this class.

Lord Rayleigh certainly deserves the thanks of all physicists and students of physics. He has rendered them a great service by what he has done hitherto. But I believe I am speaking in the name of all of them if I express the hope, that the difficulties of that which yet remains still incite him to crown his work by completing it.

H. Helmholtz

Name Index

Subject Index